Small mammals: their productivity and population dynamics

T0297125

THE INTERNATIONAL BIOLOGICAL PROGRAMME

The International Biological Programme was established by the International Council of Scientific Unions in 1964 as a counterpart of the International Geophysical Year. The subject of the IBP was defined as 'The Biological Basis of Productivity and Human Welfare', and the reason for its establishment was recognition that the rapidly increasing human population called for a better understanding of the environment as a basis for the rational management of natural resources. This could be achieved only on the basis of scientific knowledge, which in many fields of biology and in many parts of the world was felt to be inadequate. At the same time it was recognized that human activities were creating rapid and comprehensive changes in the environment. Thus, in terms of human welfare, the reason for the IBP lay in its promotion of basic knowledge relevant to the needs of man.

The IBP provided the first occasion on which biologists throughout the world were challenged to work together for a common cause. It involved an integrated and concerted examination of a wide range of problems. The Programme was co-ordinated through a series of seven sections representing the major subject areas of research. Four of these sections were concerned with the study of biological productivity on land, in freshwater, and in the seas, together with the processes of photosynthesis and nitrogen fixation. Three sections were concerned with adaptability of human populations, conservation of ecosystems and the use of biological resources.

After a decade of work, the Programme terminated in June 1974 and this series of volumes brings together, in the form of syntheses, the results of national and international activities.

INTERNATIONAL BIOLOGICAL PROGRAMME 5

Small mammals: their productivity and population dynamics

EDITED BY

F. B. Golley
Institute of Ecology, University of Georgia, USA

K. Petrusewicz
Institute of Ecology, PAS, Dziekanów Leśny, Poland

AND

L. Ryszkowski
Institute of Ecology, PAS, Dziekanów Leśny, Poland

CAMBRIDGE UNIVERSITY PRESS

CAMBRIDGE
LONDON · NEW YORK · MELBOURNE

CAMBRIDGE UNIVERSITY PRESS
Cambridge, New York, Melbourne, Madrid, Cape Town, Singapore, São Paulo, Delhi

Cambridge University Press
The Edinburgh Building, Cambridge CB2 8RU, UK

Published in the United States of America by Cambridge University Press, New York

www.cambridge.org
Information on this title: www.cambridge.org/9780521116060

First published 1975
This digitally printed version 2009

A catalogue record for this publication is available from the British Library

Library of Congress Catalogue Card Number: 74–25658

ISBN 978-0-521-20601-3 hardback
ISBN 978-0-521-11606-0 paperback

Contents

Contents

Table des matières

Table des matières

Содержание

Содержание

Contenido

Contenido

Contributors

Alibbhai, S. K.   Department of Zoology, Animal Ecology Research C. P. Oxford, England.

Arata, A. A.   Vector Biology and Control, WHO, Geneva, Switzerland.

Ashby, K. R.   University of Durham, Durham, England.

Batzli, G. O.   University of Illinois at Urbana-Champaign, Urbana, Illinois, USA.

Bernard, J.   Ministre de l'Agriculture, Station de Zoologie Appliquée de l'Etat, 5800 Gembloux, Belgium.

Bikhovski,   All Union Institute of Plant Protection, Leningrad, USSR.

Bobek, B.   Institute of Zoology, Jagiellonian University, Kraków, Poland.

Bolszakov, W. I.   Institute of Plant and Animal Ecology, Sverdlovsk, USSR.

Bourlière, F.   Laboratoire de Physiologie, Paris, France.

Bujalska, G.   Institute of Ecology, Polish Academy of Sciences, Dziekanów Leśny near Warsaw, Poland.

Briese, L. A.   Savannah River Ecology Laboratory, Aiken, South Carolina, USA.

Chew, R. M.   University of Southern California, Los Angeles, California, USA.

Christov, B.   Zoological Institute Museum, Sofia, Bulgaria.

Delany, M. J.   Department of Zoology of the University, Southampton, England.

Denisov, W. P.   University, Department of Biology, Saratov, USSR.

Drożdż, A.   Institute of Zoology, Jagiellonian University, Kraków, Poland.

Fleming, T. H.   University of Missouri, St Louis, Missouri, USA.

French, N. R.   Colorado State University, Fort Collins, Colorado 80523, USA.

xiv

Masaa M. Al-Jumaily   Royal Holloway College, Englefield Green, Surrey, England.

Mermod, C.   Institute de Zoologie, Neuchatel, Switzerland.

Meylan, A.   Station Fédérale de Recherches, Nyon, Switzerland.

Mezhzherin, V. A.   Kiev-3, Pushkinskaya 5-3, USSR.

Myllymäki, A.   Agricultural Research Center, Tikkurila, Finland.

Naumov, N. P.   Moscow State University, Moscow, USSR.

Obretl, R.   Czechoslovak Academy of Sciences, Brno, Czechoslovakia.

O'Farrell, M. H.   Savannah River Ecology Laboratory, Aiken, South Carolina, USA.

Pelikan, J.   Czechoslovak Academy of Sciences, Brno, Czechoslovakia.

Perrin, M. R.   Makerere University, Kampala, Uganda.

Petrusewicz, K.   Institute of Ecology, Polish Academy of Sciences, Dziekanów Leśny, near Warsaw, Poland.

Piastolowa, O. A.   Institute of Plant and Animal Ecology, Sverdlovsk, USSR.

Polakov, I. J.   All Union Institute of Plant Protection, Leningrad, USSR.

Popov, W. A.   Kazań 15, Malaja Krasnaja 14 kw. 24, USSR.

Pucek, Z.   Mammals Research Institute, Polish Academy of Sciences, Bialowieża, Poland.

Rowe, F. P.   Ministry of Agriculture, Fisheries and Food, Pest Infestation Control Laboratory, Tolworth, Surbiton, Surrey, England.

Ryszkowski, L.   Insititute of Ecology, Polish Academy of Sciences, Dziekanów Leśny, near Warsaw, Poland.

Schilov, J.   Moscow State University, Moscow, USSR.

Schvecov, J. G.   Institute of Agriculture, Irkutsk, USSR.

Shvarts, S. S.   Institute of Plant and Animal Ecology, Academy of Science, Sverdlovsk, USSR.

Smith, M. H.   Savannah River Ecology Laboratory, Aiken, South Carolina, USA.

Smirnov, V. C.	Institute of Plant and Animal Ecology, Sverdlovsk, USSR.
Sokolov, V. E.	Academy of Sciences, Moscow, Leninskij Prospekt 33, USSR.
Sokur, J. T.	Academy of Sciences, Kiev, USSR.
Stenseth, N. C.	University of Oslo, Blindern, Oslo 3, Norway.
Stoddart, D. M.	University of London King's College, London WC2, England.
Tahon, J.	Ministre de l'Agriculture, Gembloux, Belgium.
Tertil, R.	Institute of Zoology, Jagiellonian University, Kraków, Poland.
Wagner, C. K.	Southwestern at Memphis, Memphis, Tennessee, USA.
Wagner, S.	Southwestern at Memphis, Memphis, Tennessee, USA.
Walkowa, W.	Institute of Ecology, Polish Academy of Sciences, Dziekanów Leśny, near Warsaw, Poland.
Wiener, J. G.	Savannah River Ecology Laboratory, Aiken, South Carolina, USA.
Wunder, B. R.	Colorado State University, Fort Collins, Colorado 80523, USA.

Foreword

In September 1966, the IBP section concerned with the productivity of terrestrial communities (PT), in collaboration with the Polish Academy of Sciences, Institute of Ecology, held a technical meeting at Jablonna, Poland. There, under the chairmanship of Professor François Bourlière, at that time Convener of Section PT, general agreement was reached on the broad outlines of a programme for studying production in terrestrial ecosystems.

Of the many important decisions taken at Jablonna, one was to base the PT programme on biome and theme studies. In the latter category, emphasis was to be given to animal groups which played an important role in ecosystem dynamics. An immediate result of those discussions was the setting up of a working group on small mammals under the chairmanship of Professor Dr K. Petrusewicz with Dr L. Ryszkowski as secretary. Within a short time, scientists from twenty-seven countries were associated in studies which have culminated in the publication of this volume.

The element of co-operation which has characterized the work of the Small Mammals Group grew out of the common aims shared by all of its members and it was encouraged by the publication of a newsletter and the holding of various symposia. The first number of the Newsletter, for example, demonstrated very clearly the wide geographical coverage of the group. It included articles from Bulgaria, Canada, Czechoslovakia, France, Great Britain, Poland, Rumania, the USA and the USSR. As this volume goes to press, the newsletter continues to exist and is mailed to over 200 scientists and research laboratories.

The symposium at Jablonna was the first of many important meetings of workers on small mammals. The second meeting was held in Oxford, England, in July 1968, and a third in Helsinki, Finland, in August 1970. These meetings were reported in the following publications: *Secondary Productivity in Terrestrial Ecosystems, Energy Flow through Small Mammal Populations*, and the Proceedings of the IBP meeting on secondary productivity in small mammal populations, in *Annales Zoological Fennici*. These publications made the results of research by the working group available to the larger IBP community and to the scientific world in general. In addition, research programmes on small mammals have figured prominently in many national IBP projects.

Foreword

Two research programmes were organized by the working group: a *standard or minimum programme* designed to provide information on the impact of small mammal populations on primary production; and, secondly a *maximum programme* designed to develop research methods and to investigate special problems associated with measuring energy flow in mammal populations. The *minimum programme* has been carried out in field and forest environments, mainly in the north temperate regions, and has focused on the determination of density, sex ratio, body weight distribution, age distribution, and sexual activity of populations. The *maximum programme* has been expecially concerned with developing methods for determining density of mammals; with food habits; and with metabolism and energy assimilation in populations.

All of these research efforts culminated in a series of summary papers which were prepared at the request of the working group and discussed at a synthesis meeting in Dziekanów Leśny, Poland, on 6–10 November 1973. These papers have been rewritten in the light of the discussions at that meeting and this volume represents a summary of research by scientists working on small mammal projects in IBP over six years. While many participants made brief comments regarding the papers, others examined them carefully as referees or provided written comment. We have listed all the participants as contributors to the volume.

We are grateful to the Polish Academy of Sciences, Institute of Ecology, for providing support to the working group and for being the host for the final synthesis meeting. Editorial assistance has been provided by Susan Wagner and Kenyon Wagner in Poland. Priscilla Golley and Nancy Payne edited and prepared the typescript for the printer. Support for the completion of the volume has been provided by the IBP Central Office and the Institute of Ecology, University of Georgia.

F. B. GOLLEY

K. PETRUSEWICZ

L. RYSZKOWSKI

Preface

Our object here is to evaluate the results obtained by the IBP/PT working group on small mammals. The working group was organized in 1966 during the meeting at Jablonna, Poland, on secondary productivity of terrestrial ecosystems. The program was focused mainly at the population level. Two approaches were developed. First, a program using simple standard methods was designed to provide extensive information on density, sex ratio, body weight distributions and sexual activity of populations in forest, grasslands, and fields. By use of bioenergetic parameters, approximations of energy flow at population levels could then be obtained. Second, an advanced program was especially concerned with the development of more exact methods for determining the density of mammals in populations along with their food habits, metabolism and energy assimilation.

At the III International Meeting at Helsinki, Finland, in 1970, the working group re-examined its objectives in order to evaluate the role of small mammals in ecosystem function. The emphasis of the studies was changed from the population level to the evaluation of interactions of small mammals with other elements within the ecosystem and especially as they influence the processes of production, energy flow and mineral cycling. It was recognized that small mammals interacting with other consumers are important control components in ecosystems. In this context, the new approach was applied to problems such as small mammals as pests on crops and forests. History is replete with evidence that as man's impact on the environment is increased there is an increase in mammalian and other pests. While control measures can be instituted, they are seldom successful over long periods of time. In contrast, an ecosystem approach which considers all aspects of the system may provide the basis for successful pest management.

To summarize briefly, the activities of the group dealt with the structure and function of small mammal populations, their role in various ecosystems, and with the application of control measures.

Most results were obtained by the group studying field and forest environments in the north temperate regions. Tropical and arctic habitats were not covered intensively. Also, more studies were made of rodents than of bats, lagomorphs, insectivores, mustelids and other small mammals. During the IV International Meeting held in 1973 at

Dziekanów Leśny, Poland, efforts were made to cover these gaps in information. Even so, this synthesis of the activities of the group relies mainly on rodent studies carried out in the temperate zone.

Considerable progress has been made toward developing sound methods for taking censuses of mammal populations. The estimate of density is one of the most basic parameters in understanding the role of any population and since practically all other calculations concerning population activities rely on this parameter, much effort was devoted to this problem. To obtain reliable estimates of small mammal densities by trapping, one has to obtain information simultaneously on the number of animals present within a given area as well as on the rate of movement across the boundaries of a trapping plot. Three reliable trapping methods were developed for use with some groups of rodents and insectivores (see Chapter 2). The confidence limits of the estimates were calculated so that the results can be compared with estimates determined by other methods. Because of international cooperation it was possible to compare the densities of small mammals in different habitats of the temperate zone for the first time. This cooperation made possible an objective survey of rodent densities in different ecosystems. So far, little progress has been achieved concerning estimates of density in subterranean and arboreal species or bats and mustelids. This lack of information prevents a reliable evaluation of the role of small mammals in the tropics, where many arboreal rodent species and other poorly sampled groups of small mammals, e.g., bats, exist.

The studies on demographic parameters and bioenergetics are summarized by Bourlière (Chapter 1.1), Petrusewicz & Hansson (Chapter 7), French *et al.* (Chapter 4), Pucek & Lowe (Chapter 3), and Grodziński & Wunder (Chapter 8). Due to the disparity of studies on different systematic groups the functional characteristics of small mammals given below is pertinent mainly to rodents.

Small mammals, especially rodents, adapt to any terrestrial habitat from the arctic to the tropics provided they can obtain food. Being homeothermal animals, they can be active even under quite severe climatic conditions; for example house mice were found close to the Pole in food stores left by the Scott expedition. Breeding rodent populations have also been found in habitats created by man, such as deep-freeze meat stores, mines, steel mills and so on. Therefore, they can be described as a very adaptative group of animals responsive to any opportunity. In spite of accepted nutritional types, such as grass-eaters, seed-eaters and so on, hard and fast lines do not exist and most small mammals are omnivorous and take advantage of a variety of plant foods as well as insects, other invertebrates, fungi and even vertebrate flesh if available. In comparison, with, for example, ungulates, the herbivorous small mammals have high digestibility coefficients. This characteristic combined with the high ratio of respiration to consumption in small mammals (insects have a much lower ratio) indicates that they are more efficient in mineralization of the organic matter than insects or ungulates, if the groups are compared on the basis of an equal unit of biomass. The overall energetic efficiency of production is low in small mammals (about 2 per cent of the food consumed). Their impact on the environment relies on a relatively high reproductive rate of populations in comparison with large mammals. Analysis of the pattern of reproduction and survival within small mammal populations has shown an inverse relationship between those two demographic parameters. Within small mammal populations one can find two types of strategies for adaptation to living conditions: first, high reproduction and rapid growth to maturity combined with low survival; and second, effective survival with low reproduction. Outbreaks in numbers of small mammals having economic importance were reported mainly for those characterized by the first strategy.

Many vertebrate predators rely on rodents as key supplies of food due to their high turnover of biomass. Because of the nonselective food habits of rodents, one can consider them as a link between many primary producers and vertebrate predators of the ecosystem. This is especially evident in habitats like tundra or deserts where rodents are key vertebrate species.

On the basis of the demographic and bioenergetic features described one can characterize small mammals as a component of the ecosystem having: (*a*) high turnover rate; (*b*) very high energy cost of production; and (*c*) small dependance on climatic conditions due to homeothermy (with the exception of hibernating specics).

These characteristics determine the adaptability of small mammals as well as their output and efficiency in terms of energy flow and mineral cycling in the ecosystem. Generally speaking, output of any component per unit of time can be defined as the product of mass (standing crop) times turnover. It seems that small mammals, generally speaking, are more flexible in their level of standing crop than in turnover.

The standing crop is the result of reproduction and survival, and since turnover is determined by survival, one reaches the conclusion that the variation in the ecosystem output of small mammals is mainly influenced by variation in the reproductive processes of populations. It seems that among the different parameters of reproduction the length of breeding periods, intervals between successive pregnancies in the same female and the percentage of pregnant females vary more than litter size or pregnancy duration. There is a suggestion that the more flexible parameters of reproduction are food dependent. Many of the morpho-physiological indices of population 'welfare' which can be used to check this suggestion are discussed by Shvarts in Chapter 6.

One has to keep in mind that the standing crop of population is influenced by dispersion (Lidicker, Chapter 5) and that this factor modifies the results of reproduction in a given habitat. In terms of high energetic production cost, small mammals can be defined as poor accumulators of energy and elements.

Because of poor knowledge on element cycling in mammal populations it is premature to establish their full role in mineral cycling in ecosystems. Gentry *et al.* (Chapter 9) made an enormous effort to gather the scanty information on this topic. It seems on the basis of their studies that differences among small mammal species are relatively small compared to those found among plants. Therefore, one can assume that small mammals as a whole relatively homogeneous in terms of mineral cycling. The role of small mammals in arctic, temperate, tropic and desert ecosystems was analyzed by Batzli (Chapter 11), Golley, Ryszkowski & Sokur (Chapter 10), Fleming (Chapter 12) and Naumov (Chapter 13) respectively. In studies on the impact of small mammals on vegetation it was shown that while the quality and quantity of available food varies greatly between biomes and genera, the percentage of the available primary production consumed by rodents in the majority of cases seems to be relatively trivial (usually less than 2 per cent). Still, it is clearly evident that ecosystems with a higher amount of plant food available have higher rodent densities. Rich forests have high rodent densities, but grassland ecosystems are characterized by densities at least one power higher than forest ecosystems. Although the main bulk of data to support this conclusion was obtained in the temperate zone, the scanty information that we have from the tropics also supports it. Rodents in grasslands, together with other grazers and fires, are essential factors for removing (through mineralization) large quantities of seasonally shed above-ground primary production. Without these mineralization factors, the regrowth of the new season's vegetation would be greatly impaired because of the accumulation of plant detritus. The outbreaks of rodents in grasslands, when the consumption of primary production is higher than 2 per cent, are frequent phenomena in comparison to forest ecosystems. Outbreaks of rodents in grasslands seem to play an important role in the mineral cycling of these ecosystems.

The direct influence of small mammals on energy flow is measured by their consumption, which is rather small (except during outbreaks), in the overall picture of ecosystem energetics. The outbreaks show the potentially destructive role of small mammals.

The role of small mammals, especially rodents, in mineral cycling is not limited to consumption. Their burrowing activities may play an important role in: (a) the transport of elements across the soil profile, especially in dry habitats with no upward movement of water from deep ground water tables, and (b) the change of physico-chemical conditions of the upper soil layer; by that means indirectly influencing decay of organic matter, water storage in the soil and so on. There is a great need for studies on the evaluation of these phenomena. Nevertheless, it can be assumed that the burrowing activities greatly magnify the role of small mammals in mineral cycling beyond the limits set by consumption.

The other subtle role of small mammals is their impact on the diversity of the ecosystem. Clumped distribution of excretion spots of feces and urine, feeding habits and burrowing activity influence diversity of plant cover. Burrows provide a shelter place for many invertebrates and vertebrates. For example, in desert ecosystems, many invertebrate species can exist provided there are burrows made by colonies of rodents. The selection of seeds for food by rodents probably has a higher impact on plant diversity in forests than in grasslands; however, there is only scanty information published on this topic.

There are many other impacts of small mammals on ecosystem functioning which in turn have secondary and tertiary effects, but their quantitative evaluation is unknown. For example, there exists a rich small mammal fauna in the tropics (with more species of bats, rodents, marsupials and primates than in temperate communities), which is involved in pollination as well as in seed dispersion. Other than qualitative statements, such as that bats appear to be one of the important pollination agents in the tropics, there is little quantitative data to evaluate their role in this respect.

Due to enormous efforts within the framework of IBP, quantitative data on populations, bioenergetics, consumption and to a lesser degree on mineral cycling were gathered for small mammals. These studies have lead to several generalizations which require examination and testing in the future. For example, in the tundra rodents such as lemmings are the key species in the economy of the ecosystem. Their densities fluctuate violently and their destructive effects on plants are magnified by the slower recovery rates of vegetation. In the tundra the role of small mammals is much greater than in taiga, but we have relatively poor information available on small mammals in taiga ecosystems. In the temperate zone, the greatest impact of rodents on ecosystem economy is probably in the grasslands. Here small mammals speed up mineralization of organic matter and their burrowing activity has an effect on mineral cycling. In cultivated fields, rodents are competitors to man, but as the intensity of agriculture increases, their role decreases. In very intensive agriculture, man's control of the environment is so complete that there is no room for small mammals. Along the gradient from reforested areas to more complicated and mature forests the role of small mammals decreases although their diversity increases.

The same general picture can be sketched for the tropics where in tropical grasslands and deserts the role of small mammals, especially rodents, is the greatest. Tropical forests are characterized by very stable and diverse populations of small mammals, but their role in the forest economy is obscure.

The other aspect of small mammals to be especially stressed is connected with their role as crop pests and disease vectors (see chapters by Arata, 14.3; Naumov, 13; Myllymäki, 14.1; Rowe, 14.2). In this respect, the evaluation of the role of small mammals is oriented not to ecosystem functioning but to human welfare or economy. In terms of the ecosystem this means that man, managing for high productivity, stability, and sanitary conditions, has to provide energy subsidies from outside the system to control competitors in order to obtain the desired goals. A discussion of the effect of such actions on ecosystems is outside the scope of this chapter.

Concerning the theoretical evaluation of the measures of small mammal control, as will be pointed out several times in this volume, more efforts should be paid to control of reproduction processes than has been done to date, especially when dealing with small mammals having a very short lifespan. This approach might be especially fruitful for effective control procedures.

The evaluation of the role of small mammals in different ecosystems, as well as in various climatic zones, relies on the information gathered by the authors in the general discussion held during the IV International

Meeting of the IBP Working Group of Small Mammals. The editors wish to acknowledge the contribution of the participants at that meeting to the formulation of the ideas developed in this book.

F. B. GOLLEY
K. PETRUSEWICZ
L. RYSZKOWSKI

March, 1974

1. Introduction

1.1. *Mammals, small and large: the ecological implications of size*

F. BOURLIÈRE

The choice of small mammals as a trophic group to receive special attention within the Terrestrial Productivity section of IBP was not made haphazardly, but deliberately. There were, in fact, strong economic and methodological motivations for this choice. Their most numerous representatives, the rodents, are major vertebrate 'pests'. Some species are responsible for the loss of a large proportion of the agricultural production in a large number of developing countries, whereas others are reservoirs or carriers of lethal pathogens. Consequently the physiology, pathology, behaviour and ecology of rodents have been studied more intensively during recent decades than in most other mammalian orders.

However important it is for the improvement of 'human welfare' to reduce competition with vertebrate pests and thus to increase 'biological productivity', there were more cogent reasons – ecologically speaking – for concentrating on small rather than large mammals. First, sizable natural populations of small mammals are to be found almost everywhere, while undisturbed populations of large mammal species are rare where IBP research teams are mainly situated. Furthermore, size itself is an important, albeit often underestimated, parameter in considering the relationships of homeothermic vertebrates to their natural environments. This point perhaps requires some clarification.

What is a small mammal?

Small or large mammals do not constitute definite taxonomic entities. The adult weight of contemporary terrestrial animals belonging to class Mammalia ranges from 1.7 grams (the shrews *Suncus etruscus* and *Sorex tscherskii*) to over 5 tonnes (the African elephant). The various species are not, however, evenly distributed along a linear weight gradient: most of them cluster between a few grams and a few kilograms

1

on the one side, and between a few tens and a few hundreds of kilograms on the other. On the whole, 'middle-sized' species are far less numerous than the small- and large-sized ones. Such a bimodal weight distribution is not due to mere chance. These two major size categories of mammals probably correspond to the two alternative strategies which are open to populations of terrestrial, warm-blooded vertebrates for the exploitation of environmental resources.

Although small- and large-sized genera or species are (or were) to be found in many living or fossil mammal orders – for instance, rodents as large as a hippopotamus (the extinct genus *Eumegamys*) and ungulates as small as a hare (the contemporary chevrotains) – most orders comprise a majority of representatives belonging to one or the other of the two major size categories. The distribution of these size categories among contemporary representatives of terrestrial mammals is shown in Table 1.1 together with an indication of the predominant food habits of most species within each order. Small mammals have been considered here to include any species whose adult live weight ranges from less than 2 grams to about 5 kilograms. This latter figure has been decided upon, more or less arbitrarily, by the IBP Small Mammals Working Group so as to include lagomorphs but to exclude middle-sized animals such as badgers and wild canids.

A cursory glance at Table 1.1 clearly reveals that the overwhelming majority of contemporary species of terrestrial mammals are of a small size. Furthermore, most of them eat a fairly large variety of vegetal material (grazers, browsers, seed-eaters, nectarivores), although a reasonable minority is insectivorous or carnivorous. On the other hand, large mammals are almost exclusively vegetarians, the only striking exceptions being the large canids, felids and bears. It should also be added that all fossorial species are of a small size, together with most scansorial and flying forms. Large size is indeed a handicap for burrowing in soil, as well as climbing trees or running on branches, except for monkeys and apes which can use their hands, feet and (sometimes) tail to grasp their supports.

It is worthwhile to note at this point that the bimodal distribution of body weights among terrestrial mammals has no counterpart among swimming or flying warm-blooded vertebrates. All intermediates in size exist between the otter and the blue whale, as well as between the Cuban hummingbird and the mute swan. The lower energy cost of locomotion in water and air, as compared with running (Schmidt-Nielsen, 1972*a*), probably allows greater flexibility in the quest for food or mates, and does not, in aquatic mammals and birds, exert a similarly strong pressure for selection between two alternative types of environmental adaptation.

These comments on small mammals do not include the bats, which while small are quite different from the ground or tree living forms. Bats as flying mammals experience a different set of constraints compared to their fellows of similar body size. The IBP studies have largely ignored the bats, although recent studies in the tropics where the bat fauna is large, diverse and important have begun to rectify this bias.

Table 1.1. *Number of terrestrial genera and species, food habits and size category of contemporary terrestrial mammals*

Orders	Number of contemporary genera*	Number of contemporary species*	Food habits of most species†	Size category of most species‡
Monotremata	3	6	A	S
Marsupialia	81	242	A/V	S
Insectivora	77	406	A	S
Dermoptera	1	2	V	S
Chiroptera	173	875	A	S
Primata	47	166	V	L
Edentata	14	31	A/V	L
Pholidota	1	8	A	S
Lagomorpha	9	63	V	S
Rodentia	354	1687	V	S
Carnivora	96	253	A	S
Tubulidentata	1	1	A	L
Proboscidea	2	2	V	L
Hyracoidea	3	11	V	S
Perissodactyla	6	16	V	L
Artiodactyla	75	171	V	L

* After Anderson & Jones (1967).
† The following broad categories have been distinguished: V, predominantly vegetarians; A, predominantly animal eaters.
‡ Two broad size categories are recognized: S, small, i.e., adult body weight less than 3 kg; L, large, i.e., adult body weight over 5 kg.

The advantages of being a small mammal

The most apparent, but not necessarily the most important, advantage of small size for a terrestrial mammal is to allow relatively easy concealment from predators hunting by sight. The smaller a ground living animal is, the easier it is for it to forage without being readily located by an aerial predator, particularly since most species are nocturnal or crepuscular in habits. Obviously such a protection is far from being absolute; some owls have evolved elaborate adaptations for locating their prey by ear and many small carnivores can track their prey by scent. Coupled often with concealing coat coloration, a small size accords a further benefit to many marsupials, insectivores, and rodents: they do not need to spend much energy in fleeing from their predators or in fighting with them. Furthermore, many have developed defensive behavior patterns such as rolling themselves into a ball, erecting their quills, playing dead or losing their tails by autotomy in the mouth of a somewhat bewildered predator. These are all efficient energy sparing adaptations.

The second advantage of small size for a terrestrial mammal is to provide easy access to a number of food sources which are only sparsely exploited by other vertebrates: fresh leaves and buds in the lower grass layer, fallen seeds and nuts, invertebrates in litter layers, arthropods on tree trunks and branches, etc. A small size also pre-adapts ground living Insectivora and Rodentia to explore and take advantage of the food resources of the soil, both plant (rhizomes, bulbs) and animal (earthworms, many arthropods) in nature. Burrowing has probably led to permanent subterranean life, a mode of life achieved independently by a few marsupials, insectivores and rodents.

Small size also enables rodents and other mammals to take full advantage of the microclimates which are to be found in most environments, whether they be within the leaf-litter of a forest, under the grass cover of a savannah or below the deep snow layer of an arctic tundra in winter. A number of species even enlarge or build up, at a low energy cost, temporary or permanent shelters in which they can take refuge when life becomes impossible 'outdoors' during part of the day or at some season of the year. Many desert, high altitude or subarctic rodents provide good examples of this use of artificial microclimates, which large mammals are seldom able to take advantage of, even less able to create by themselves.

A further advantage of many small mammals is the polyestrous pattern of reproduction, in which short periods of estrus and progestation alternate cyclically, so that a female can conceive at a second or subsequent estrus if the first is infertile. Bats, however, do not follow this pattern. In temperate regions most are monestrous. In contrast to most small mammals, many large mammals have, each year or even at longer intervals, a single and relatively long period of estrus when the tract is receptive to spermatozoa, followed, either spontaneously or in response to copulation, by a progestational phase; in this case the female may remain infertile for the rest of the year if she does not conceive.

The high rates of conception, coupled with large litter sizes and short gestation and growth periods, provide many small mammals with a production potential unequalled among other homeothermic vertebrates. The physiological reasons which allow the weight of the litter at birth to reach 50 per cent of the mother's weight in some rodents (compared with 5.7 per cent in man and 0.2 per cent in the polar bear) are not yet clear. However, such high fertility enables many insectivore and rodent populations to cope efficiently with drastic changes in their environment, and to recover quickly from 'ecological disasters' which cause local extinction of populations in most large terrestrial mammals.

On a long-term basis small mammals enjoy further advantages over large species. Not only do their populations have a short generation time and a high turnover rate, they can also, when necessary, split into a number of local populations well adapted to peculiar local conditions (Blair, 1950; Baker, 1968; Dice, 1968).

Furthermore, the short life span of small mammals, when associated with polymorphism, also facilitates a marked change in genetic composition during a cycle in the size of a population. For instance, Krebs and coworkers (1973) have found evidence of large changes in gene frequency at two loci (transferrin and leucine aminopeptidase) in wild populations of *Microtus pennsylvanicus* and *M. ochrogaster* in southern Indiana, in association with population changes. This type of observation supports the hypothesis that demographic events in *Microtus* are genetically selective and that losses are not distributed equally over all genotypes. A key factor in this case might be the unequal dispersal tendency of the various genotypes. At all events, such a mechanism provides small mammals with a further possibility of adapting their populations to short-term changes within the population and/or its environment.

The constraints of being a small mammal

Small size also has its disadvantages. The most obvious is the high energy cost of homeothermy. For all animals there is an important relationship between body size and metabolic rate. However, this relationship is much more pronounced for homeotherms than for poikilotherms. A small mammal has proportionately a much greater surface area than a large one; this imposes a greater body demand in rates of heat transfer and evaporative water loss.

A greater energy loss must be balanced by an increased metabolic rate if body temperature is to be kept constant (Kleiber, 1961). As pointed out by Schmidt-Nielsen (1972b), oxygen and nutrients must be supplied to the cells of the smallest mammal at rates that are 100-fold those in the largest mammal. These differences between species of different sizes may even be greater in carnivores than in herbivores and in species with different habits. Iversen (1972) has recently found that the metabolic rate of weasels is 100 per cent higher than expected from the mammalian standard curve; however, in fossorial heteromyids the metabolic rate is lower than predicted from the well-known Kleiber formula.

Correlatively, most of the life processes are speeded up in small mammals, excluding the bats. Gestation and growth period, as well as life span are shortened, ageing processes accelerated, and under field conditions many species have an average life expectation of less than one year.

A high energy expenditure of the body requires a high daily rate of food intake. Some adult palearctic shrews consume much more than the equivalent of their own body weight each day, and many rodents and lagomorphs 30 to 70 per cent (Davis & Golley, 1963). Consequently, the trophic impact upon a given ecosystem of the same biomass of small or large mammals will be very different. Two hippopotami will sustain themselves with $\simeq$ 200 kg of fresh herbage, whereas the same biomass of *Microtus agrestis* will eat ten times as much plant material. In this perspective, increase in size within a given mammalian order can be viewed as an energy sparing mechanism which slows down the energy flow within the community.

The loss of body heat being relatively smaller in large warm-blooded vertebrates, the larger mammals have an advantage in cooler regions (the so-called Bergmann's Rule). However, in equatorial climates small mammals are less at a disadvantage. In hot–humid environments with ambient temperatures constantly in the upper part of the thermoneutral range, a reduction in metabolic rate is made possible, and has in fact been demonstrated by Hunkeler & Hunkeler (1970) and by Hildwein (1972). On the other hand, in strongly seasonal environments in both temperate and tropical zones, many small mammals can only maintain their populations by periodically entering a state of reduced metabolism (variously labelled hibernation, estivation or seasonal torpor), by eating concentrated food (seeds, nectar) during the favorable season, and by storing energy in their body fat deposits (and/or in their burrows)

6

when food is abundant. Some species display daily fluctuations in body temperature and basal metabolism in response to food shortage. Others, which avoid high daytime temperatures and consequent water loss by entering a burrow, have 'obligatory' periods of daily torpor. Most small mammals, however, cannot survive for more than a very small number of days in the absence of food.

Another handicap of small terrestrial mammals, overlooked until very recently, is the high energetic cost of their locomotion. The experiments conducted by Schmidt-Nielsen (1972*a*, *b*) and his coworkers have clearly shown that the cost of running a given distance is a great deal higher for small mammals than for large ones. The fact that a horse can move one gram of its body weight over one kilometre more cheaply than a mouse is another evolutionary advantage of a large body size.

Behaviorally speaking, a short life span can be a serious drawback for small mammal species. It prevents elaborate socialization of the young, limits the opportunities for learning from the experience of adults and does not facilitate the development of social traditions. While a complicated social structure may somewhat balance this disadvantage (Naumov, 1967), the long-term consequences of a short life span often confine small mammals to rather conservative ways of life and to rigidity in behavior, forcing them to rely mainly upon genetically-fixed skills. Indeed, it generally requires a long period for the usual selection processes to build up new behavioral adaptations. On the contrary, a long life expectancy allows for a long period of socialization among larger animals. It is now well established, at least for mammals, that not only social organization, but social life itself is almost entirely dependent upon the individual's early rearing environment. Furthermore, as clearly illustrated by Poirier (1972), it is through socialization that a group of large mammals passes its social traditions and ways of life to succeeding generations. In this way, socialization ensures that adaptive behavior will not have to be 'rediscovered' anew in each generation. It is probably not an accident therefore, that the elaborate social systems and flexible behavioral patterns are often found among elephants, dolphins, whales, wolves, African wild dogs, monkeys and apes, and less often among shrews, mice, voles or even weasels.

Summary

The small mammal strategy of exploitation of environmental resources, based upon a rapid adaptation to short-term changes in population and/

or environmental modifications, as well as to a variety of local conditions, implies a strong trophic impact upon natural resources. The role played by small mammal consumers in the flow of energy within natural and man-modified ecosystems is therefore much more important than that of the larger long-lived species, at least in temperate and subarctic communities. The much smaller populations of ungulates, carnivores and primates have to rely more on mobility, learning processes and social organization, and relative independence from local or seasonal conditions, to ensure their survival. For them morphological and physiological evolution progressively yields precedence to behavioral evolution.

1.2. *Productivity investigation in ecology*

K. PETRUSEWICZ

Productivity investigations began in hydrobiology in the thirties and forties (Winberg, 1934; Juday & Schomer, 1935; Hutchinson, 1938; Borutzky, 1939; Juday, 1940; Lindeman, 1942; Ivlev, 1945), but only much later was the idea of biological productivity popularized in ecology by Odum (1959). In the sixties, through the activity of IBP, productivity has been made the central object of field and laboratory investigations in terrestrial and aquatic environments throughout the world. In this way a new field of study, generally known as productivity investigation, was created.

What are productivity investigations? Some biologists and ecologists confuse productivity investigations with community or population studies; for instance, prey–predator relationships, regulation of numbers, or the numerical dynamics of the populations. In common parlance productivity studies are considered as studies of the energy content of organisms which tend to replace by energy units the numbers of individuals or quality of biomass which are the common units of population or community dynamics. Of course, there is nothing more misleading. Productivity investigation is a qualitatively different field of ecological research. Briefly, it can be defined as the study of the processes of organic matter production and its transformation in the biosphere. This study has led to the discovery and development of a number of concepts. In particular it introduced to ecology a kinetic approach and has enabled us to consider ecological units as open energetic systems. All this permitted ecologists to:

(1) investigate processes which run continuously in time and not just compare states at different moments in time;

(2) distinguish the concept of standing crop from the production process, and to introduce into ecology a number of parameters which are cumulative in time, and characterize or express such processes as production, consumption, energy flow, etc.;

(3) pay attention to and introduce into ecological practice not only the numbers and biomass of organisms but also their metabolic rate;

9

(4) determine the function of the biosphere units (ecosystems, trophic levels, populations) as energy systems.

It is difficult to say what is original and what is derivative in science. It is probably not important to decide whether the study in productivity investigations of organic matter production and its transformation has been original, producing a number of new concepts, or whether these ideas were originally developed elsewhere. In either case, productivity investigations have resulted in numerous new data and a deeper understanding of ecological processes.

Standing crop versus production and turnover

In any ecological research, and especially in productivity investigations, we must clearly distinguish between production and other processes which occur continuously in time (such as consumption, assimilation, etc.), and states occurring at a particular moment in time such as biomass and numbers.

In the majority of ecological studies and in all productivity research we start from the numbers and/or biomass of a population. This gives us the quantitative state at any moment per unit of space. This state, usually called the standing crop, can be expressed in terms of number of animals (numbers, N), weight units (biomass, B) or in energy units. Nevertheless, it is only a state at a definite point in time.

The changes of state with time are also studied, and are expressed as differences in numbers (ΔN) or biomass (ΔB) between two points in time.

The investigations of standing crop and its changes in time are extremely important. They were developed into a large and active field of ecological investigation, commonly called population dynamics investigation, which has contributed significantly to the development of ecological knowledge. As a result of them it has been possible to determine the shape and regularities of population growth, and the seasonal and annual changes in population dynamics. However, these studies still focus on standing crops and changes in standing crops; they do not study continuous processes in time. Even the most accurate knowledge of standing crops and their changes in time provide little or no information on the amount of organic matter (energy) which passes through the population. To give an example: it was found (Petrusewicz, 1970) that in 1966 the average standing crop of hares was 103 kg/100 ha in an area of Poland; in the same year this population produced 164 kg of biomass. The average standing crop gives no hint about the production of the hare population. Standing crops can even give misleading information. Elton's pyramid of numbers (expressed in biomass) is sometimes reversed, meaning that there are more predators than prey – which is, of course, impossible. Information on production of the populations can explain this apparently unrealistic pattern.

In order to base analysis and comparisons upon comparable units, we measure production either in weight units – fresh weight (in practice mostly after having killed the animal), dry weight, ash-free weight – or in terms of energy content of tissues of the body (see Chapter 8).

The concept of the production of a population, an individual, or a trophic level is the total amount of organic matter or energy produced by the unit through reproduction and body growth, and not lost by metabolic processes, whether or not this matter survives to the end of the period under question, or was eliminated through consumption, death or emigration.

This concept is complex and it can be considered in different ways. From the point of view of energy flow, production may be defined as a net balance of food transferred to the tissue of the population or individual during a definite time period, i.e., a net balance between assimilation and respiration.

$$P = A - R = C - FU - R, \qquad (1.1)$$

where C is consumption, FU are rejecta (feces and urine), R is respiration (cost of maintenance), and A is assimilation or energy flow (Petrusewicz, 1967*a*, *b*; Petrusewicz & Macfadyen, 1970; see also Fig. 1.1).

This scheme is simplified. It is set up from the point of view of ecological investigations, especially in the field. The concept of rejecta (FU) containing the feces (F) and urine (U) may be objected to by physiologists. From their point of view, urine should be included with assimilation and production, as it is first assimilated and then ejected. The ecologist may argue that urine remains within the organism for a very short period of time and there is no way nor much sense to separate it from feces when long-term ecological investigations are concerned. Certainly, the notion of 'short time period' is subjective, yet the concept of rejecta (FU) is convenient in ecological studies since it refers to that part of the consumption which is not utilized by the population (organism) but enters and leaves as matter containing energy. If more detailed investigations are required, e.g., in the productivity of an individual or cycling of matter, we can separate urine and introduce the concept

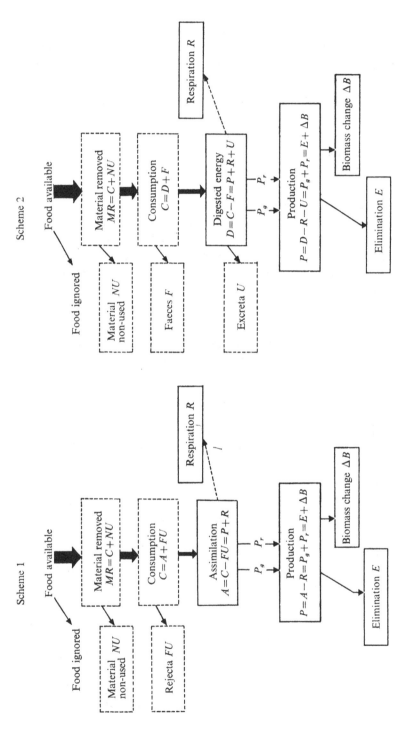

Fig. 1.1. Schemes of energy flow through a population (from Petrusewicz, 1967b).

of digested matter (energy) (see Fig. 1.1). A detailed discussion of these concepts can be found in Petrusewicz (1967a, b) and Petrusewicz & Macfadyen (1970).

From the point of view of the fate of the produced biomass (energy), production is the amount of biomass accumulated by a population during a given time (ΔB), plus matter (energy) eliminated or lost in other ways than by respiration:

$$P = \Delta B + E, \tag{1.2}$$

where E stands for elimination. E should always be added in (1.2), for in spite of the fact that elimination reduces population standing crop, the matter that is eliminated was in the population and should be included with production. In contrast, ΔB should be used with the appropriate sign, since it can have a positive or negative value (Fig. 1.2).

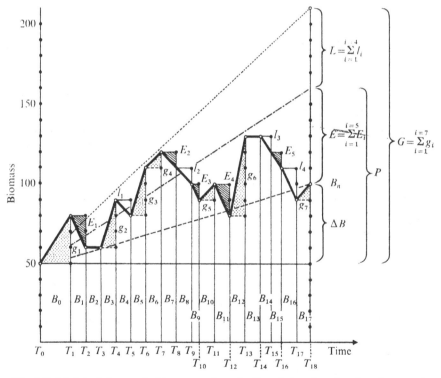

Fig. 1.2. Model of changes in biomass and production. Let us imagine that the curve is based on an ideal set of information so that all changes of biomass have been recorded and presented graphically (i.e., frequency of records was high enough so that in intervals between them there were only increases or decreases). P = production, E = total elimination (death losses, emigration), L = weight losses through time T_0–T_{18}; g_i = consecutive biomass increases; l_i = consecutive weight losses (from Petrusewicz, 1967a).

If we consider the ways in which biomass is produced, production may be divided into that due to reproduction (P_r) and to body growth (P_g):

$$P = P_r + P_g. \tag{1.3}$$

As we have pointed out above, the differences in standing crop ($\Delta B = B_2 - B_1$) between two moments in time tells us little or nothing about production during time period $\Delta T = T_2 - T_1$. The relationships expressed in (1.2) indicate that if there is no elimination during a given period, the difference between standing crops (ΔB) represents production (P). If elimination is balanced by production, the difference between standing crops is zero ($\Delta B = P - E \approx 0$), and then the difference in states tells us nothing about production. The lack of elimination is not imaginable in any population. Instead, it often happens that in well balanced populations elimination equals production over a long time period.

Odum (1959) has emphasized in ecology the concept of turnover (θ), which represents a relationship between standing crop and production:

$$\theta_{(T)} = P_{(T)}/\bar{B}_{(T)}. \tag{1.4}$$

This index has a considerable cognitive significance. It expresses the rate of biomass change during a definite time period (T). More detailed discussion of the turnover concept is presented in Chapter 7. Here we want only to note one aspect of this concept. When the relationship between plants and herbivore or prey and predator is studied, even the most accurate knowledge of biomass dynamics or the calculation of an average biomass does not provide us with a knowledge of the possibilities of the prey to serve as food for predators.

Considering the above concepts several further comments are necessary:

First, (1.2) indicates that there is an essential distinction between production and differences in standing crop ($\Delta B_{(T)}$) during any time period. Production contains the total biomass (energy) produced over a given time period, whether or not it survives to the end of that period. Not only a change in standing crop but also the total elimination is included in production. Elimination can involve the whole organism (dead, killed, emigrated) or other losses from organisms (excluding loss of weight), as for instance in moult.

Changes in the standing crop in a balanced population approximate to zero, or at least are extremely low as compared with elimination over a long time period such as a year or more. Petrusewicz (1970) found

14

that over five years the average of the changes in the standing crop of a hare population on January 1 was only about 0.02 per cent of elimination. In this case (1.2) may be modified as follows:

$$P \approx E. \tag{1.5}$$

Second, according to (1.1), biomass used for maintenance (R) is not included in the production. Even the organic matter incorporated into the tissues of organisms but later used in life processes is not included. Let us take as an example an animal such as a marmot hibernating in the winter or dry season. The weight of this animal increases from spring to autumn as it fattens and/or is growing. Since it hibernates and does not take any food in winter, energy input ($A = P + R$, (1.1)) is zero, but respiration occurs all the time the animal is alive, although at a very low rate. Therefore, the animal loses weight and production is negative. Consequently production will be larger for the spring–autumn period than for the spring–spring period. It should be remembered then that the weight losses during the periods when $A < R$ should be included in production or at least accounted for and made explicit. These relations are presented in Fig. 1.2 and 1.3. Periods when all or a majority of

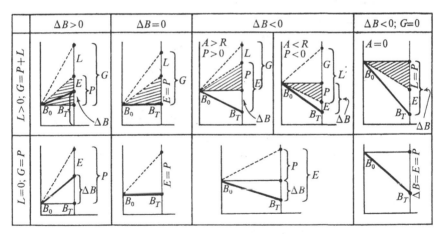

Fig. 1.3 Relations between the total biomass growth (G), production (P), elimination (E), and weight loss (L), when $B \lessgtr 0$, and when $R \lessgtr A$. $B =$ biomass change, $R =$ respiration, $A =$ assimilation (from Petrusewicz, 1967*a*).

individuals in the population lose weight are not rare in nature. For instance, they may occur during a cold or dry season, or during the mating season for males. These phenomena should be remembered in productivity studies.

Third, production may be due to reproduction and body growth ($P = P_r + P_g$). This relationship is considered in detail in Chapter 7. Here we only want to point out that production due to reproduction represents a considerable percentage of the total production, and in precocial small mammals it may reach 60 per cent of the total (Petrusewicz *et al.*, 1968). Accurate knowledge of P_r is, therefore, extremely important in productivity investigations.

Numbers versus number of discrete individuals

Investigation of the dynamics of populations is a very important research subject in ecology. These investigations have resulted in a large body of data and numerous theoretical concepts; they are also (especially in small mammal investigations) frequently a starting point for other types of ecological studies. But the dynamics of numbers, even if most accurately known, says little about the processes occurring between two moments in time.

To give an example: Petrusewicz *et al.* (1969), found that there were 152 individuals of the red bank vole on an island of 4 ha on June 16, and 304 individuals on July 31. Therefore the difference between standing crops was 152.

$$\Delta N = N_{T_2} - N_{T_1}. \tag{1.6}$$

But the number of individuals born (γ_r) from June 16 to July 31 was 306. Thus, the input of new individuals to the population during the six weeks in question was 306 and not 152. This may also be expressed in another way. In the population during the study period there were 458 (152 + 306) discrete individuals (γ in Fig. 1.4).

The total budget of individuals in the population can be written:

$$N_{T_2} = N_{T_1} + \gamma_{r(T)} - E_{N(\Delta T)} \tag{1.7}$$
$$304 = 152 + 306 - 154.$$

We are interested not only in number of animals added to the population (γ_r) and eliminated (E_N) but also in the total number of discrete individuals which were present in a population during the time period ΔT:

$$\gamma = N_{T_1} + \gamma_r. \tag{1.8}$$

The above reasoning contains two categories of concepts: on the one hand, two standing crops (N_{T_1} and N_{T_2}) at two time moments (T_1 and

T_2) and the differences between them ($\Delta N = N_{T_2} - N_{T_1}$), on the other hand the number of discrete individuals which were added (γ_r) or eliminated (E) or were present (γ) in the population throughout the time period (ΔT).

The number of newborn and animals eliminated from the population can be calculated for a time period, such as spring, summer or growing

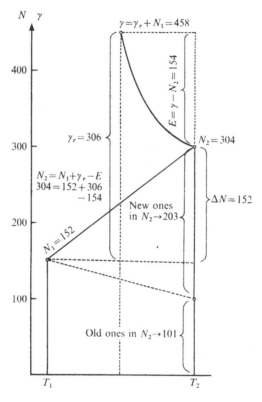

Fig. 1.4. Numbers (N) versus number of discrete individuals (γ). N_1 and N_2 = numbers at times T_1 and T_2; γ_r = number of newborn, γ = number of animals which were present and E = number of animals eliminated during time period T_2–T_1.

season, and then summed for a whole year. These sums, on the condition that ΔT is known, make sense and provide useful information on the process occurring in the population. On the other hand, summing up the standing crops makes no sense. During a year 6, 10 or 12 censuses can be taken. Of course, the sum of 6 and 12 censuses will be different, but this difference reflects only the method used. The relationship between the population dynamics and real changes in the number of discrete individuals is presented in Fig. 1.5.

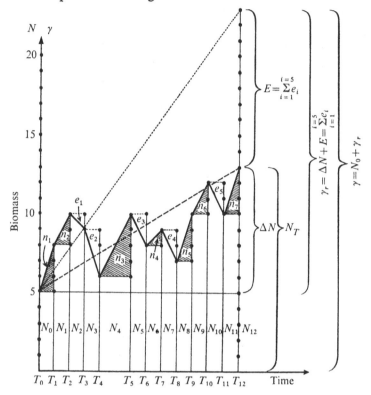

Fig. 1.5. Dynamics of population due to birth and eliminations. Let us imagine that the curve is based on an ideal set of information so that all changes of biomass have been recorded and presented graphically (the frequency of records was high enough that in the intervals between them there were only increases or decreases). N_i = standing crops (numbers), n_i = positive and e_i = negative changes in numbers during particular time intervals; N = change in numbers during the period T_1–T_{12}; E = elimination (sum of e_i) and γ_r = number of newborn (sum of n_i) during T_1–T_{12}, γ = number of discrete animals present in population during T_1–T_{12}.

In productivity investigations the concept of turnover of individuals (θ_N) is:

$$\theta_{N(T)} = \gamma_{(T)}/\bar{N}, \qquad (1.9)$$

where γ is the total number of discrete individuals present during time T and $\bar{N}$ is the average number within the same time period.

Turnover of individuals is similar to turnover of biomass but not identical. Turnover of biomass ($\theta = P/B$) refers to P, that is to the input of biomass to the population; turnover of numbers refers to total number of discrete individuals in a population during the time period in question, not only to an increase of numbers. By analogy, this means that $\theta = P/B$ would be $\theta_N = \gamma_r/\bar{N}$ and not the commonly used $\theta_N = \gamma/\bar{N}$.

Petrusewicz (1966*a*) has shown that:

$$\gamma_{(T)} = \bar{N} \cdot T/\bar{t}, \tag{1.10}$$

where $\bar{t}$, is the average length of residency of individuals in a population during the time of study (T) i.e., the average length of life of an individual but only within time period T.

It can be readily seen from (1.9) and (1.10) that:

$$\theta_{N(T)} = T/\bar{t}, \tag{1.11}$$

or, if we consider T as a unit (i.e., one year) and use it to measure $\bar{t}$, then:

$$\theta_N = 1/\bar{t}. \tag{1.12}$$

A more detailed discussion of the θ_N concept, its relation to the turnover of biomass (θ) and practical application to small mammal investigations, will be presented in Chapter 7.

Energy flow and metabolism

Assimilation is the energy inflow to the population, i.e., the energy which is incorporated in the tissue of individuals and not just passed through their intestinal tracts.

Formula 1.1 ($A = P + R$) tells us that the energy flow to the population can be accumulated as production (P) or used for maintenance, expressed as respiration (R). The energy accumulated in the population can be stored as biomass, augmenting the retention of highly organized, energy rich organic matter in the ecosystem, but after a period of time all production is eliminated and returns to the ecosystem as food for higher trophic levels or to decomposers. The major share of elimination usually goes directly to decomposers. The energy used for maintenance is lost from the ecosystem as heat.

Perception of these events has directed the attention of ecologists towards the importance of metabolism in ecological investigations, and has stimulated the development of bioenergetics studied both at the individual level (ecological physiology) and at the population level (physiological ecology). A detailed theoretical discussion of these problems, methods of study and the results obtained in this field will be presented in Chapter 8.

We have used the term productivity several times. This term is sometimes used as a synonym of production, and in some publications it denotes the production of young. Based on our discussions at the meeting of the IBP terrestrial productivity group at Jablonna in 1966 (see Petrusewicz, 1967*b*), we shall use the term productivity as a general concept to denote all processes involved in production; that is consumption, rejection, respiration, etc.

The introduction into ecology of the parameters of productivity, representing not only standing crops but also processes, and expressed quantitatively in comparable units, has resulted in a deeper insight into ecological investigation and initiated a new stage in the development of ecology concerned with ecological systems.

Types of productivity investigations

Productivity investigations may be concerned with both individuals and ecological units, such as a population, trophic level or ecosystem. Productivity of individuals is a subject of physiology, but because of its importance for ecology such studies are often conducted by ecologists (see also Chapter 8). These data, together with population parameters such as numbers, sex ratio, and age structure are used as basic empirical data for productivity investigations. From these data we can obtain productivity values through reasoning based on theoretical assumptions and models. In the field we can rarely determine productivity empirically and then only for some processes such as consumption (C) and egestion (F).

As was mentioned at the beginning of this subchapter, productivity studies are a very young branch of biology and as far as I know the first productivity approach to the investigation of small mammals was presented in 1960 (Golley, 1960). Recently, due to the activity of IBP, a large quantity of data has been gathered and a new branch of science has been developed. Although productivity studies are young, they are already very diversified. It is probably not an exaggeration to say that each ecological research center has slightly different objectives and its own preferred methods. In general, however, three main types of research can be distinguished.

First, we can be interested in the energy budget of a population. In such a case we approach the population as an energetically open system. We make efforts to obtain detailed data on the population inputs and outputs, the retention of organic matter (energy) in the population, its turnover rate, the rates of the population biomass output such as the amount of organic matter (energy) transferred to other parts of the ecosystem or dissipated in the form of heat (entropy), or proportioned among different components of the population energy budget (intra-population efficiency indices). This enables us to compare the energy requirements of populations of different species, geographically different populations of the same species, or the same population in different years or phenological periods.

Intrapopulation indices of efficiency are very useful and help characterize the type of population function. The following indices are most commonly used: assimilation/consumption efficiency (A/C), production/assimilation efficiency (P/A), commonly known in hydrobiological investigation as K_2 score (Winberg, 1971), and production/consumption, or 'ecological efficiency' (P/C), in hydrobiology K_1 score.

Ratios such as P/R, P_r/P, P_g/P and P_r/P_g are also useful. A deeper insight into the character of system function can be gained from the turnover rate ($P/\bar{B}$).

The second approach consists of analyses of population function from the point of view of the ecosystem, these being mostly investigations of particular food chains. They give a deeper, quantitative and comparable determination of the niche of a population or of a trophic link in ecologically similar populations and result in a detailed determination of the ecological role or impact of the population on the ecosystem. They also permit us to evaluate the economic significance of the population and provide the scientific basis for the control activities of man. Further, they represent a necessary link in the research of eco-system function and permit us to identify and quantitatively determine feedback and other mechanisms of ecosystem regulation (see also Chapter 10).

In investigations of this kind we may be interested in the effects of the population on producer or predator components of the food chain.

Knowledge of the impact of populations of small mammals on primary production is very important from the economic point of view, studies being chiefly concerned with determining ecological pressures on vegetation. The best measure of this is the amount or percentage of material removed (MR, see Fig. 1.1), but it is a very poorly known quantity and investigations of this kind should be intensified both from the practical and cognitive points of view. Since we do not usually know the amount of material removed but not utilized, we base our estimates upon consumption (C). In some cases the differences between MR and

C may be insignificant, e.g., for granivores or insectivores, but for grasseaters or barkeaters they may be large; for example, the beaver removes a whole tree from the forest and ingests only the bark from small, young branches. From the point of view of the function of the ecosystem the interpopulation efficiency indices may be very useful in this regard. For instance, the ratio $MR\lambda/P\lambda - 1$ represents the proportion of material (energy) retained after the transfer from the lower to higher trophic level.

From the practical point of view we may be interested in the amount of yield 'stolen', destroyed or changed. Here our competitors are either feeding on food directly used by man, such as grain crops, vegetables and fruits, or else they may be in competition with domestic animals, such as rabbits and voles versus sheep or cattle.

When we consider the relation of small mammals to their predators, we are interested in the amount of organic matter that a small mammal population can 'offer' to the higher trophic level. In that case the most useful parameters of productivity are production and elimination and also the indices of efficiency, such as predator consumption/prey production ratio ($C\lambda_2/P\lambda_1$). It is also of interest to know the number of pathways of matter (energy) from small mammals to the predators. This knowledge makes it possible to determine more precisely the ecological role of small mammal populations, and obtain a deeper insight into the regulatory role of small rodents in the ecosystem.

Finally, the third approach to productivity investigation consists of the analysis of energy flow and matter cycling in the ecosystem. In these types of investigations a new set of concepts can often be involved, e.g., respiration of the ecosystem (which is a sum of the respiration of all living components of ecosystem, both autotroph and heterotroph), as compared to the primary production which is newly created in the ecosystem. While we can sum respiration of different trophic levels, summing production or assimilation of different trophic levels is not a correct procedure. For example, adding primary production to production of herbivores sums the total value of primary production including its part which has accumulated in the animals.

These kinds of investigations permit us to gain an insight into different functional types of ecosystems. They are extremely important and frequently represent a final objective of ecology. A more detailed discussion of this approach can be found in Petrusewicz & Macfadyen (1970).

I would like to thank several of my colleagues who, during the IV Small Mammals Working Group Meeting, made remarks, additions and criticisms and would like to thank especially Drs Chew, Drożdż, and Linn for their help in improving the chapter.

2. Density estimations of small mammal populations

M. H. SMITH, R. H. GARDNER, J. B. GENTRY,

D. W. KAUFMAM & M. H. O'FARRELL

Accurate estimation of population density is perhaps the most important objective of IBP-related research. Not only is it worth studying on its own merit, but it is used in all calculations of standing crops and transfer coefficients between functional units of ecosystems. Inaccurate estimates of density decrease the precision of predictions about ecosystems. Many other variables in field ecology can be totally or partially studied in the laboratory and the results extrapolated to field conditions. Density must be measured *in situ* with many uncontrolled variables introducing biases into the estimates. Most problems associated with density measures in small mammal populations stem from the failure to recognize their spatial and temporal dynamic nature and the complexity of behavioral responses to census techniques.

Our objective in this paper is to discuss the available methods for obtaining density estimates in small mammal populations and to make recommendations concerning future work in this area. Literature citations have been selected to illustrate key points and not to supply an exhaustive list of all appropriate papers. The general outline of the paper follows the approximate sequence of decisions and actions taken during the conduct of the census procedure. As in actual field studies, some issues presented later directly modify earlier decisions discussed in the article. We realize that there is not just one way to take a census of small mammal populations, but there are certain key decisions that modify the subsequent usefulness of the data. Frequently the imposition of secondary objectives, such as elucidating community structure, confound the results beyond practical limits of interpretation.

Statement of objectives

The formulation of the objectives should precede the beginning of the work and not be an afterthought that results from an attempt to write a cohesive paper from a set of data collected in an almost random manner. For our purposes we will assume that we are discussing the

Small mammals

design of a study whose primary objective is as follows: to obtain a reliable estimate of the density of various species represented in some typical small mammal community. This estimate must be calculated with appropriate confidence intervals and some method of checking the model of calculation must be available to give credibility to the estimate. We realize there are several vague terms in our statement of the objective, but given the general nature of this paper these are unavoidable. As will become obvious, the design of specific studies depends not only upon the situation in question and the specific objective but also upon the secondary objectives. Rather than set specific secondary objectives at this time, we will present a series of conditional cases as we discuss typical decisions that determine the experimental design.

Precensus decisions concerning classification variables

Many variables affect the probability of capture. Some of these are discrete or classification variables (e.g., trap type) and decisions regarding them are usually made prior to the beginning of the census period. Others are uncontrolled and in many cases continuous variables (e.g., weather) and are normally recorded during the census period. Of course other variables fall into both of these categories (e.g., season) and the order of discussion in no way implies anything about their relative importance.

Census methods and trap types

Population density of small mammals has been estimated using a variety of traps and techniques. The three main categories of methods used for detection of small mammals are non-trapping, removal trapping, and non-removal trapping. The method used in a census should be based upon the species to be studied, the duration of time of the study (either a point in time estimate or sequential estimations through time) and the secondary objectives.

Several types of non-trapping techniques are available for small mammal detection. Density estimation using direct observation is possible for some small mammals using a removal plot technique (Hanson, 1968). Indirect observation using tracking boards (Lord *et al.*, 1970), dropping boards (Emlen *et al.*, 1957), sand transects (Bider, 1968) and runway counts (Lidicker, 1973) can be used to detect the relative but not the actual numbers of small mammals. Recently,

26

Marten (1972) described a method for small populations of rodents using live trapping and marking with subsequent estimation using tracking boards. It is doubtful that this technique can be used for a wide variety of species.

In certain situations, excavation of burrows is an effective technique for density estimation. This has been done in an open field (Smith, 1968a) and inside a temporary enclosure (Dieterlen, 1967c) or coupled with poison (Gromadzki & Trojan, 1971) to prevent escape while digging. Digging out burrows of *Peromyscus polionotus* gave more accurate estimates than trapping(Smith, 1968a); however, this technique is not practical for most small mammals. Mice have been captured by hand from flooded burrows (Andrzejewski & Gliwicz, 1969; Ryszkowski *et al.*, 1971), from under shocks of grain (Linduska, 1942), or from nest boxes (Brant, 1962) but these methods are probably not practical for most small mammals.

Removal census methods find their greatest use in density estimation for a point in time (Hayne, 1949; Grodziński, Pucek & Ryszkowski, 1966; M. H. Smith *et al.*, 1971). Three basic types of traps used in removal studies are the kill traps, live traps, and pitfall traps. The most efficient trap for most rodents is the kill trap (for a review see Wiener & Smith, 1972). Small snap traps are probably more efficient than live traps for some insectivores (Kale, 1972). For insectivores and certain rodents pitfall traps have a high capture efficiency (Edwards, 1952; Aulak, 1967; Brown, 1967; Pucek, 1969; Hamer *et al.*, 1971). Density estimates using removal procedures can be made using live traps when animals are needed for other reasons.

Non-removal census methods are more time consuming but provide sequential density estimates for the intact population and allow the analysis of other population parameters (Brant, 1962; Tanton, 1969; Tanaka, 1972). Live traps and pitfalls are commonly used with this technique. Pitfalls are more efficient in studies of certain small mammals (Andrzejewski & Wroclawek, 1963; Chelkowska, 1967; Pucek, 1969; Dub, 1971a). Differences in the efficiency of live traps of different sizes are debatable (Quast & Howard, 1953; Kisiel, 1972); obviously large traps can be insensitive to tripping by smaller species (Grant, 1970). In addition, other aspects of trap construction alter species-specific probabilities of capture (Brant, 1962; Hansson, 1967).

A problem common to both removal and non-removal census techniques is the determination of numbers of traps per station. Multiple traps per station avoid saturation of trapping stations and result in a higher capture rate for the population, but too many traps at a station lowers the probability of capture per trap (Andrzejewski *et al.*, 1966). Combinations of trap types at each station may also offset inclement weather bias due to trap sensitivity (G. C. Smith *et al.*, 1971; Wiener & Smith, 1972) or species-specific preference for different traps (Aulak, 1967). Hansson (1967) found that two or more single-capture live traps per station were more successful than a single multiple-capture live trap in reducing biases caused by behavioral interactions. These considerations are particularly important in high density situations (Tanaka, 1963*a*).

Bait

The presence or absence of bait as well as the type of bait explains part of the variability in trap response. Certain baits elicit a higher trap response than others, although the response may change among different age and sex classes or among the different habitats (Fowle & Edwards, 1954; Carley & Knowlton, 1968; Patric, 1970). Bait acceptability also varies seasonally (Fitch, 1954) and is a function of the availability of food in the habitat (Smith & Blessing, 1969). In the worst situation failure of traps to capture certain animals may actually be explained by ineffectiveness of the bait. Specific baits can be used to capture a variety of mammals including shrews and rodents (Beer, 1964), and these should be used when studying small mammal communities. Animals also respond to traps without bait, but in most instances their response is enhanced by its presence (Andrzejewski & Wroclawek, 1963; Carley & Knowlton, 1968; Balph, 1968; Buchalczyk & Pucek, 1968).

Bait is sometimes placed at each station for one to seven days before the census period begins (Grodziński, Pucek & Ryszkowski, 1966; Tanton, 1969; Zejda & Holisova, 1971). There is usually good acceptance of the prebait but this varies according to species and possibly reproductive class (Holisova, 1968; Myllymäki, 1969–70; Myllymäki, Paasikallio & Häkkinen, 1971). One reason for prebaiting is to increase the initial probability of capture by conditioning the animals to the trapping situation. Prebaiting accelerates the rate of capture, decreases its variance (Gentry, Golley & Smith, 1971) and increases the probability that the data will fit a regression model for population estimation (Hayne, 1949; Zippin, 1956; Tanaka & Kanamori, 1969). The technique has also been used to mark animals prior to the census to measure the edge effect (Ryszkowski, 1971*a*).

Under certain circumstances prebaiting can allow population estimates to be made in less time (Tanton, 1969). In addition, it is absolutely essential for some methods of population estimation (Smith *et al.*, 1969–70), but there are instances where it does not work (Grodziński, Pucek & Ryskowski, 1966; Gentry *et al.*, 1968). Other disadvantages include the probable concentration of animals on the edge of the sampling area resulting in an inflated density estimate (Zejda & Holisova, 1970; Gentry, Smith & Beyers, 1971*a*) and a disruption of the spatial organization of the populations. This would prevent the use of the data for studying social organization (Calhoun, 1964; Gentry, Smith & Beyers, 1971*b*). If it is possible to study social organization from trapping results, then it would seem best to avoid prebaiting unless it is needed to mark animals for calculation of the edge effect. In non-removal studies prebaiting adds additional food and may thus affect subsequent population dynamics (Gentry, 1968; Smith, 1971). Prebaiting is another complicating variable whose simple effects and interactions with the myriad of other variables that we must deal with are mostly unknown.

Trapping configurations

The spatial arrangement of the traps is usually determined by the shape and amount of habitat available and the intensity of effort required for an adequate sample. The basic configurations are transects, grids, or some modification thereof. Transect or line trapping is most often used in studies of short duration when the primary objective is the collection of specimens and secondarily the estimation of relative density (Calhoun & Arata, 1950–7). A census line alone cannot be used to determine absolute density unless habitat and species-specific calibration factors are known (Stickel, 1948; Hansson, 1967).

The simplest use of line trapping for density estimation can be accomplished by trapping on a census line followed by trapping on an assessment line (cross line method; Fig. 2.1*a*). Animals captured on the census and assessment lines can be marked and released or removed. The assessment line should be placed at an acute angle to the census line and trapping conducted for a short period (M. H. Smith *et al.*, 1971). The angle would depend upon the expected area of effect around the census line and the spacing of the trap stations on the assessment line, thus optimizing the number of stations in each area. In the ideal situation there would be a minimum of ten stations in the area of effect and a like number in the area unaffected by the trapping at either end of the assessment line. The primary advantage of the cross line is the relatively small amount of effort and habitat required for its operation. Unfortunately, the use of this simple design will be complicated by the variance in the spatial locations of animals responding to differences in

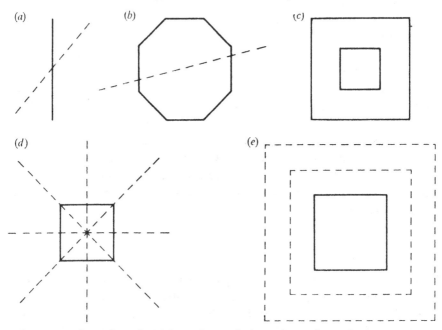

Fig. 2.1 Sample configurations of trapping methods used to estimate density. Configurations are not drawn to scales. Solid and dashed lines represent census and assessment lines respectively. (*a*) Cross lines; (*b*) octagon with one assessment line; (*c*) grid with inner square indicated; (*d*) grid with eight assessment lines; and (*e*) grid with outer concentric squares or assessment lines. Trapping is usually conducted on grid or census line first, followed by trapping on assessment lines.

habitat (Olszewski, 1968). Little is known about the dispersion of small mammals in a relatively homogeneous habitat (Opuszynski & Trojan, 1963; Wallin, 1971).

The problems associated with the cross line method can be overcome by using several sites and the pooled results to calculate density. The lines may be scattered or connected to form a continuous line, either straight or circular. Assessment lines must be arranged so they do not modify the capture rates on other lines near or in the area of effect (Kaufman *et al.*, 1971). In practice the circular census line is easier to locate in the habitat if it is in the shape of an octagon (Wheeler &

Calhoun, 1968; Fig. 2.1*b*). It also does not require a long linear stretch of homogeneous habitat.

Two basic octagon designs have been used, although the success of this method is not completely demonstrated (Gentry, Smith & Chelton, 1971; Kaufman *et al.*, 1971). In theory there should be a lower rate of capture along the assessment line as it goes through the area of effect around the census line. This phenomenon refers solely to unmarked animals if those captured on the census line were marked and released. In one study, assessment line trapping revealed an unexpected increase in the number of captures around the census line (Gentry, Smith & Chelton, 1971). Variation in the probability of capture along the assessment lines due to prior trapping on census lines violates one of the basic assumptions of the method (Hansson, 1974). More work is needed to check this and other assumptions, but it is possible that the assessment lines were not long enough to extend into the unaffected area.

The length of the assessment lines depends upon the distance animals are likely to move during the census period. Adamczyk & Ryszkowski (1968) found that *Clethrionomys glareolus* and *Apodemus flavicollis* move an average of 157 and 180 m, respectively and long range movements have also been reported by Faust *et al.* (1971) and Tanton (1969). It is now apparent that our initial view of the distribution of small mammals in a relatively static spatial pattern was erroneous.

A grid is a series of parallel lines with some common distance between them. It is frequently used in studies of relatively long duration where sequential trapping is used to give information concerning temporal (usually seasonal) fluctuations in population characteristics. Grids are usually rectangular and located in the middle of relatively homogeneous habitat. They should be large enough to easily include the average distance moved by the small mammals being studied. One of the primary biases may be a sampling area that is too small to adequately study density-related population processes.

Trapping can be used to estimate population number which can then be divided by the area of the grid to give an estimate of density. This approach usually results in an overestimate of density because it fails to recognize that the sampling area is larger than the grid (M. H. Smith *et al.*, 1971). Data gathered from grids can be used in two ways to calculate the area sampled. The first compares the number of captures at outer stations to those at inner stations (Hansson, 1969*a*; Pelikan, 1970; Smith *et al.*, 1969–70). The other approach involves use of prebait to mark animals as residents on the grid (Ryszkowski, 1971;

Fig. 2.1*c*). These inner grid techniques have certain disadvantages (Gentry, Smith & Beyers, 1971*b*; M. H. Smith *et al.*, 1971), and this has resulted in modifications of the grid configuration.

Two basic modifications exist: one uses assessment lines extending from the grid (M. H. Smith *et al.*, 1971; Fig. 2.1*d*), and the other uses outer concentric squares of traps removed from the edge of the grid (H. D. Smith *et al.*, 1972; Fig. 2.1*e*). Both techniques attempt to measure movement relative to the censusing activity on the grid by examining the spatial location of animals in the border zone. Published results utilizing these configurations are few and need replication in a variety of environments. We know of a series of such tests for the grid and assessment lines, but the results are as yet unpublished. Preliminary analyses indicate the usefulness of this technique in habitats ranging from deserts to lowland hardwood forest.

There is a need to combine a series of configurations and techniques into one study to compare the resulting density estimates. Ryszkowski, Gentry & Smith (1971) proposed a comparison of the method of using prebaiting to determine density from the trapping records on the grid (Ryszkowski, 1971*a*) with that utilizing assessment lines (M. H. Smith *et al.*, 1971). Four such tests have been conducted, and the preliminary analysis indicates good agreement in most cases (i.e., density estimates ± two standard errors overlap for the two methods).

Grid size and trap spacing are also important variables. As a grid is increased in size the relative bias of the edge effect should be reduced until it is negligible (Faust *et al.*, 1971), although this requires a grid that is larger than those commonly used. Small grids may give approximately the same results as larger ones (Myllymäki, Paasikallio & Häkkinen, 1971; Pelikan, 1971; Zejda & Holisova, 1971), but these studies have not calculated density with confidence intervals so it is difficult to evaluate the results. Grid size is also a function of trap spacing which is critical to the measurement of other population characteristics, such as apparent home range (Hayne, 1950).

Trap spacing is important in modifying density estimates (Tanaka, 1966) and should be based on the animals' ability to detect traps at some distance defined here as the 'recognition distance'. In a preliminary analysis of species in a lowland hardwood forest of South Carolina, the recognition distance was 2.9 m which is approximately 1/5 of the common trap spacing (15 m) used in this area (Gentry, Golley & Smith, 1971) or 1/30 of the home range diameter. Calhoun has suggested spacing traps at 1/6 the average diameter of the home range of the dominant species (Smith *et al.*, 1969–70). Wheeler & Calhoun (1967) introduced the concept of 'perception swath' in discussing trap spacing, which is approximately twice the recognition distance. Theoretically trap spacing should be based on the recognition distance but more work is needed to empirically define this value. A trap spacing of 15 m may be a good compromise for most of the common species studied throughout the world; this certainly appears to be the case in the southeastern United States (Smith *et al.*, 1969–70).

Habitat

Small mammal species have become adapted to certain habitat types where their needs for food, shelter, moisture, etc., are satisfied. The habitat may be vertically stratified allowing more species to co-exist (Getz & Ginsberg, 1968; Rosenzweig, 1973). Certain habitat features (e.g., uprooted trees) influence movement patterns and the intensity of contacts between individuals (Olszewski, 1968). Arboreal species represent a special problem and may not be adequately sampled unless traps are placed in trees (Gentry *et al.*, 1968). Fossorial and semi-fossorial species present a similar problem. These considerations suggest that density may be more appropriately calculated as a function of volume rather than area. Density should also be calculated on a habitat-specific basis where study areas include more than one habitat type (Blair, 1941). Therefore, the type of habitat, its degree of heterogeneity, and special features must be considered when undertaking a population census.

Season

Small mammals exhibit annual cycles in numerous population and individual variables (Smith, 1974). These cycles have a bearing on the approach to census studies both directly and with respect to the achievement of secondary objectives. The choice of season(s) and techniques of censusing, therefore, should be partly based upon whether the information needs to be collected once, periodically during the year, or over the entire annual cycle.

Two major variables, density (Briese & Smith, 1974) and trap response (Hansson, 1967; Pucek, 1969), undergo annual fluctuations. Animals that hibernate or estivate have a probability of capture of zero at certain times. Therefore, population densities in the same or different habitats should be estimated during the same time of the year. The

Small mammals

reproductive season and period of peak density are two important times for such studies (Ryszkowski, 1969*b*). In southern latitudes where breeding seasons sometimes extend throughout the year (Davenport, 1964), sampling must be more frequent. Changes in trap response throughout the year require census techniques that account for changes in the probability of capture, in order that annual changes of density are not primarily a reflection of fluctuations in trap response.

Models

Selection of the appropriate statistical model should be one of the early considerations in planning a population census. Concern must be given to assumptions in the selected model, inclusion of checks on the validity of these assumptions, sample sizes required by the model, and secondary objectives. Generally, the more assumptions that can be met the more efficient the population estimate will be (Cormack, 1968).

Although estimates of population size may not be seriously affected by an inadequate model (Junge, 1963) the variance estimate will be biased (Cormack, 1972) and may be imprecise even when the assumptions appear to be met (Roff, 1973*a*). Therefore, empirical estimates of the variance should be used (Paulik & Robson, 1969; Roff, 1973*a*).

Duration of the census

The duration of the study will be partly determined by the sample size necessary to provide a density estimate with a specified level of precision. In determining the duration of a study the factors to be considered are: the model for analysis, likelihood of violations of the assumptions in the model, desired level of precision for estimates, sample size (n), and intensity of sampling, as well as the secondary objectives.

Precision is the probability that $\hat{N}$ (estimated population size) does not differ from the true population size (N) by more than pN, where p is the level of accuracy (see (2.1); Robson & Regier, 1964),

$$1 - \alpha \leqslant P\left(-p < \frac{\hat{N} - N}{N} < p\right) \tag{2.1}$$

where α is the selected probability level and $1 - \alpha$ is the precision. The level of accuracy suggested for population estimates was given as 0.10.

Roff (1973*a*) studied sampling intensities (n/N) necessary to achieve a coefficient of variation of 0.05 (accuracy = 0.10) for Petersen and

Jolly–Seber estimators and found that as the population increases sampling intensity may be allowed to decrease, although sample size increases. Formulae given by Roff were used to estimate sampling intensity and sample size at the 10 per cent level of accuracy (Table 2.1).

Table 2.1. *Sample sizes* (n)† *necessary to obtain* 10 % *level of accuracy of the population size* (N)

Petersen method			Jolly (3 sample)		
N	n/N	n	N	n/N	n
5	0.987	5	5	0.923	5
25	0.944	24	25	0.832	21
50	0.899	45	50	0.767	38
75	0.861	65	75	0.719	54
100	0.828	83	100	0.682	68
125	0.799	100	125	0.649	81
150	0.775	116	150	0.621	93
175	0.752	132	175	0.596	104
200	0.732	146	200	0.573	115
225	0.714	161	225	0.553	124
250	0.697	174	250	0.533	133
275	0.681	187	275	0.516	142
300	0.667	200	300	0.499	150
325	0.653	212	325	0.485	158
350	0.641	224	350	0.471	165
375	0.629	236	375	0.458	172
400	0.618	247	400	0.445	178
425	0.608	258	425	0.433	184
450	0.597	269	450	0.422	190
475	0.589	280	475	0.412	196
500	0.579	290	500	0.402	201

† The tabulated values were obtained by selecting N, setting the coefficients of variation at 0.05 (Roff, 1973a) and solving for n/N and n. Petersen CV $\sqrt{[(1 - P)/(NP^2)]}$. Jolly–Seber CV $= \{[(1 - P)^2/N](1 + 1/P)\}^{\frac{1}{2}}$. $P = n/N$ or sampling intensity.

A practical suggestion for mark-recapture techniques is that the product of the number marked and the number examined for marks should equal four times the population size ($nM = 4N$; Robson & Regier, 1964). When populations are small ($N < 15$), Boguslavsky (1956) provides a means for terminating sampling by examination of sequential captures.

If regression formulae are used to calculate $\hat{N}$ for removal studies then accuracy increases with the duration of sampling (i.e., the number

35

of nights traps are set; M. H. Smith *et al.*, 1971). Accuracy is not greatly affected by the duration of study when multinomial formulae are used, but is primarily determined by the sampling intensity. Tables for sampling intensities necessary to obtain desired levels of accuracy are available for this method (Zippin, 1958).

The final solution to the determination of duration must result from optimization of all factors so that assumptions of the selected model are met, desired levels of precision obtained, and requirements of special sampling schemes satisfied.

Additional variables affecting probability of capture

Weather

Activity patterns of most vertebrates are affected by changes in weather (Briese & Smith, 1974; Gibbons & Bennett, 1974). For example, terrestrial animals show increased activity on warm, cloudy nights with even greater activity when these conditions are accompanied by rain (Bider, 1968), but responses of these animals to weather changes may be delayed for up to four days (Sidorowicz, 1960). In addition, weather affects abundance (Smith, Gentry & Pinder, 1974) and 'trappability' of small mammals (Sidorowicz, 1960; Mystkowska & Sidorowicz, 1961).

Usually the effects of weather on small mammal activity are discussed in general terms or in respect to the most obvious weather variable. For example, Gentry & Odum (1957) considered weather and winter activity of old-field rodents, while Blair (1951) and Caldwell & Connell (1968) studied the relation between activity of the old-field mouse (*Peromyscus polionotus*) and clear moonlit nights. However, attempts to statistically evaluate individual weather variables on small mammal activity are few (Gentry *et al.*, 1966; Marten, 1973). These studies show that weather effects are difficult to isolate due to the correlation and interactions between rainfall, humidity, temperature, moonlight, barometric pressure, season, habitat, species, etc., although a possible approach to this problem might involve principal component analysis (Hinds & Rickard, 1973).

Increases in probability of capture with changes in weather seem to result from greater distances moved rather than changes in trap response, Several lines of evidence indicate this is true. Pitfall methods, not involving trap response in the classic sense, reveal that movement is influenced by weather changes (Briese & Smith, 1974), and rainy weather produces an increase in captures on the edges of grids during removal trapping (Gentry, Golley & Smith, 1971). Further, reinvasion of an area is correlated with mean monthly air temperature (Smith, 1968*b*). It should follow that the area of removal or range of movement is also a function of weather. The effect of weather on density estimations can be minimized by selecting a model which does not require a constant probability of capture over the duration of the study (Kaufman *et al.*, 1971).

Behavioral response to traps

Differential trappability will result from variability in exploratory behavior and response to novel objects (Balph, 1968). Species that are neophilic are caught at a faster rate than those that are more neophobic (Faust *et al.*, 1971) and individuals within a species also show a wide range of response to traps (Young *et al.*, 1952; Petrusewicz & Andrzejewski, 1962; Sheppe, 1966*a*; Balph, 1968). When considering this behavioral response, several interrelated factors (age, sex, reproductive condition) must also be considered.

Age affects trappability because of the animal's prior experience and social rank. Older animals frequently rank higher in social stature and may be caught first and more often than young individuals. This has been shown for both removal (Carley & Knowlton, 1968; Gentry *et al.*, 1968) and non-removal trapping (Andrzejewski & Rajska, 1972; Summerlin & Wolfe, 1973). Variability in trap response within populations is suggestive of social effects (Petrusewicz & Andrzejewski, 1962; Gliwicz, 1970), while intracohort variability is also associated with sex and breeding conditions (Gliwicz, 1970). In removal studies, reproductive males are taken at a higher rate (Myllymäki, 1969–70; Myllymäki, Paasikallio, Pankakoski & Kanervo, 1971) and capture rates are higher during the breeding season (Chelkowska, 1967; Gliwicz, 1970).

Olfactory or auditory stimulation may alter the probability of capture (Calhoun, 1964). In multiple-capture traps, the presence of *Apodemus flavicollis* inhibited the trappability of other *Apodemus* and *Clethrionomys glareolus* (Kalinowska, 1971). Additionally the presence of *Clethrionomys* had no effect on *Apodemus* and enhanced the capture of other *Clethrionomys*. Subordinate *Sigmodon hispidus* tended to avoid traps when treated with conspecific scent, but dominant individuals did not (Summerlin & Wolfe, 1973).

Small mammals

Non-removal trapping presents a problem since there is usually a change in probability of capture between initial and subsequent captures. Animals are placed in two groups, trap-prone and trap-shy, on the basis of frequency of recapture (Tanton, 1965; Andrzejewski et al., 1971). This may be equivalent to the neophobic–neophilic division discussed earlier. Tanaka (1956) denotes three types of populations: Type I, initial capture rate < recapture rate; Type II, initial capture rate > recapture rate; Type III, initial capture rate = recapture rate. Most populations and species of rodents tend to be Type I while few are Type III (Tanaka, 1963b). Prebaiting fails to change the trap response of the population from Types I or II to Type III (Tanaka, 1970). In reality, trap response is probably a continuously distributed variable but the shape of the function for the inter- and intra-specific cases needs documentation.

Movement patterns

Trapping alters the behavior of small mammals, their responses being further complicated by age, sex and breeding condition. For example, males traverse larger areas than females, thereby increasing the probability of encountering a trap (Howard, 1960; Brown, 1969). Among invading animals, males constitute the highest proportion of captures, females and subadults being intermediate and juveniles the lowest (Myllymäki, 1969–70; Dub, 1971b; Myllymäki, Paasikallio & Häkkinen, 1971). Evaluation of differential movement requires a comparison of the numbers of animals in each sex–age category in the surrounding populations and under these conditions differential movement may be shown to be apparent rather than real (Briese & Smith, 1974).

If the area of the border zone is not taken into account, movements of animals on to the edge of sampling areas bias density estimates upwards. This edge effect can be the result of normal movements of animals living partially on the grid (Ryszkowski, 1969b) or animals may also move into vacant areas upon the removal of residents (Chelkowska & Ryszkowski, 1967). The difference between removal rates of individuals marked by prebaiting and those that are unmarked (Pucek & Olszewski, 1971; Gentry, Smith & Beyers, 1971a) is due to delayed movement onto the grid (M. H. Smith et al., 1971). Recognition of the spatial dynamic nature of the small mammal community is, therefore, essential to the proper design of census studies.

Interspecific removal rates

Up to this point we have been mainly concerned with interactions of intraspecific behavior with census techniques. Calhoun (1964) observed interspecific differences in small mammal capture rates and suggested that differences were related to social organization of the small mammal community. Differences in the removal rates (summarized in Table 2.2)

Table 2.2. *Relative removal rates of some small mammal species*

Species	Fast†	Intermediate†	Slow†
Rodents:			
Apodemus agrarius		12	12
Apodemus flavicollis	7, 9	12	12
Clethrionomys gapperi	8, 13		
Clethrionomys glareolus	9, 12	7	
Micromys minutus	12		
Microtus pennsylvanicus		11	
Microtus pinetorum			2, 8
Mus musculus			10
Napaeozapus insignis		13	
Ochrotomys nuttalli	13	5, 6	1, 2, 3, 4, 13
Peromyscus gossypinus	1, 2, 3, 4, 5, 6		
Peromyscus leucopus	8, 13	13	
Peromyscus maniculatus	13	13	
Rattus exulans		10	
Rattus rattus	10		
Reithrodontomys humulis			14
Sigmodon hispidus	14		
Insectivores:			
Blarina brevicauda	1	8, 11	2, 3, 4, 5, 6, 13
Sorex araneus			7, 9
Sorex cinereus			8
Sorex longirostris	1		1
Sorex minutus			9
Suncus murinus	10		

† 1. Gentry, Golley & Smith (1968); 2. M. H. Smith *et al.* (1971); 3. Kaufman *et al.* (1971); 4. Gentry, Smith & Chelton (1971); 5. Faust, Smith & Wray (1971); 6. Gentry, Golley & Smith (1971); 7. Pucek & Olszewski (1971); 8. Calhoun (1964); 9. Aulak (1967); 10. Barbehenn (1969); 11. Barbehenn (1974); 12. Grodziński, Pucek & Ryszkowski (1966); 13. Nabholz (1973); 14. Briese & Smith (1974).

can create serious problems in assessing densities. *Peromyscus gossypinus* is removed from grids at a faster rate than *Ochrotomys nuttalli* or *Blarina brevicauda* (Gentry, Golley & Smith, 1971). In one case *O. nuttalli* did not appear until day 6, and in another, *B. brevicauda* did not appear until day 12; these species and perhaps others would not have been detected by the Polish standard minimum grid (Table 2.2; Grodziński, Pucek & Ryszkowski, 1966). Polish investigators observe small inter-specific differences in rodent capture rates (Pucek & Olszewski, 1971). Trapping periods are short (3–5 days) and prebaiting is routinely used (Grodziński, Pucek & Ryszkowski, 1966); insectivores are removed by pitfall traps (Aulak, 1967).

Calhoun's (1964) hypothesis on species-specific removal rates is based on the spatial arrangement and movement patterns within the small mammal community. From this hypothesis the dominant species should be captured at a more rapid rate and have larger home ranges than the subordinate species. In contrast, Faust *et al.* (1971) found that *B. brevicauda*, a species removed at a slower rate, had the largest home range. One of Calhoun's assumptions is that the distribution of species in the small mammal community is not clumped, but this assumption is rarely met and different species have different removal rates even when spatially separated (Gentry, Golley & Smith, 1971; Kaufman *et al.*, 1971).

An alternative to Calhoun's hypothesis is that species differ in their reaction to traps (neophobia versus neophilia; Faust *et al.*, 1971). Also, initial trap response may be different from subsequent responses (Tanaka, 1963*b*; Balph, 1968; Faust *et al.*, 1971). Regardless of the reasons for different removal rates, they do create problems in estimating densities, but any model which does not require a constant probability of capture over the duration of the study will lessen the effect of differential removal rates. Continued application and development of these models should be encouraged.

Statistical methods for density estimation

Population estimation by non-removal techniques

Since 1889 when Petersen suggested a method of population estimation using the ratio of marked to unmarked captures (LeCren, 1965), the investigation of theoretical models of population estimation has been a dynamic field. We will consider some general models for mark–recapture and removal studies (see Cormack, 1968; Hanson, 1967 for reviews) followed by some special methods to solve some of the problems encountered in population estimation.

Assumptions commonly made by mark–recapture models are:
(1) The animals do not lose their marks.
(2) The captures are correctly recorded as marked or not marked.

(3) Marking does not affect the probability of survival.

(4) The population is either open or closed, therefore

    (*i*) no gain or loss of members during sampling, or

    (*ii*) there is recruitment and immigration but death and emigration affect marked and unmarked animals equally (in this case survival rates are assumed to be constant, i.e., not correlated with age), or

    (*iii*) knowledge is available from other sources which permits an allowance to be made for migration, birth and death prior to the analysis of the data.

(5) The population is randomly sampled so that either

    (*i*) every animal has the same probability of capture, or,

    (*ii*) if there exist strata within the population which have different probabilities of capture then the marked animals are proportionally distributed through these strata.

With the Petersen method (Lincoln index) one period of marking is followed by one period of recapture. This method requires that assumptions (1), (2), (3), (4.*i*) and (5.*i*) hold. The population estimation is made by

$$\hat{N} = \frac{Mn}{m} \qquad (2.2)$$

where:

M = number of marked animals released from the first sample;

n = number of animals in the second sample;

m = number of marked animals in the second sample.

Because $\hat{N}$ overestimates N by $1/m$ the equation $\hat{N} = m(n + 1)/(m + 1)$ with bias of e^{-m} has been suggested by Bailey (1952) for small populations.

If M_i (the total number of marked animals in the population before the *i*th sample) is known, then the Schnabel (1938) estimate

$$\hat{N} = \Sigma n_i \cdot M_i / \Sigma m_i \qquad (2.3)$$

where:

n_i = number in the *i*th sample;

m_i = number in the *i*th sample that are marked;

may be used with sequential sampling and is identical to a series of Petersen estimates (Cormack, 1968).

Bailey's (1951) triple-catch method gives a deterministic estimate of birth and death rates (Cormack, 1968) but has been shown to be less efficient than the Jolly–Seber estimates (Roff, 1973a).

The Jolly–Seber stochastic model sequentially estimates population size and requires assumptions (1), (2), (3), (4.*ii*) and (5.*i*) above. This more general method uses more information from the population, and allows for the removal of animals during the study and the estimation of survival, immigration and probability of captures. The population size at time *i* is found by

$$N_i = \frac{n_i}{m_i} \qquad \hat{M}_i = n_i + \frac{n_i Z_i S_i}{m_i \, r_i} \qquad (i = 2, \ldots, k - 1) \qquad (2.4)$$

where samples are taken on k occasions and the population size is estimated at time i and where:

$\hat{M}_i$ = estimated number of marked animals in the population at time i;

Z_i = number marked before the ith sample which are not caught in the ith samples but are caught subsequently;

S_i = number released from the ith sample;

r_i = number of the S_i that are caught subsequently.

Other formulae are given by Jolly (1965). Biases resulting from unequal catchability, birth and survival rates have been discussed by Cormack (1972).

A recent model suggested by Manly & Parr (1968) is a special case of the Jolly–Seber model (2.3). Its usefulness comes from a modification of assumption (4.*ii*) so that mortality rates may be age related (Manly, 1970). Population size at time *i* is found by

$$\hat{N}_i = \frac{n_i \cdot m'_i}{r'_i} \qquad (2.5)$$

where:

m'_i = total number of animals marked and known to be alive at time i (in class Z_i);

r'_i = the number of m_i caught at time t_i.

The mathematical models which these methods are based on must be shown to hold before an estimator may be justifiably used. Assumptions (1), (2), and (3) are required by all mark–recapture models and can be met by careful marking and recording procedures. Assumption (4) can be checked by experimental techniques and/or an alternate model chosen or formulated. However, the most frequently violated assumption is that of equal catchability (assumption (5). This is a major weakness of mark–recapture methods because changes in trappability cannot readily be detected by statistical methods (Roff, 1973*b*) or compensated for by statistical refinements (Hanson, 1967). A possible solution for heterogeneity of trapping may be provided by using different trapping techniques for marking and recapturing (Junge, 1963). Eberhardt (1969) and Barbehenn (1974) recommend that traps should be shifted at least once during the sampling period. Every technique, both biological and statistical, must be used to check and countercheck the validity of the assumptions (Cormack, 1968).

The estimation of the variance by empirical techniques has been recommended because biases in both the estimated population size and the variance result when the requirements of the model are not met (Cormack, 1972). Another reason is that the variance is so highly correlated with the number of recaptures that the confidence limits sometimes exclude the true value of N. Empirical confidence limits may be calculated by selecting before the study begins the sampling intensity for which the coefficient of variation equals 0.05 and then obtaining the required sample size (Table 2.1). The confidence limits around $\hat{N}$ are then equal to $\pm\ 0.10\ \hat{N}$ (Roff, 1973*a*).

New models and modifications of old ones for special applications continue to be suggested. Hanson (1967) and Rupp (1966) have shown that mark–recapture and removal methods are special cases of change-in-ratio estimators. Chapman & Junge (1956) have suggested a Petersen model which substitutes assumption (5.*ii*) for (5.*i*). Seber (1962) discussed modifications to the Jolly–Seber model when the number of recaptures are small. Tanton (1965*b*) proposed a population estimator for cases of unequal catchability when the frequency of recaptures follows the negative binomial distribution, but the method will work with other distributional functions. He extrapolates the function for the zero class and adds the animals in all classes to estimate N. If Tanton's method is valid, it represents a new approach, but we must devise ways to check both the accuracy and assumptions of his technique.

Population estimation by removal techniques

Population size may be estimated using two basic approaches (1) total enumeration or (2) sampling (Hanson, 1967). Enumeration of all individuals (N) in a population from a known area (A) yields a true density value (D). Using this approach on a small island or in an enclosed area, trapping is directed toward total removal without the complication of migration. If the population is distributed over a greater area, it becomes necessary to subsample to estimate density. Total enumeration within sample plots randomly placed in the habitat can be used to estimate mean number of animals per plot as well as the standard error (Hanson, 1967). This method requires numerous temporary enclosures and is not practical for small mammals. When enclosed sample plots and removal trapping are used for the estimation of density ($\hat{D}$), there are two values that need to be estimated, population size ($\hat{N}$) and the area from which that population was removed ($\hat{A}$).

The simplest estimation of N is the total number of animals removed from an unenclosed plot. This technique assumes that nearly all animals in the population are caught. An index calculated from the mean difference between other estimates of numbers and the total number of individuals caught can be used to test this assumption (Grodziński, Pucek & Ryszkowski, 1966). This technique may be useful for studies in which a relative density is sufficient.

The most commonly used technique for estimation of populations for removal studies is the linear regression technique relating daily captures (Y) to previous number captured (X) (Hayne, 1949; Zippin, 1956). This technique assumes that (1) births, deaths, immigration and emigration do not occur or, if they do, their effects cancel each other out, (2) probability of capture is constant for all individuals, (3) probability of capture remains constant throughout the census period and (4) trapping effort during successive removals remains the same (Hanson, 1967). In this situation

$$Y = a_1 + a_2 X \qquad (2.6)$$

where a_2 is the probability of capture, (P_c), $a_1 = P_c N$ and $\hat{N} = X$ when $Y = 0$. A curvilinear regression technique that could be used for removal trapping is described by Tanaka & Kanamori (1967).

Using these assumptions, the number of animals captured on any day is the product of the probability of capture and the number of animals on the plot, but this approach is not applicable in most studies since the probability of capture does not remain constant (Gentry, Golley & Smith, 1971; Janion *et al.*, 1968). Variation in P_c can be reduced by shortening the period of removal (Buchalczyk & Pucek, 1968) – in certain situations some species on the area are not censused during short periods of removal (Gentry *et al.*, 1968) – and prebaiting has also been used to increase probability of capture as well as its constancy (Grodziński, Pucek & Ryszkowski, 1966; Babinska & Bock, 1969; Gentry, Golley & Smith, 1971). Large daily variations in P_c decrease the correlation coefficient (r) and make resulting population estimates questionable (M. H. Smith *et al.*, 1971). Because of the non-independence of consecutive daily captures and the cumulative nature of X, r is biased upward (Acton, 1966). As a minimum, if the fit of the line is non-significant at $P \leqslant 0.05$, then this method of estimation can not be used. For short trapping periods the significance value of r is large (e.g., $r_{(0.05)} = 0.997$ for 3 days and $r_{(0.05)} = 0.950$ for 4 days) (Steel & Torrie, 1960) and confidence intervals are very large (M. H. Smith *et al.*, 1971). Many trapping studies conducted over a few days have probably yielded statistically nonsignificant population estimates.

Estimation of population size when probability of capture does not remain constant is possible using a maximum likelihood method and assuming a geometrical distribution (form specified) of captures across a time scale (Janion *et al.*, 1968). The mean of the conditional distribution of time of capture for the animals is used in conjunction with published tables to yield $\bar{P}_c$ and $\hat{N}$. Their formulae cannot be used without their tables, and thus are not included here. Procedures for estimation of confidence intervals were not provided by the authors. In addition no way is provided to check the assumption of the specified geometrical distribution; we feel that calculation of confidence intervals and application of the technique should depend at least partly on the fit of the data to the model. The technique can be used for data gathered by both removal and non-removal trapping.

Estimation of population size can also be made using the assessment line technique (Kaufman *et al.*, 1971; M. H. Smith *et al.*, 1971). This procedure does not require a constant P_c for all individuals through time. Movement in the border zone can occur without violating the assumptions of the method, since area is estimated from captures along assessment lines. The technique does assume that (1) for the habitat being studied there is relatively constant average rate of capture over distance and (2) P_c in and out of the area of effect is equal. It is necessary that density be high enough for the slope of the regression line in the unaffected area to be statistically different from zero. Since estimation of number by this technique is intimately associated with the estimation of density and area sampled, we will give the detailed method in a later section.

Area and density estimation

The simplest method of estimating density ($\hat{D}$) is by the formula

$$\hat{D} = \hat{N}/A_g \qquad (2.7)$$

where A_g is the area of the grid and associated confidence intervals for $\hat{D}$ are calculated with $\hat{N}$. In this section, we will assume $\hat{N}$ is estimated by some method previously described and will not specify or reference the method here. The arbitrary border zone method simply adds a border of a certain distance from all sides of the grid. This distance is usually some multiple of the trap spacing or is based on information about average distances moved by species (Stickel, 1954; Pelikan, 1967). In this case

$$\hat{D} = \hat{N}/(A_g + A_b) \qquad (2.8)$$

where A_b is the area of the border zone. Both of these approaches make unreasonable or unsubstantiated assumptions about the area of sampling.

Two techniques exist for calculating the border zone without pre-baiting (Hansson, 1969; Smith *et al.*, 1969–70). Both of these rely on the distribution of captures from outer to inner trap stations. The edge effect is demonstrated by the increased number of captures at outer traps compared to those at traps near the center of the grid. Smith and co-workers plotted probability of capture per station (P_{cs}) against the distance from the edge of the grid moving inward and P_{cs} was then calculated from the cumulative number of captures/cumulative number of traps for each concentric belt of stations from the outside inwards. Simultaneous solution of two linear functions provided an estimate of the width of the border zone (W_b). Thus

$$\hat{A} = (W_g)^2 + 4W_g \cdot W_b + \pi W_b^2 \qquad (2.9)$$

where $\hat{A} = A_g + A_b$ and W_g is the width of the grid. Density is calculated as in (2.7). W_b in this method is probably equal to r in Hansson's technique (1969). For consistency we will use W_b instead or r and W_g in place of a in the discussion of his approach. Hansson gives

$$\hat{N}_b/b^2 = \hat{N}_a/(W_g^2 + 4W_g \cdot W_b + \pi W_b^2) \qquad (2.10)$$

where $\hat{N}_b$ is the estimated number from an inner homogeneous square, b is the width of this inner square and $\hat{N}_a$ is the estimated population in the area sampled. W_b and b are calculated by inspection of the data and testing the distribution of captures with a chi-squared heterogeneity

46

test. Basically the inner square includes as much area as possible without the chi-squared test showing spatial heterogeneity. A series of tests should be conducted and the resulting statistics compared. One bias in the technique is that these tests are not independent and therefore the level of probability increases as a function of the number of tests ($P >$ 0.05 with more than one test). In both methods W_b is supposed to approximate to the radius of an average home range for the species captured. Density in Hansson's technique is calculated as in (2.7) with $\hat{N}_a = \hat{N}$. Neither Hansson nor Smith *et al.* give a method for calculating the confidence intervals for the sampling area, but it should be possible to use the variance in captures per station in the outer and inner areas to make these calculations.

M. H. Smith *et al.* (1971) discuss another method, the 'inner square' method, which has been extensively used (Aulak, 1967; Buchalczyk & Pucek, 1968; Adamczyk & Ryszkowski, 1968; Pelikan, 1970). In this method

$$\hat{D} = \hat{N}_b/A_b \tag{2.11}$$

where A_b is the area of an inner square of sampling stations and is equal to b^2 in (2.9). No attempt is made to estimate the width of the border zone; associated confidence intervals for $\hat{D}$ are calculated with $\hat{N}_b$. This method assumes that immigrants do not penetrate the inner square, but this assumption may not be justified (M. H. Smith *et al.*, 1971).

Bias due to the penetration of immigrants to the inner square may be reduced by shortening the census period or in using a prebait to mark the residents on the grid. In the latter method (Ryszkowski, 1971a), it is assumed

$$\hat{D}_b = \hat{D}_s \tag{2.12}$$

where $\hat{D}_b = \hat{D}$ in (2.17) and $\hat{D}_s$ is the density in the total sampling area. In addition

$$\hat{D}_b = \hat{N}_b/A_b \tag{2.13}$$

with

$$\hat{A}_b = 75^2 + 4(75W_b) + \pi W_b^2 \tag{2.14}$$

and $\hat{N}_b$ is calculated from those animals marked or captured in the inner square.

For the total area

$$\hat{D}_s = \hat{N}_a/\hat{A} \tag{2.15}$$

and

$$\hat{A} = 225^2 + 4(225W_b) + \pi W_b^2 \tag{2.16}$$

47

for a 16 × 16 grid with a 15 m interstation interval. From (2.11), (2.12) and (2.16)

$$\hat{N}_b/\hat{A}_b = \hat{N}_a/\hat{A} \tag{2.17}$$

and from (2.13), (2.15) and (2.14)

$$\hat{N}_b/(75^2 + 4(75W_b) + \pi W_b^2) = \hat{N}_a/(225^2 + 4(225W_b) + \pi W_b^2) \tag{2.18}$$

Since $\hat{N}_b$ and $\hat{N}_a$ can be solved by other techniques, (2.17) can be solved for the width of the border zone (W_b). Then, substituting W_b into (2.13) and solving for $\hat{A}_b$ or into (2.15) and solving for $\hat{A}$, (2.10) and (2.14) can be solved for density. A bias could be introduced in this method if the width of the border zone for the inner area is not equal to that for the total area or if the densities in the two areas are not equal. Because of the interaction between prebaited stations in the middle of the grid and the lower intensity of such interaction on the edge of the grid, this assumption may not be justified and needs to be checked (Gentry, Smith & Beyers, 1971*b*). In addition, no method for the calculation of confidence intervals is given for $\hat{A}$.

Another technique that uses outer belts of traps (dense line) around the grid to assess the border zone was described by H. D. Smith *et al.* (1972). In a sense the total grid is the inner square in this method. Application of the technique has been confined to a single mark–release study but by prebaiting on the dense line and later removal on the grid, the technique can be modified with appropriate mathematical formulae to work for either approach. Placement of the dense line is extremely important and currently has to be determined by trial and error. A method of calculation for the confidence intervals is given. This technique needs to be tested in a variety of habitats and some general rules for the placement of the dense line derived from the data.

The last method to be discussed is the assessment line technique first elaborated by Wheeler & Calhoun (1967). The lines pass through areas influenced by the removal trapping and areas not affected. Absence of animals captured during the census is detected by a decrease in the rate of capture over distance. The technique can be applied using many unique combinations of census lines, census grids and assessment lines (Gentry, Smith & Chelton, 1971; Kaufman *et al.*, 1971; M. H. Smith *et al.*, 1971). Modifications of this technique can be used in mark–release studies (Nabholz, 1973). Calculations for both $\hat{N}$ and $\hat{A}$ are incorporated into this method.

We will illustrate the method of calculation for a census grid and eight assessment lines (M. H. Smith *et al.*, 1971). Accumulating captures along a trap line results in a straight line (Gentry, Smith & Chelton, 1971; Kaufman *et al.*, 1971; M. H. Smith *et al.*, 1971) which can be calculated by the least squares method (Steel & Torrie, 1960). The slope (*B*) of this line represents the ambient rate of capture per unit

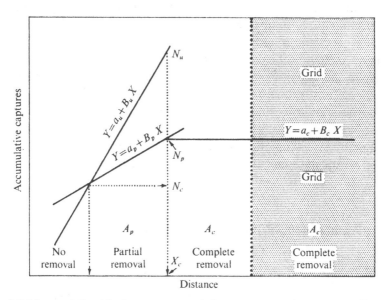

Fig. 2.2 Linear relationships between accumulative captures on the assessment lines as a function of distance from the outer ends of the assessment lines to the ends inside the grid. X_c denotes the edge of the area (A_c) from which all mammals had been removed. A_p is the area of partial removal. N_u equals the number of animals that would have been caught along the assessment lines up to X_c if A_p were equal to 0, N_p the accumulative number of animals actually caught at X_c, and N_c the number of animals that would have been caught if A_c had included the entire area of A_p (modified from figure in M. H. Smith *et al.*, 1971).

distance for a given number of trapping nights; *B* is dependent upon the density of small mammals and environmental factors responsible for varying P_o during the trapping period. If all of the mammals have been removed from the sampling area, *B* will abruptly change at the edge of the sampling area and will equal 0 inside this area (Fig. 2.2). The exact point of change in *B* can be calculated by the simultaneous solution of the two equations for the straight lines,

$$Y = a_c + B_cX \tag{2.19}$$

and

$$Y = a_u + B_uX. \tag{2.20}$$

Y is equal to the accumulative number of captures, X is distance in meters and a is the intercept of the line. The subscripts refer to the area of complete removal (c) and the unaffected area (u). Normally B_u would be statistically greater than B_c and can be tested by a one-tailed t test between the number of captures at each station in the two areas. Differences in the rate of capture should not be tested by a direct comparison of the slopes since the associated errors are biased because of the cumulative nature of Y and the non-random character of X. Another problem is associated with the decision concerning which points to include in the calculations of (2.19) and (2.20). Normally the change in the rate of capture in the two areas is obvious, but the inclusion of the ambiguous points into the calculations of either (2.19) or (2.20) results in little difference in the point of intersection of the lines. Similar considerations apply if there is an area of partial removal.

The point (X_c) along the assessment lines representing the edge of the area of complete removal ($\hat{A}_c$) can be calculated by

$$X_c = \frac{a_c - a_u}{B_u - B_c}. \qquad (2.21)$$

The width in meters (W_c) of $\hat{A}_c$ outside the grid is

$$W_c = L_a - X_c. \qquad (2.22)$$

L_a is the length of the assessment lines outside the grid or 120 m for the 16×16 grid used by M. H. Smith *et al.* (1971). If the captures were accumulated with the end of the assessment lines farthest from the grid as the starting point, then the size of the sampling area

$$\hat{A}_c = (2W_c + W_g)^2 \qquad (2.23)$$

with W_g equal to the length of the side of the grid. Dividing the actual number of mammals caught on the grid (N_g) by $\hat{A}_c$ yields density ($\hat{D}$).

In addition to $\hat{A}_c$, there is usually an area of partial removal ($\hat{A}_p$; Fig. 2.2). The width (W_p) of $\hat{A}_p$ can be calculated in a manner similar to that in which W_c was calculated. An equivalent area of complete removal ($\hat{A}_{cp}$) must be calculated for $\hat{A}_p$. The number of animals (N_u) that would have been captured on the assessment lines if there had not been an area of partial removal is equal to Y_u, and

$$Y_u = a_u + B_u X_c \qquad (2.24)$$

The number actually caught up to the edge of the area of complete removal (N_p) can be calculated by

$$Y_p = a_p + B_p X_c; \qquad (2.25)$$

$N_p = Y_p$ at this point. $\hat{N}_c$ is equal to the number of animals that would have been caught on the assessment lines if the area of complete removal had extended up to the edge of the area of partial removal.

$$\hat{A}_{cp} = \frac{\hat{N}_u - N_p}{\hat{N}_u N_c} \cdot \hat{A}_p \qquad (2.26)$$

In other words, $\hat{A}_{cp}$ equals the ratio of animals that should have been caught on the assessment lines if there had been no area of partial effect to those actually caught in this area times the size of the area of partial effect. This assumes $\bar{P}_c$ is equal for animals of the same species in the unaffected area and in $\hat{A}_p$. Interspecific individual or daily variation in $\bar{P}_c$ does not invalidate the method of calculation. Since trapping in these areas is accomplished along the assessment lines on the same nights, environmental effects on $\bar{P}_c$ in the two areas should be approximately the same.

From the above it follows that

$$\hat{D} = \frac{N_g}{A_c + A_{cp}} = \frac{N_g}{A_s} \qquad (2.27)$$

Confidence intervals must be calculated for the density ($\hat{D}$). N_g is assumed to be approximately equal to N, the real number of animals in $\hat{A}_c + \hat{A}_{cp}$. Thus N_g does not have a confidence interval. $\hat{D}$ will be in error by an amount equal to

$$D - \hat{D} = \frac{N - N_g}{\hat{A}_c + \hat{A}_{cp}}. \qquad (2.28)$$

This error is probably negligible, and there is no known way to estimate size of the error since P_c is not constant over time.

Both $\hat{A}_c$ and $\hat{A}_{cp}$ have confidence intervals. Normally the errors associated with the straight line could be used to place confidence intervals about a given point (Steel & Torrie, 1960). However, the trap stations are not picked or spaced at random and accumulative captures cannot decrease with distance, so the sampling errors associated with X and Y are biased.

If the data for each assessment line were treated separately, then there would be eight independent area estimates for $\hat{A}_c$ and eight for $\hat{A}_{cp}$. Standard errors could be calculated in the usual way (e.g., with

$$S_{\bar{A}} = \sqrt{\left(\frac{\Sigma\{\hat{A}_c{}^2 - [(\Sigma\hat{A}_c)^2/n]\}/n}{n-1}\right)} \qquad (2.29)$$

51

$n = 8$ for the 8 assessment lines). However, there are usually insufficient captures on single assessment lines to use this method of calculation.

Three of the techniques discussed appear to be promising for density estimation. These are the assessment line technique (Kaufman *et al.*, 1971; M. H. Smith *et al.*, 1971) dense line technique (H. D. Smith *et al.*, 1972) and prebaiting technique (Ryszkowski, 1971*a*). Each needs to be tested in a variety of habitats and experimental validation via comparative studies is needed (e.g., Ryszkowski *et al.*, 1966).

The discussion thus far has been limited to estimating density (D) at one location. A more important question is that of obtaining an average density ($\hat{D}$) for a habitat, region or some other geographical or ecological unit. For example, Spitz (1972) has shown that samples from small areas of a few hectares may poorly represent the dynamics of the population living in an area of several hundred hectares. Whatever the sampling scheme, it must be designed to account for the vast habitat heterogeneity encountered and be simple enough (e.g., cross line method) to enable sufficient replication. The number of samples required for a given level of confidence in the density estimate may be calculated from the range in density and the mean density across sampling plots (Table 2.3). Replication needed to attain a given level of confidence decreases

Table 2.3. *Number of samples* (N_s)† *needed to estimate the mean density* ($\bar{D}$) *with a standard error equal to* 10 % *of the mean given different ranges* (R_d) *in* $\hat{D}$ *across samples.* $N_s = \{(R_d/6)/(0{\cdot}10\hat{D})\}$

Density range (R_d)	Mean density $\bar{D}$									
	1	2	3	4	5	10	25	50	100	500
1	3	1	1	1	1	1	1	1	1	1
5	69	17	8	4	3	1	1	1	1	1
10	278	69	31	17	11	3	1	1	1	1
15	625	156	69	39	25	6	1	1	1	1
20	1 111	277	123	69	44	11	2	1	1	1
25	1 736	434	193	109	70	17	3	1	1	1
50	6 939	1 736	771	434	278	69	11	3	1	1
100	27 778	6 944	3 088	1 736	1 112	278	44	11	3	1
500	694 444	173 611	77 154	43 399	27 778	6 944	1111	278	69	3
1000	2 777 778	694 444	308 654	173 618	111 116	27 778	4445	1111	278	11

† Samples of one cannot be used to calculate a standard error and should be used only as an indication of the low number of replicates required.

with an increase in the mean density and a decrease in the range. The amount of effort required to estimate the mean density with appropriate confidence intervals is generally quite large because most situations are characterized by a relatively small mean and large range in density.

Recommendations

The large number of variables which affect density estimates require that appropriate experimental designs be used to validate existing techniques and develop new ones. Critical variables must be controlled, manipulated or held constant in both open and closed populations to provide a wide data base from which conclusions can be drawn. Such data for known populations are particularly needed.

Long-term trapping studies are needed for a variety of habitats and species. Extended trapping periods allow a more critical view of the factors which influence probability of capture. Data of this type are necessary to understand the role of social interactions, movement patterns, reproductive cycles and environmental factors influencing the probability of capture.

A single method for estimating densities may not be applicable to the variety of situations encountered in small mammal studies. Future studies should not be devoted to just a consideration of one or two standard census methods but should be designed to ensure accurate density estimations. When large amounts of data have been gathered in this manner, broad comparisons between species, habitats and geographical areas will be possible.

We thank Sarah Collie, Glennis Kaufman, Paul Ramsey, Jean Turner, Peggie Whitlock and Gloria Wiener for help in preparing the manuscript and Jo Gipps, Nil Stenseth and Lennart Hansson for offering constructive comments on the manuscript. This effort was supported by contracts AT(38-1)-310 and AT(38-1)-819 between the US Atomic Energy Commission and the University of Georgia.

3. Age criteria in small mammals

Z. PUCEK & V. P. W. LOWE

It is often necessary to establish the age of animals in ecological studies. For example, unless the ages of individuals or generations in a population can be ascertained with appropriate accuracy it is not possible to make a detailed analysis of population dynamics, calculate the growth rate of individuals, fit growth curves, estimate maximum and average longevity, age of sexual maturity or puberty or to determine age specific fertility, natality or mortality. It is, therefore, almost always essential for an ecologist engaged in measuring energy flow in mammal populations, in the practical management of populations used for economic purposes or in pest control, to be able to determine the ages of the animals he is studying. In order to choose the most appropriate methods and to evaluate the results properly, it is important to have a knowledge of the life history of the species and to be familiar with the environment, particularly with those factors likely to influence or vary the characters being used as a basis for age determination.

There is a large and constantly increasing number of publications on methods for determining age in mammals, and, of course, it will never be possible to give a complete, let alone an up-to-date bibliography of the subject. Here, we have endeavoured to include the most important papers describing the different methods, significant improvements or modifications and their application in defining or estimating age. The authors' task has been facilitated to some extent by several recently published reviews. A short review of methods for determining age in mammals was given by Klevezal & Kleinenberg (1967) along with a detailed description of their studies on determining age from the layered structure of bones and teeth. Friend (1967a, 1968) analyzed eye lens weight in Lagomorpha, Carnivora, Ungulata, Cetacea, and certain birds as a criterion of age. Laws (1962) reviewed methods for determining age in Pinnipedia, with particular emphasis on annual incremental lines in tooth cement, while Sergeant (1967) compared the first applications of this technique to land mammals. Literature (up to 1966) on methods for determining age in Cetacea and Pinnipedia is listed by Jonsgård (1969), and there is a comprehensive account by Habermehl (1961) of methods used for domestic, fur-producing and game animals. There is also a wide-ranging chapter on determination of age by Taber (1969).

Small mammals

Methods for determining age by body measurements, the genital system, the state of integumentary features and by skeletal development in selected orders of land animals have been described in general outline by Rybář (1970). Recently, a very comprehensive and detailed review has been published by P. A. Morris (1972) on methods of age determination in all groups of mammals but, although the author refers to more than 150 different publications, no assessment is made of their usefulness. The author has also limited himself chiefly, though not entirely, to English-language papers and certain journals, and has consequently omitted reports, which, in our opinion, are important.

While no purpose may be served by another review, a new compilation and realistic assessment is required for the present synthesis volume. We have not aimed at giving exact descriptions of the techniques for estimating age in the different species, but at examining the methods currently applied and assessing their suitability for use with small mammals. Hopefully, this chapter will enable the reader to choose the appropriate method for age determination of various small mammal species.

Age determination methods based upon the growth process

Increase in body size

Generally speaking, body size in animals is directly correlated with age. Also, body weight or a particular set of body measurements can be used as indices of age. Unfortunately, these usually enable one to make only crude age classifications of individuals in a population, dividing them into relatively few, broad age classes. Even then, there is often overlap between juveniles and adults and such criteria cannot be used when detailed studies of age structure or age specific properties of a population are required. Several studies of differential growth in rodents have related the different generations to the seasons of their birth (Shvarts *et al.*, 1964; Martinet, 1967; Zejda, 1971; Adamczewska-Andrzejewska, 1971, 1973*b*). Similarly, seasonal changes in the growth rates of shrews have been reviewed by Pucek (1970). These studies have shown that the seasonal effect causes individual animals of different generations to reach the same body weight at different ages, and for this reason body weight cannot usually be used as an index of age.

Skeletal growth and development

Skull growth and ossification of sutures. Although the skull in most small mammals continues to grow throughout life, its growth tends to be

56

seasonal. The most reliable variables appear to be measurements of skull length (condylobasal length), mandibular length or length of the maxillary tooth row.

In areas with base-rich soils, the degree of ossification of the skull features may be used as a general expression of age in large and medium sized mammals, but where animals are restricted to acid heaths, the sutures of the skull in many species of mammal never ossify completely. In small rodents, the development of crests and ridges has been found to be more useful; with increasing age and muscular development both tending to become more pronounced (Wasilewski, 1952, 1956a, b; Rossolimo, 1958; Gebczyńska, 1964, 1967). However, these features may be no better guides to age than is body weight unless the results of such studies have been related to specimens of known age.

Fusion of epiphyses. For determining age of medium to large size mammals, fusion of the epiphyses, especially of the long and tail bones, has been found to be useful, particularly during their early development as juveniles. X-rays give more reliable results than observations made on skinned and cleaned bones or when assessments are made by palpation.

This technique has not been used very often for small mammals (Petrides, 1951; Kirkpatrick & Barnett, 1957). Carson (1961) found it useful in determining the age of squirrels (*Sciurus carolinensis* and *Sciurus niger*) and Morris (1971) used it on hedgehogs (*Erinaceus europaeus*). Rybář (1969, 1971) also used this method for bats (*Myotis myotis, Rhinolophus hipposideros*), and obtained reasonable accuracy up to about fifteen weeks old in the former and six weeks of age in the latter. After these ages, the skeletons became completely ossified.

One of the most sophisticated studies using the X-ray process was undertaken by Tarasov (1966), to determine the age of *Microtus gregalis* and *Lagurus lagurus*. He was able to identify seven age classes up to one year. Of these he thought the first three were the most reliable. Hagen (1955, 1956) by visual inspection of the ossification of the epiphyses of the tail vertebrae in the common vole (*Microtus arvalis*) claimed an accuracy of aging of between one and two months. However, when Hansson tried this method, he found it unsatisfactory (Askaner & Hansson, 1967).

Not enough is known about the variation in skeletal development and ossification of the epiphyses, especially in the small mammals, but it seems likely that these processes also may be affected by season, and, therefore, will be inconsistent between generations.

The baculum and sexual maturation. The baculum is present only in certain mammals (Carnivora, Pinnipedia, Chiroptera and some rodents and primates). It is a cartilaginous element which supports the male genital organ. It continues to increase in size until puberty is reached, these size changes indicating physiological rather than chronological age, and in species where puberty is reached at the same time by every individual within a cohort or generation, the baculum can be useful in separating the juveniles from the adults. It is used in this way most frequently in studies involving the medium sized carnivores (Mustelidae, Canidae), beavers and Phocidae. There are very few references to its use for determining age of small mammals. Differences in size and shape of bacula distinguish juveniles from adults in species of *Perognathus* and *Dipodomys* (Burt, 1936), while Elder & Shanks (1962) and Chipman (1965) found similar results with *Ondatra zibethicus* and *Sigmodon hispidus*, although the classes were not altogether free from overlap. Hamilton (1946) and Anderson (1960), however, found considerable variation between individuals of the same age in both *Clethrionomys gapperi* and *Microtus pennsylvanicus*. Degn (1973) found that the weight of the baculum could be used to separate juveniles from adults of *Sciurus vulgaris* during the summer months and after August–September it could still be useful if combined with the weight of the eye lens.

On the other hand, in *Microtus montanus, Clethrionomys glareolus* and *Myopus schisticolor*, Arata *et al.* (1965) and Artimo (1964, 1969) found that the size and degree of ossification of the baculum separated the mature from immature individuals without difficulty, regardless of season or age. However, they found no correlation between size or development of the baculum after puberty.

Many authors have shown that puberty is not age specific (Adamczewska, 1961; Shvarts *et al.*, 1964). Also, it is well known that whether or not an individual attains puberty by a certain age often depends on other population processes (Pucek, 1960; Stein, 1961; Bujalska, 1970, 1973). For instance, in many species of rodents the breeding season may be extended even into the winter months (Zejda, 1962; Newson, 1963; Kryl'cov, 1964; Kubik, 1965; Haitlinger, 1965; Smyth, 1966). Under these circumstances, there will be almost no correlation between physiological and chronological age. Clearly the size and degree of ossification of bacula is of only limited value in estimating the ages of individuals.

Incremental layers in bones. Descriptions of layered structures within and around the hard elements of the skeleton of mammals date from the middle of the nineteenth century when Owen (1840–5) revealed the existence of such structures in the dentine and cementum surrounding the roots of teeth in marine mammals. Kler (1927) demonstrated the presence of appositional layers in the external compact bone tissue, and suggested that they were homologous with similar structures to be found in the bones of poikilothermal vertebrates and could be used to determine the age of individuals. This suggestion was developed by Brjuzgin (1939) for reptiles and by Chapskii (1952*a*), who demonstrated the possibility of age determination in *Phoca groenlandica* on the basis of the layers within the mandibles. Other authors have demonstrated layers in the periosteum zone of bones in various species of marine mammals (Laws, 1960; Nishiwaki *et al.*, 1961; Kleinenberg & Klevezal, 1962; Tikhomirov & Klevezal, 1964) and land mammals such as *Citellus pygmaeus* (Mejer, 1957). Mainly due to Klevezal and coworkers, the layered structure of the periosteum in various bones, including the long bones, has now been demonstrated in more than one hundred species of mammals belonging to nine orders (for a detailed review, see Klevezal & Kleinenberg, 1967; P. A. Morris, 1970; Klevezal, 1970, 1972, 1973; Klevezal & Mitchell, 1971; Ivanter, 1973).

The basic character of a bone depends on a continual rebuilding of its parts during the individual's life, and this is partly achieved by reinforcement with dense appositional material which is deposited around the bone from the periosteal sheath. The rate of deposit of this type of bone differs from season to season and according to the physiological condition of the animal. Each winter, with a reduction, followed by a cessation of growth, a narrow layer of compact bony tissue is deposited. In the summer, with an increased rate of growth, a broader layer of less dense bone is laid down. This succession of layers of bony tissue allows calculation of the number of winters an individual has survived, and thus its age in years.

The technique consists of preparing microsections or, more usually, histological sections from bones which have previously been decalcified. The sections are then stained by standard methods. As the periosteal zone varies in depth and clarity between different parts of a bone, it is usually necessary to obtain several samples of material before attempting to count the layers. In rodents, parts of the mandible just behind the alveolus of the incisor have proved better for this purpose than sections taken elsewhere. Recently, Klevezal (1973) has suggested that the phalanges of the digits might be used, which would allow one to age animals from populations being marked by toe-clipping alive.

Small mammals

It is not always easy to estimate the age from the number of layers present because additional layers sometimes are formed. The number of layers, however, generally agrees with those found in the tooth cement and dentine. Sometimes, in species of mammals having a life-expectancy of several years, the layers, laid down in the bones in the early years of life, are reabsorbed later, and cannot be used for age determination purposes. Incremental lines are best defined in populations living in areas with a contrasting seasonal or continental climate (Klevezal, 1973). In most small mammals, which rarely survive more than one winter, this method is obviously of only limited value.

Tooth succession and development

The sequence of loss and replacement of deciduous teeth by permanent teeth has been used to determine the age of individuals of many species of mammals, but in most small mammals the permanent teeth are acquired either before birth or while still in the nest. However, in some species such as *Apodemus flavicollis*, the very young may be separable from older individuals in this way (Adamczewska-Andrzejewska, 1967).

Microtines have open-rooted molars which continue to grow as the occlusal surfaces are worn down. The bank voles (*Clethrionomys*) and muskrats (*Ondatra*) are exceptional in that their teeth are open-rooted to begin with, but later partially close leaving two quasi-roots, which then continue to grow throughout life as in the other species. The length of these roots has been used as an index of age in these species, particularly the roots of the first mandibular molar (Prychodko, 1951; Wasilewski, 1952; Zejda, 1961; Pucek & Zejda, 1968), but sometimes those of the second maxillary molar (Koshkina, 1955; Tupikova *et al.*, 1968).

Zimmerman (1937), Koshkina (1955) and Tupikova *et al.* (1968), expressed the length of the roots as a fraction of the length of the tooth, $\frac{1}{4}$, $\frac{1}{3}$, $\frac{1}{2}$, etc. Age classes based on this method were claimed to be accurate to within 2 to 3 months.

Vernier calipers or a measuring microscope have been used for measuring molar roots exposed with a fine dental drill (Sealander, 1972) or the whole teeth may be extracted after boiling (Pucek & Zejda, 1968). X-ray technique was used to make measurements quicker and less time-consuming (Semizorova & Popper, 1971).

The age at which the roots of the molars are formed varies with the time of year when the animal was born and the species. Roots are developed at about 3 months in *Clethrionomys glareolus* and *C. rutilus*

60

(Wasilewski, 1952; Koshkina, 1955; Shaw *et al.*, 1959; Zejda, 1961; Mazák, 1963; Tupikova *et al.*, 1968, 1970) at about 7 to 8 months in *C. rufocanus* (Koshkina, 1955) ($6\frac{1}{2} \pm 3$ months, Vitala, 1971) and $2\frac{1}{2}$ to 3 months in *Ondatra zibethicus* (Cygankov, 1955; Trnková, 1966). The age at which roots are developed, however, varies with the season of birth. In *C. glareolus*, Zejda (1971) observed that roots were developed by 2 months of age in individuals born in the spring (March to May) but not until 3 months of age in individuals born in the autumn (September and later). Lowe (1971) observed a still wider range in root development time between the first and last born cohorts, ranging from 8 weeks in April/May to 18 weeks in late October. Differences in the growth rate of the roots related to season also have been observed by Zejda (1961, 1971) and Haitlinger (1965) and measured by Lowe (1971), who found that growth rates varied within an overall range of 0.05 to 0.55 mm/month, depending on the cohort and the season of the year.

Incremental layers in tooth cement and dentine

The number of incremental layers in tooth cement and dentine has been found to be a useful criterion of age in many species of mammals, but only when the mammal is long-lived, such as whales, seals, ungulates and carnivores. There are relatively few publications on this method of age determination for small mammals. Klevezal & Kleinenberg (1967) have reported on *Myotis myotis*, *Nyctalus noctula*, *Spermophilopsis leptodactulus*, *Marmota baibacina*, *Apodemus agrarius*, *Rattus norvegicus*, *Meriones tamariscinus*, *Cricetus cricetus* and *Ondatra zibethicus*. Cementum layers were also found in *Spermophilus beecheyi* by Adams & Watkins (1967), in *Sorex araneus* by Kleinenberg & Klevezal (1966), in the Uinta ground squirrel (*Spermophilus armatus*) by Montgomery *et al.* (1971), and in the vampire bat (*Desmodus rotundus*) by Linhart (1973).

It is clear, however, that this technique, which essentially measures only the number of winters the individual has survived, is of little use for small mammals who rarely survive more than one year.

Weight of the eye lens

The basic concept of using the eye lens as an indicator of age is that the lens growth is continuous throughout an animal's life (Smith, 1883). Donaldson & King (1937) and Leopold & Calkins (1951) showed that the weight of the eye lens in rats (*Rattus norvegicus*) increased throughout life due to material of ectodermal origin being deposited continuously, thus increasing the size and mass of the lens with age. Lord (1959) was the first to use the dry lens weight as an indicator of age in cottontail rabbits (*Sylvilagus floridanus*), the method being quickly adopted by many workers and applied to a variety of species, mainly medium to large mammals, e.g., lagomorphs, carnivores and ungulates (Friend, 1967*a*, *b*, 1968; P. A. Morris, 1972). The method was also used with some success on small mammals, chiefly rodents.

The technique is relatively simple. It consists of fixing the eyeballs (or lenses in the larger animals) in formalin, excising the lenses, cleaning them thoroughly and drying them in an oven (or by chemical means – Berry & Truslove, 1968), and then weighing them singly or in pairs. Though simple, the method is far from easy if the results are to be always comparable, and various authors (Friend, 1968; Caboń-Raczyńska & Raczyński, 1972; P. A. Morris, 1972) have emphasized the need for strict standardization.

One of the misunderstandings that has arisen is over the concentration required for the fixative. Some authors using Romejs (1953) basic histological techniques, have interpreted the directions for preparing a 10 per cent formalin solution for fixing the eye lenses, to mean 1:3 parts commercial 40 per cent formaldehyde in water. Others have presumed this 10 per cent dilution to be a 1:9 parts formaldehyde in water, which results in a 4 per cent solution, a concentration commonly used for fixing tissues. It is, however, well known that the strength and kind of the fixative used may have a considerable effect on the weight of an organ or tissue (Pucek, 1967). This was clearly demonstrated by Broekhuizen (1973), who used both 1:3 and 1:9 dilutions of formalin to dry eye lenses from the brown hare (*Lepus europaeus*) with very different results. It should also be noted that the actual weighing procedure should not vary; eye lenses, being highly hygroscopic, rapidly increase in weight after removal from a drying oven if left in a damp room. The consequent error may exceed 1 per cent, which, as P. A. Morris (1972) suggests, is too high, particularly where comparable data are wanted from several populations of small mammals.

It is also inadvisable to use material from animals which have been frozen. Östbye & Semb-Johansson (1970) showed considerable variation in the weight of eye lenses within samples of Norway lemmings (*Lemmus lemmus*) which had been frozen for a long time.

With increasing age, errors from these sources are likely to become more serious since at first there is a marked increase in the weight of the eye lens but later the increase is slower and generally independent of the current body weight (Friend, 1967a; Östbye & Semb-Johansson, 1970; Karpukhin & Karpukhina, 1971). Confidence intervals increase as the older age classes are reached, and therefore this method becomes less accurate with the older generations. Adamczewska-Andrzejewska (1973a) has suggested that age estimates may vary by 2 weeks in common voles (*Microtus arvalis*) 20 weeks of age, 3 weeks in individuals of 21 to 30 weeks of age and by as much as 14 weeks in animals of 58 weeks of age.

Habitat and genetic factors may also affect the rate of growth of the eye lens. The influence of diet, laboratory conditions and differences between natural habitats on the weight of the eye lens is given by Friend (1967b, 1968) and P. A. Morris (1972). In certain small rodents Shvarts *et al.* (1964), Dobrinskij & Mihalev (1966) and others found that the eye lens weight increased more rapidly in individuals born in spring than those born in autumn. Adamczewska-Andrzejewska (1973a) noticed that *M. arvalis* of the spring cohort had a rate of lens growth nearly double that of the autumn-born animals, though the differences were not statistically significant. In the laboratory, Östbye & Semb-Johansson (1970) found great variation in eye lens weight between lemmings of the same age, reared under the same conditions. At ages between 140 and 160 days, the rate of lens growth was found to have slowed considerably, but it then increased again in the older animals, the changes being apparently correlated with the inhibited growth in body weight.

Chambers (1962) suggested that the weight of the eye lens might be affected by genetic heterogeneity in a population and Berry & Truslove (1968) have supplied evidence of this effect in a laboratory population of house mice (*Mus musculus*) of different strains. They concluded that this method of age determination could not be used because of differences in eye lens weight between genetic strains of mice in the population.

Despite these objections, the data obtained by the majority of authors indicate that the weight of the eye lens is the most useful indicator of age, particularly, if used with animals having a restricted breeding season so the current year's young can be clearly distinguished from the older animals on the basis of a frequency distribution (Rieck, 1962; Caboń-Raczyńska & Raczyński, 1972; Degn, 1973). Moreover, during the first period of growth cohorts can also be distinguished by this method (Beale, 1962).

Although there is often greater variation in age specific weight of the eye lenses in small mammals than in the lagomorphs, the standard

63

Small mammals

deviation is far smaller with this method than any other using body or skull measurements and it can be recommended as the best method for determining the age of voles, lemmings, mice, dormice, squirrels and bats (Martinet, 1966; Dobrinskij & Mihalev, 1966; Askaner & Hansson, 1967; Perry & Herreid, 1969; Fisher & Perry, 1970; Louarn, 1971; Adamczewska-Andrzejewska, 1971, 1973a, b).

Changes in body organs during growth (*biochemical methods*)

There are several published accounts relating changes occurring in tissues and organs during development to temporal age. Pirie & Heyningen (1956) and Dische *et al.* (1965), have described such changes in the eye lenses of rats, cattle and humans. Water-soluble protein is constantly being added to the lens during its growth and is then converted to insoluble protein. It has proved possible to measure this process by determining the quantity of one of the amino acids, tyrosine. The quantity of tyrosine present has been found to be proportional to the protein content and in rats increases with age up to 900 days (Dische *et al.*, 1965); in laboratory mice (*Mus musculus*) up to 9 months (Dapson *et al.*, 1968); and in *Peromyscus polionotus* up to 750 days (Dapson & Irland, 1972); in each case, independently of body weight. This analysis has proved a better way of predicting age than any of the other methods; for instance, in *Peromyscus polionotus* the 95 per cent confidence limits at 100 days old were 96 to 107 days, and at 400 days old were 390 to 426 days (Dapson & Irland, 1972). At the same time, it should be remembered that, as Otero & Dapson (1972) have emphasized, this technique requires rather more complicated apparatus and analyzes than other methods. It does, however, have another advantage in that the total nitrogen content of the eye lens appears to be unaffected by malnutrition. This was first demonstrated by Kaufman & Norton (1966) and Kaufman *et al.* (1967) using pigs (*Sus scrofa*) up to the age of 320 days. In other species, the method may be less useful; for instance, Grau *et al.* (1970) found that in *Procyon lotor* the increase in nitrogen content of the eye lens declined abruptly after 12 months, and the method could be used to discriminate only between the young of the year and adults.

Fedyk (1974a) demonstrated a significant correlation between the total protein and total water content in the bank vole, and suggested that this relationship could be used as a method for determining age. The best correlation with age was given by the protein content, but again there were significant differences between cohorts depending on

64

the season of their birth, even during the first few days of life (Fedyk, 1974*b*). Since measuring the water content of rodents is a relatively simple procedure, Fedyk also used this parameter to determine their ages. For voles up to 3 months of age the range of accuracy lay between ±2 to ±10 days, if born during the autumn period; for spring born cohorts the range lay between ±7 to ±14 days.

Since many species have no obvious morphological features which might be used to determine the age of individuals accurately, biochemical methods, such as those described, should be given greater consideration, particularly in routine ecological studies involving small mammals. Other biochemical criteria may also prove worthy of investigation (Rockstein & Hrachovec, 1963), especially where the age determination technique used cannot be allowed to damage the specimens.

Age determination methods based upon other structures and structural degradation

Moult succession and type of pelage

Type of pelage and moult succession may be a good indicator of age, at least during a limited period of an animals' life. The use of pelage as an age indicator is based on the cyclical character of hair growth, with its alternating periods of activity and rest. In young animals, juvenile hair cycles occur at definite ages, irrespective of season.

It was found that the first three cycles of moult are remarkable constant in the laboratory mouse, *Mus musculus* (Chase & Eaton, 1959) and the rat, *Rattus norvegicus* (Johnson, 1958*a, b*) and in some wild rodents (*Microtus californicus*, Ecke & Kinney, 1956; *Ochrotomys nuttalli*, Linzey & Linzey, 1967). Koponen (1964, 1970) found that hair growth expresses quite accurately the age of the Norwegian lemming, *Lemmus lemmus*, up to 5 to 6 weeks. It also is useful for older lemmings, except breeding females, up to 2 months, and allows five age classes of chronological age to be distinguished. Chipman (1965) has stated that succession of moult is useful additional criterion of age in *Sigmodon hispidus* and Barrier & Barkalow (1967) were able to distinguish four age classes in the gray squirrel, *Sciurus carolinensis*, taking into consideration color and texture of the winter pelage.

There is a lack of similar studies in other rodents and the method itself has rather limited value, because any age classification is possible only during the first weeks or months of postnatal development.

Small mammals

Dehnel (1949) and Dunajeva (1955) found clear differences in the pelage of young of the year and overwintered shrews of the genera *Sorex* and *Neomys*. Characteristic of overwintered shrews was more or less worn fur on ears, tail and feet. By combining the features of the pelage with its length and wear of teeth there is no problem in distinguishing two basic age groups in these genera.

Tooth wear

Tooth wear, when combined with tooth replacement, has been long used for determining age in mammals (Habermehl, 1961). The general assumption is that wear of teeth is proportional to the age of an individual and is a most traditional and simple method which gives quite satisfactory results if relative age groups are needed. However, much more correct data are expected when age-reference models based on animals of known age are available for comparison. The usual technique then is to allocate the specimens to previously separated age groups based on 'type' individuals. Unfortunately such 'type' individuals are rarely selected from a sample of individuals of known age, and in most cases they represent different stages of tooth wear only.

There are numerous studies on tooth wear in small mammals in which only subjective age classes have been distinguished (Naumov, 1934; Varshavskij & Krylova, 1948; Kubik, 1952; Deparma, 1954; Lozan, 1961; Skoczén, 1966; Hoslett & Imaizumi, 1966; Adamczewska-Andrzejewska, 1967, 1973b, and many others). Most of the previous works concerning the tooth wear technique, including those by Russian authors, have been summarized by Tupikova (1964). Descriptions of age classes accompanied by tables to distinguish ages, and sketches of tooth wear patterns for a number of the most common rodents and moles as well as shrews, are available.

The main drawback of this method is the subjectivity of the particular author in assessing the degree of tooth wear. A good example of this problem is given by Keiss (1969) who invited eleven biologists to estimate the age of a series of elk of known age, and noticed errors in their estimations of up to 7 years. Similar although not so extensive errors have been stated by Szabik (1973) for *Capreolus capreolus*. In a comparable experiment with red deer, similar errors could also have occurred because of the considerable variation in the size of the teeth, but if the stage of eruption reached by the third molar was also taken into account in each jaw examined, few animals were aged incorrectly by more than one year (Lowe, 1967). No similar tests have been tried on small mammals with the exception of the mole (*Talpa europaea*), where the results published by Deparma (1954) and Skoczeń (1966) were supported by Grulich (1967), who assessed tooth wear by measuring the height of certain tooth cusps. Measurement of the height of selected cusps or whole teeth and calculation of tooth-wear indices, although more laborious, does enable much more objective classification of the material into age groups and appears to be a preferred method by many authors (Connaway (1952), Viktorov (1967) and Dapson (1968) in shrews; Okhotina (1966), Usuki (1966) and Grulich (1967) in moles).

Two techniques for measuring the area of exposed dentine have been proposed for herbivorous and omnivorous mammals, whose tooth crowns are relatively flat. Shorten (1954) measured the area of wear in *Sciurus carolinensis* by the amount of light penetrating through the negative of a sketch made under a drawing apparatus. Tanaka (1968) measured the total size of occlusal area of the upper molars in the rat and calculated the molar wear index as a ratio of total worn surface to tooth row length times molar breadth. The index was calculated to eliminate the effect of individual variation due to differences in molar size. Such methods seem to be rather complicated when adopted to serial material of some hundreds or thousands of specimens.

Unfortunately, most authors do not have at their disposal wild individuals of known age, and comparisons with laboratory reared animals are completely valueless for such studies.

Individual variation in tooth wear is not usually investigated and is probably minimized when the material being studied is allocated to previously selected tooth patterns or age classes. A very good example of such variation was found in an exhaustive study of *Apodemus agrarius* by Adamczewska-Andrzejewska (1973*b*), who discovered broad ranges of tooth wear classes among individuals of the same lens weight class when comparisons of the age estimations by these two methods were made.

The rate of tooth wear, especially in connection with the differential hardness of the enamel and dentine and the changing size of the wearing surface, has not been carefully studied either. Geographical differences in the rate of tooth wear connected with the hardness of food or type of soil can be expected (Szabik, 1973), the latter effect being especially important in the digging mammals. There are also seasonal changes in the hardness of the enamel as described in *Sorex araneus* by Adamczewska-Andrzejewska (1966) and different rates of tooth wear in

Small mammals

Meriones erythrourus born in different seasons were observed by Gintlis (1959). Finally, the possibility of genetic variations in dental structure and hardness was suggested by Berry & Truslove (1969).

Tooth wear is not usually used as the only guide to age, but in combination with other characters (despite the objections mentioned above) it seems that rather satisfactory results might be obtained with this method in many species of small mammals.

Other methods of age determination

As we have been concentrating on the determination of age in small mammals, special techniques developed for other mammal groups will be only briefly discussed in the following paragraphs. It seems worthwhile to do this because a knowledge of these may encourage some workers to develop new techniques or to adapt those already used for other mammals. It is suggested that the reader look at the reviews mentioned at the beginning of the chapter or other special literature for more detailed discussions of these methods.

Some keratinized and continuously growing organs, such as nails and claws, show clear yearly rings or layers when studied externally or sectioned, and have been used as indicators of age in Pinnipedia (Plehanov, 1933; Chapskii, 1952*b*; Tikhomirov & Klevezal, 1964). It is clear that such determinations of age are only possible in those mammals, marine species for example, which do not wear down their claws. Claw wear is a serious drawback with land mammals, but it seems that not all claws wear to the same extent, and some effort in this direction might be fruitful. As yet, no one has tried the technique on small mammals.

Other keratinized organs such as horns or baleen plate, have been used for age determination in Bovidae and Mysticeti whales (P. A. Morris, 1972). Also, the laminated structure of an earplug and the number of corpora albicantia remaining in the ovary have recently been studied extensively in Mysticeti whales and were found to be good criteria of age in these long-living mammals (Utrecht-Cock, 1965; Roe, 1967; P. A. Morris, 1972).

Gerontologists have found many biochemical changes in the mammalian organism. Some examples were mentioned earlier but there are probably many more possibilities to explore using these changes as indicators of chronological age in wild mammals. It has been suggested, for example, that degenerative changes in collagen fibres, such as loss of elasticity with advancing age, might be used as a guide to age (P. A. Morris, 1972).

Bone marrow, which is red in young animals, is replaced by fat and changes color to yellow as the animal ages. A very simple technique was suggested by Habermehl (1961) for age determination of cattle using this character. The method needs some detailed study if it is to be adopted for free-living wild mammals.

General discussion and conclusions

The statistical accuracy of any method used for determining age in mammals should be estimated, as for example by Lidicker & MacLean (1969) or Adamczewska-Andrzejewska (1973*b*). Since processes of growth and development are variable and many methods are based upon those variable characters, it is very important to know the errors and confidence intervals of such estimations as well as the relative effectiveness of different techniques.

One of the methods for testing age determination procedures is a comparison of estimates obtained by different methods. Examples of such studies, mainly in larger mammals were reviewed by P. A. Morris (1972). Some others are described below.

Bujalska *et al.* (1965) compared relative-age estimates of *Lepus europaeus* obtained by four different methods; ossification of the skull, pelvis, radius and ulna, and eye lens weight. It is clear from their tables that the greatest degree of correlation between methods was obtained when ossification of the skull and pelvic sutures were compared (89 per cent). When the degree of ossification of the pelvis or skull and ulna notch (epiphysis) were compared, only 72 and 73 per cent agreement respectively was observed. The agreement of lens weight and epiphysis techniques, used simultaneously, for separating young of the year and older hares was the highest (96 per cent). In this case, as in Dolgov & Rossolimo (1966), relative-age classes delineated by one method do not completely correspond with those obtained by other techniques. Thus, it is not possible to say which method is better but one may decide which of those that are highly correlated it is easiest to use.

More valuable results are obtained when a method's accuracy is tested on samples of individuals of known age. Grau *et al.* (1970), comparing four different methods of determining age of *Procyon lotor* found an average agreement of 83 per cent between tooth wear classes and actual age. The agreement varied between 69 and 100 per cent with different age groups.

Perry & Herreid (1969) found in a small sample ($N = 48$) of known-age *Tadarida brasiliensis* that 73 per cent of all bats were placed in the same age categories by tooth wear and lens weight methods. However, in a larger sample of 482 bats of unknown age, only 49 per cent were aged the same by both methods and 15.5 per cent showed disagreements by two age classes. Agreements of a similar order were found in each of five cave populations of bats that were studied. The authors' conclusion is that the accuracy of both methods is very similar.

When comparing epiphysial notches and the eye lens weight in *Tamiasciurus hudsonicus*, Davis & Sealander (1971) found 82 per cent agreement in distinguishing two basic age groups of animals – under 16 months of age or older. They observed the most discrepancies in animals of 12 to 16 months, i.e., about the time of epiphysial fusion.

A comparative study of age determination methods in *Apodemus agrarius* of known age was recently presented by Adamczewska-Andrzejewska (1973*b*). Growth parameters together with age variability in a complex of morphological characters commonly used in determination of the age of small mammals were analyzed. These included body and skull measurements, tooth wear and eye lens weight. The errors of age estimation were calculated by comparison with animals of known age. Although very high correlation was found between tooth wear or eye lens weight and actual age, the level of agreement between these two methods was rather unexpected. It appeared that tooth wear trends do not coincide exactly with age or eye lens weight, and there were individuals with very worn teeth in low lens weight classes. These observations suggest either large individual variation in tooth wear or that both tooth wear and lens weight accumulated variability at comparable rates. The examples mentioned thus show clearly how great the discrepancy can be between different methods.

The general conclusion of various authors is that for more correct estimation of age several methods should be used simultaneously, though this is rather difficult in routine studies of growth curves, age structure or other population parameters. Nevertheless, we stress that the main sources of errors between different methods may be connected with individual variability of particular characters, and nonparallel growth rates of the body parts or nonuniform rates of ossification of different parts of the skeleton. It is clear that in testing a method it is necessary to do so against a series of animals of known age rather than against other methods of age classification.

It is sometimes necessary to use several techniques to divide material into clear-cut age classes during the whole life span of the species as different methods may operate satisfactorily up to a defined age only

(Zimmerman, 1972). On the other hand, simultaneous comparison of different estimates, especially based on the linearly-changing characters, enables possible separation of age groups from a scatter diagram, even when there is some overlap between the classes of one or other method (Bree *et al.* 1966; Mead, 1967; Degn, 1973).

Calculation of confidence limits is very important when estimating age by any method. Unfortunately, this is done in few studies and if calculated, it is not always clear how the statistic was derived (Connolly *et al.*, 1969; Myers & Gilbert, 1968). It is necessary to note when calculating confidence limits about a regression line that the x-axis (age) is a nonrandom variable, selected arbitrarily by the investigator (Simpson *et al.*, 1960). Age of unknown animals is a random variable when predicted from the regression equation. Consequently, confidence limits based on a nonrandom variable cannot be used accurately and would be smaller than the true value. Special formulae are necessary, therefore, in order to calculate the confidence limits (Dapson & Irland, 1972). Further, it is known that the confidence limits for individual estimates are always larger than for mean estimates (Simpson *et al.*, 1960). More accurate limits are obtained when variability in the intercept and slope is taken into consideration. Such limits are asymmetrical and broaden with advancing age. Confidence limits of this type are shown in papers by Dudzinski & Mykytowicz (1961), Rongstad (1966), Keith *et al.* (1968) and Dapson & Irland (1972). The practical consequence of adopting asymmetrical confidence limits is decreasing accuracy of age estimations for older age classes (Dapson & Irland, 1972; Adamczewska-Andrzejewska, 1973*a*).

There are few small mammal species for which accurate methods for determining age have been developed. Among the most promising methods are tooth wear, lens weight and biochemical changes. However, we cannot determine the age of most mammals with satisfactory accuracy with any of the methods tested to date. Age determination of small mammals during the very early stages of their ontogenesis, that is during the development in the nest, has been reasonably successful and external characters are usually sufficient to tell the correct chronological age of an individual in this case. However, since this period of small mammal life is of little interest in ecological studies of a trappable population, it was not emphasized in this chapter.

From the preceding text it is clear that the determination of mammalian age is based upon growth and developmental changes in the organism. The rate of growth is one of the distinguishing factors between

71

age classes. When it is rapid, one obtains discontinuity between age groups in a sample; if it is slow, differences are obscured or tend to disappear. Thus, in the early stages of development, some rapidly changing characters may serve as good indicators of age, but usually not in older age classes when the phase of rapid growth is over.

In older individuals, variability of characters increases due to differing conditions during the development of particular individuals, different habitats, geographical and population situations. Nonuniform growth rates even appear among individuals of the same litter. Thus, there is a great need for further study of the various methods and their application to small mammals. For the purpose of estimating accuracy, more comparative studies on known-age samples are also needed.

Little is known about the influence of genetic factors on the variability of characters as possible guides to age. Some data indicating the importance of these factors have been presented, but we want to stress that genetic factors should be considered in the studies on age determination of mammals.

It is very important to standardize techniques. Examples of differences caused by changes in procedure of age estimation are shown in the section on eye lens weight. Possibly similar differences can be found when working with other methods. More careful descriptions of methods, including all technical details used in a scientific report, may help to standardize methods, since many researchers often change previously accepted procedures making comparisons impossible.

4. Patterns of demography in small mammal populations

N. R. FRENCH, D. M. STODDART & B. BOBEK

Objectives

The population consequences of life history phenomena have been well established both theoretically and experimentally. Still, few studies have evaluated the interaction of mechanisms under natural conditions. It has been pointed out that rates of increase as determined under laboratory conditions represent the maximal rate of increase for the environment of the experiment. In reality, populations are exposed to a spectrum of conditions, not the least of which is the intermittent breeding season. Small mammals optimize various life history traits to achieve suitable mechanisms for maintaining their populations through time. Increased reproductive capacity is such a trait, as is increased survival.

Our purpose in this report is to survey the recent literature on small mammals for demographic statistics, to collate these, and to compare them in a search for general trends of the means by which small mammal populations have developed their capacity to survive, to persist, and to exploit the habitats in which they have developed. By comparison of the combinations of characteristics that occur together in the small mammal groups, we reasoned it would become apparent which sets of attributes worked most successfully for these animals. We also seek relationships of these attributes to environment and habitat.

We have attempted to organize or classify small mammals into groups according to demographic characteristics. Starting from the taxonomic organization of small mammals, we have examined the general traits of the group and then the traits of individual species. Those that failed to fit the general ecological classification of the group were classified within other groups. In this way we hoped to achieve a demographic organization of small mammal species, arranged according to life history characteristics at the population level.

Methods

The survey of the recent literature, upon which this report is based, includes articles which have appeared in the last decade. We were not restricted to the literature of this period, however, and if useful data were found in older literature, they too were included. Our attempt at a comprehensive survey was restricted to the last ten years. We were specifically interested in those categories of information known to be related to rates of population increase or those categories which might for ecological reasons be correlated with demographic parameters. It was necessary to select standard units of measurement and to make appropriate conversions of literature values. The categories of data utilized and the units of measurement are given in Table 4.1. Survival data were converted to a monthly basis by assuming exponential rates of loss in populations.

Table 4.1. *Categories of information and units of measurements*

Information	Units
Habitat	Vegetation formation, location
Reproduction	Seasonal or nonseasonal
Litters	Number per season
Young-of-the-year	Reproduction by young during the season of their birth
Litter size	Number of young or embryos per litter
Maximum prevalence of pregnancy	Maximum proportion of females breeding
Adult or young survival	Fraction of each age group surviving per month
Overwinter survival	Fraction surviving per month in the nonbreeding season
Yearlong survival	Fraction surviving per month over one year
Survival of litters by season	Fraction surviving per month for early-season or late-season litters
e_0	Mean life expectancy in months at birth
e_n	Mean life expectancy in months at age n
Home range, area	Hectare
Home range, maximum distance	Maximum distance of individual movements in metres
Density, seasonal	Variation in numbers per hectare within season
Density, interseasonal	Variation in numbers per hectare between seasons
Density, period	Number of years between peak populations

Availability of information on different taxa of small mammals varies greatly. Some taxa, such as the Muridae, Microtinae, and Cricetinae, are well represented in the literature. In each, however, there seemed to be one or a few outstanding papers. There is less information about others and a complete paucity in certain areas.

The information was summarized in tabular form for comparative purposes, organized on a demographic basis (Table 4.2) This table contains all of the relevant quantitative data uncovered in our survey. Doubtless there are important omissions, but sufficient information is included for most groups to permit generalizations regarding demographic characteristics. Although there is wide variation within groups and sometimes among different populations of the same species, the following discussion is based on mean values for the ecological groups.

Classification

Based on life history data (summarized in Table 4.3), most small mammals fall into three general types. The first of these is the group characterized by high reproductive rate, low survival rate, and high density tolerance. Comparison of survival rates is shown in Fig. 4.1.

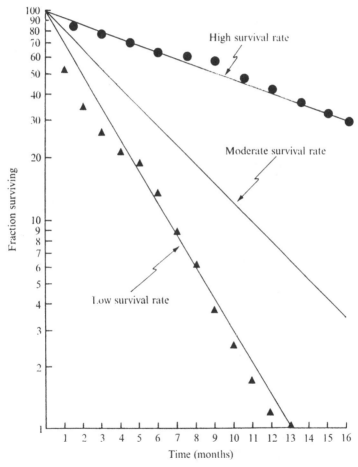

Fig. 4.1. Survival curves illustrating high (*Perognathus formosus*, French *et al.*, 1974), moderate (*Peromyscus leucopus*, Snyder, 1956), and low (*Clethrionomys glareolus*, Bobek, 1973) survival rates.

75

Table 4.2. *Demographic and home range statistics of small mammals*

Species	Habitat Location [authority]	Reproduction				
		Seasonal or non-seasonal	Number of litters	By young of year	Litter size Embryo rate	Max % pregnant
Murid type						
Akodon azarae	Grassland Argentina [1]	S		Yes		
Aplodentia rufa	Coniferous forest Washington [2]					
Conilurus penicillatus	Scrub N. America [3]				2.0	
Mus musculus	Rocky island S. England [4]	S		Yes	5.1–10.7	0.38
	Hay ricks S. Australia [5]	Non				0.93
	Wheat field S. Australia [6]			Yes		
	Reed bed S. Australia [7]	S				1.0
	Volga delta USSR [8]					
Oryzomys palustris	Island Louisiana [9]	S			3.7 high den. 6.0 low den.	
Rattus blanfordi	Mysore, India [10]					
Rattus exulans	Hawaii [11] Islands	Non	4.3		4.5	
	Pacific Ocean [12]					
Rattus norvegicus	Southern Sakhalin USSR [13]	S		Yes	8.3–9.1	
	River valley California [14]	S		Yes		
Rattus rattus	Hawaii [11]	Non	2.3		4.8–5.7	
Rattus wroughtoni	Mysore, India [10]					
Sigmodon hispidus	Grassland Kansas [15]					
	Grassland Tennessee [16]	S			7.6	0.32
Microtine type						
Alticola lemminus	Lower Lena River Yakutia [17]	S	1	No	7.4–7.7	
Alticola roylei	Pamirs Mid Asia [17]	S	2–3	Yes	4.8	
Apodemus agrarius	S. taiga USSR [17]	S	5	Yes	5.0–8.3	
Apodemus argenteus	Deciduous forest Japan [18]			Yes	3.2–4.1	1.0
Apodemus flavicollis	Mixed forest S. Sweden [19]	S				
	Deciduous forest Poland [20]					
	Beech forest Poland [21]				6.3, 5.9, 6.0	
	Forest E. Poland [22]					
	Deciduous forest S. Poland [23]				5.9	
	Mixed woodland S. Austria [24]					
	Spruce forest S. Sweden [25]					
	Lowland forest Czechoslovakia [26]					
Apodemus sylvaticus	Mixed forest S. Sweden [19]	S	4			
	Mountains USSR [17]	S	3–4	Yes	c. 8.2	
	Mixed hardwood Oxford [27]					
	Open field S. Sweden [28]					
	Spruce forest S. Sweden [25]					

	Survival (monthly)						Home range (area and/or maximum distance)	Density (ha^{-1})		
									Interseasonal	
	Adult	Young	Overwinter	Yearlong	Litters by season	e_0		Seasonal	Magnitude	Period (yrs)
[1]	0.82									
[2]							0.2–1.25 ha			
[3]										
[4]	0.8–0.85		0.1, 0.4		1st litter <0.5				1.5–50.0	
[5]								12.1/m³		
[6]								0–85		
[7]							♀ max 35 m ♂ max 24 m	0–100 11–623		
[8]										3
[9]							♂ 0.33 ha ♀ 0.21 ha			
[10]								2.65		
[11]										
[12]	0.95	0.83				$e_0 = 1.82$				
[13]										
[14]										
[11] [10] [15]								10.0 3.1–16.1		
[16]									2.3–9.1	
[17]					1st survives winter	$e_0 = 10–11$				
[17]										
[17]					1st 3–4 months					
[18]										
[19]										
[20]		0.63 high 0.70 low	0.65 high 0.75 low	0.71 high 0.76 low					5.9 high 3.7 low	
[21]						$e_1 = 2.9–3.6$			{10.9 high 2.9 med. 4.4 low	
[22]									51.2–58.0	
[23]									13.7	
[24]							♂ 2.2 ha ♀ 1.0 ha ♂ breeding 5.0 ha			
[25]	0.625							3.0, 1.0		
[26]							♂ 3.26 ha 813 m ♀ 1.31 ha 58.2 m			
[19]										
[17]										
[27]								12.5		
[28]								5.0		
[25]	0.525							9–14.0		

Table 4.2 (*continued*)

Species	Habitat Location [authority]	Reproduction				
		Seasonal or non-seasonal	Number of litters	By young of year	Litter size Embryo rate	Max % pregnant
Microtine type (*continued*)						
Apodemus sylvaticus (*continued*)	Mixed hardwood Oxford [29]	S				
	Mixed woodland S. England [30]					1.0
	Mixed hardwood Oxford [31]					
	Mixed woodland Durham, England [32]	S				0.42
	Mixed woodland Durham, England [33]					
Arvicola terrestris	Pond S. Moravia [34]					
	Pasture Cent. Switzerland [35]				4.05–4.4	
	(Review) USSR [36]	S	2–4	Yes	6.0–8.0	
	Stream N. Scotland [37]	S	1.5–2.3	No	6.4	
	Stream N. Scotland [38]					
	Stream S. Moravia [39]					
	(Review) W. Siberia [40]					
	(Review) N. USSR [41] Vychega, USSR [41]					
Clethrionomys gapperi	Coniferous forest Canada [42]		3	Yes	5.85, 5.69, 5.34, 6.0, 5.7, 5.38	
Clethrionomys glareolus	Mixed hardwood Oxford [43]					
	Taiga–Birch Swedish Lapland [44]				5.0–7.7	c. 1.0
	Mountain and plain Switzerland [45]					
	Mountain and plain Switzerland [46]	S		At 600 m Yes At 1700 m No		
	Spruce forest S. Sweden [25]					
	Mixed hardwood Oxford [47]					1.0
	Lowland forest Czechoslovakia [26]					
	Lowland forest Czechoslovakia [48]	S		Yes		
	Mixed woodland England [30]					1.0
	Mixed hardwood Oxford [31]					
	Taiga Kola P., USSR [49]	S		Yes		
	Pine forest Poland [50]	S			5.2	
	Island population E. Poland [51]	S				
	Island population E. Poland [52]	S		Yes		1.0
	Island population E. Poland [53]	S		Yes	4.43–5.15	

	Survival (monthly)						Home range (area and/or maximum distance)	Density (ha^{-1})		
									Interseasonal	
	Adult	Young	Overwinter	Yearlong	Litters by season	e_0		Seasonal	Magnitude	Period (yrs)
[29]			0.03–1.0					3.8–23.0		
								0.5–22.6		
								0.37–2.5		
[30]								8.8–37.6		
								2.6–26		
[31]								1.7–18.5		
								7.5–29.4		
								1.03–12.1		
[32]								2.6–44.7	2.6–44.7	
[33]								0.6–15.6	0.6–15.6	
									12.4–44.5	
[34]							♂ 166 m ♀ 55 m			
[35]									Up to 978	
[36]								110–116	10.0	11
[37]					1st 9 months Subs. 3 months	$e_0 = 4.5$				
[38]							♂ 112 m ♀ 76.5 m			
[39]							Sum. Wint. 170 m ♂ 500 m 135 m ♀ 80 m	4.0–7.9/ 100 m 6.3–9.3/ 100 m		
[40]										8–11
[41]								0.01		
[41]									0.01	
[42]								13		
[43]		Early 0.43 Late 0.27	0.69 0.74		♂ 0.24 ♀ 0.36					
[44]								5	40	
[45]			Low in lowland High in upland							
[46]										
[25]				0.013				3–13 1–13		
[47]			0.80–0.87 0.84–0.97 0.73–0.95 0.72–0.50		Summer 0.33–0.68 Autumn 0.71–0.86				137 247	
[26]							♂ 0.82 ha 40.8 m ♀ 0.71 ha 21.4 m			
[48]										
[30]	0.57–1.0		0.56–1.0 0.75–0.95					1.1–52.0 0.6–44.0		
[31]	0.72 0.97 0.42–0.82		>0.80					18.5–52.3 24.2–44.0 23.1–39.0		
[49]										4
[50]					0.1–0.37 1st 0.3–0.47 Subs.					
[51]					0.81, 0.79, 0.48 1st 0.33, 0.56, 0.83 2nd			14–76 17–50 17.5–99.0		
[52]			0.52		†k_1 2.12 k_2 2.62 k_3 0.79 k_4 0.89					
[53]										

Table 4.2 (*continued*)

Species	Habitat Location [authority]	Seasonal or non-seasonal	Number of litters	By young of year	Litter size Embryo rate	Max % pregnant
Microtine type (*continued*)						
Clethrionomys glareolus (*continued*)	Mixed woodland S. Sweden [19]	S		Yes		
	Mixed hardwood Oxford [54]				Summer 4.43 Winter 3.53	
	Mixed hardwood Oxford [27]					
	Deciduous forest Poland [20]	S				
	Beech forest Poland [21]	S			4.2, 6.0, 6.0	
	Pine forest Poland [50]	S			5.2	
	Southern taiga USSR [17]	S	5	Yes	5.0–8.3	
	Marijskoj. USSR [55]					0.86
	(Various) USSR [56]					
	Forest E. Poland [57]					
	Forest Poland [58]				4.8	
	Mixed woodland Durham, England [32]	S				0.36
	Mixed woodland Durham, England [33]					
	Forest E. Poland [22]					
	Deciduous forest S. Poland [23]				5.7	
Clethrionomys rufocanus	Taiga Kola P., USSR [49]			Yes		
	Mountain taiga N. Urals [17]	S	3	Yes		
	(Various) USSR [56]					
Clethrionomys rutilus	Coniferous forest Canada [42]			Yes	5.85,5.57,5.07 6.35, 5.58,5.30 6.81,6.60,4.50	
	Forest China [59]	S				
	Mountain taiga Urals [17]	S	3	Yes		
	(Various) USSR [56]					
	Taiga Novosybirsk, USSR [60]		2–4		6.76–7.25	0.75
	Taiga Sajahy, USSR [61]				5.31–7.38	
Lagurus lagurus	USSR [17]	S		Yes		
Microtus agrestis	Alpine hay meadows Swedish Lapland [44]				6.2–7.4	
	Agriculture land S. lower Saxony [62]					
	Spruce forest S. Sweden [25]					0.44
	Mixed hardwood Oxford [43]	S			4.6	
	Mixed agriculture Finland [63]					
	Pasture S. Sweden [64]					
	Mixed hardwood Oxford [65]					
Microtus arvalis	Agrocenose W. Poland [66]					

	Survival (monthly)						Home range (area and/or maximum distance)	Density (ha^{-1})		
									Interseasonal	
	Adult	Young	Overwinter	Yearlong	Litters by season	e_0		Seasonal	Magnitude	Period (yrs)
19]										
54]										
27]								12.5 forest 37.5 bracken		
20]		0.50 high 0.69 low	0.72 high 0.69 low	0.61 high 0.69 low					32.1 high 10.2 low	
21]	1st month 2–5 5–6 1–3 months c. 0.60 high 7–12	0.61 0.81 } low 0.68 } den. <0.60 } den.				$e_1 = 2.2\text{–}3.2$			15.6 high 5.8 med. 4.5 low	
50]					0.64–0.84 spring 0.74–0.86 summer			2.6–17.2		
17]					1st 3–4 months					
55] 56]									0.1–31.0	4
57]						$e_1 = 2.4\text{–}3.9$				
58]		0.62								
32]								11.6–57.7 21.9–62.5	11.6–57.7 21.9–62.5	
33]	0.63–0.82								42–66.7	4
22]									30.8–51.2	
23]									10.1	
49]										5
17]					1st 0.60 overwinter					
56]							0.06–0.42 ha		0.4–30.0	
42]								20		
59]										
17]					1st 0.60 overwinter					
56]							0.08–0.43 ha		8.2–98.0	
60]							♂ 0.19–0.67 ha ♀ 0.06–0.12 ha		11–110	4
61]										
17]										2–3
44]								5	50	
62]										3–4
25]				0.13				9–30 18–67		
43]					1st litter poor					
63]								280 trapout 308 live trap		
64]								5–60		
55]								120		
66]								10–15	Up to 175	

81

Table 4.2 (*continued*)

Species	Habitat Location [authority]	Seasonal or non-seasonal	Number of litters	By young of year	Litter size Embryo rate	Max % pregnant
Microtine type (*continued*)						
Microtus arvalis (*continued*)	Grassland France [67]	S				
	Grassland W. Poland [68]					
Microtus californicus	Grassland San Francisco Bay [69]	S				
Microtus gregalis	Forest tundra N. USSR [17]	S	4	Yes	6.6 1st 8.7 2nd 9.9 3rd & 4th	
Microtus juldaschi	Pamirs Central Asia [17]	S	2–3	Yes	3.5–3.9	
Microtus middendorffi	Forest tundra N. USSR [17]			Yes		
	Taiga Jamal/Urals [70]	S	5	Yes		
Microtus ochrogaster	Grassland Indiana [71]	Non		Yes	5.0	
Microtus oeconomus	Forests Poland [72]					
	Taiga Jamal/Urals [73]	S	5	Yes		
	Forest tundra N. USSR [17]	S	4–5	Yes		
	(Various) USSR [56]					
Microtus oregoni	Scrub Oregon [74]	S			3.2	1.0
Microtus pennsylvanicus	Wayne Co., Pennsylvania [75]	S				
	Ithaca, New York [76]					
	Michigan [77]	S				
	Grassland Indiana [71]	Non		Yes	5.0	
Microtus pinetorem	Coniferous forest Virginia [78]	S			3.2 max. 1.9 mean	0.70
Insectivora						
Blarina brevicauda	Deciduous forest New York [79]			Yes		
Neomys anomalus	Mixed forest Poland [80]	S			8.8	
Neomys fodiens	Mixed forest Poland [80]	S			4.4	
Sorex alpinus	Spruce forest [81]					
Sorex araneus	Mixed Netherlands [82]	S	3	Rarely	6.45	
	Deciduous forest Oxford [83]					
	Spruce forest [81]					
	Mixed forest Poland [80]	S			5.4	0.50–0.83
	Mixed forest Poland [84]				6.4	0.88
	Field Sweden [85]					
	Forest E. Poland [86]			<2%	6	
Sorex minutus	Spruce forest [81]					
	Mixed Netherlands [82]	S	3	Rarely	6.2	
	Mixed forest Poland [80]	S			5.7	
	Forest E. Poland [86]			<10%	6	
Sorex vagrans	Conifer–meadow Colorado [87]					

	Survival (monthly)						Home range (area and/or maximum distance)	Density (ha⁻¹)		
									Interseasonal	
	Adult	Young	Overwinter	Yearlong	Litters by season	e_0		Seasonal	Magnitude	Period (yrs)
67]					1st lit. highest 2nd lit. lowest					
68]								724–1212		
69]										2–4
17]										
17]										
17]										
70]					1st max. 3 months					
71]	0.2							334–358		
72]									97–130	
73]										
17]					1st max. 3 months					
56]							0.09–0.22 ha		0.1–45.0	
74]							♂ 0.53–0.6 ha ♀ 0.35–0.5 ha		0–15.0	
75]									290 145	
76]							0.75–1.2			
77]								140		
71]		0.1–0.12		0.76 ♂ 0.86 ♀ }Incr. 0.53 ♂ 0.64 ♀ }Decl.					136–148, 210	
78]										
79]				0.80						
80]										
80]										
81] 82]				0.96 young 0.95 adult	spring 0.94 summer 0.98	e_0 = 7–11	0.04–0.06 ha	12.3–17.7 13.3–18.5	1.1	
83]							0.28 ha	1.8–6.9		
81] 80]							0.05 ha		17.5	
84]										
85]									3–10	
86]										
81] 82]				young 0.90 adult 0.94	spring 0.90 summer 0.94	e_0 = 5.5–6.1	0.05–0.19 ha	3.9–11.1 5.3–10.5	4.6	
80]										
86]										
87]							3.0 ha 152 m			

83

Table 4.2 (*continued*)

Species	Habitat Location [authority]	Seasonal or non-seasonal	Number of litters	By young of year	Litter size Embryo rate	Max % pregnant
Cricetine type						
Allactaga elater	Fergana, USSR [88]		2			
Gerbillus dasyurus	Desert India [89]					
Hylomyscus stella	Forest Uganda [90]	S			3.2	0.71
Lophuromys flavopunctatus	Forest Uganda [90]	S			2.2	
Meriones hurrianae	Desert India [91]					
Meriones tamariscinus	Zaisan Basin USSR [92]	S	2	No (elsewhere, yes)	5.2–5.5. (1st) 4.9 (2nd lit.)	
Ochrotomys nuttalli	Mixed forest Tennessee [93] Deciduous forest Illinois [94]					
Onychomys leucogaster	Desert shrub New Mexico [95]					
Onychomys torridus	Desert shrub New Mexico [95] Desert Arizona [96]					0.80+
Peromyscus californicus	(Review) [97] (Review) [97]		3.25		1.91 1.87	
Peromyscus gossypinus	(Review) [97]					
Peromyscus leucopus	Shrub desert New Mexico [95] (Review) [97] Oak–hickory forest Michigan [98] Oak–hickory forest Michigan [99] Deciduous forest Michigan [100] Islands Canada [101]		4–5		3.4–5.5	
Peromyscus maniculatus	Coniferous forest Great Slave Lake [42] (Review) [97] Coniferous forest British Columbia [102] Coniferous and clear-cut Washington [103] Shrub to alpine White Mts., California [104]	S S S S	2 4 Fewer (high) More (low)	Some Yes	old 5.36–6.15 all 5.45–6.11 5.06 4.63 high elev. 4.01 }med. 3.64 4.00 low	0.75
	(Review) [97] Shrub desert New Mexico [95] Coniferous forest British Columbia [105] Coniferous forest British Columbia [106] Farm land Minnesota [107	S	4		4.0–5.1 4.5	0.70
Peromyscus polionotus	(Review) [97]				3.1	
Peromyscus truei	(Review) [97] (Review) [97]		3.4		3.52 2.84	
Praomys morio	Forest Uganda [108]	Non			3.3	
Reithrodontomys fulvescens	Pine savannah E. Texas [109]					
Reithrodontomys halicoetes	[110]	S				0.36
Reithrodontomys megalotis	[110]	S				0.37
Reithrodontomys raviventris	[110]	S				0.48

	Survival (monthly)						Home range (area and/or maximum distance)	Density (ha^{-1})		
									Interseasonal	
	Adult	Young	Overwinter	Yearlong	Litters by season	e_0		Seasonal	Magnitude	Period (yrs)
88]										
89]				0.72		$e_1 = 0.9$				
90]										
90]										
91]				0.70		$e_1 = 1.4$	♂ 0.009 ha			
92]							♀ 0.01 ha			
93]							♂ 0.26 ha 31.4 m	0.61–1.46	0.1–1.5	
94]							♀ 0.24 ha 31.7 m / 0.49 ha 73.2 m		0.3–12.1	
95]							2.7–4.6 ha		0.19	
95]							2.7 ha		0.01–0.39	
96]									0.61–3.3	
97]	c. 0.19								c. 8.0	
97]										
97]								3.7–6.7	up to 11.8	
95]							1.26–1.82 ha		0.12–0.32	
97]								7.4–16.8 max.	0.99–26.8	
98]			0.85		$e_1 = 1.3$ (spr.–sum.)	$e_1 = 6.5$		2.7–5.9	2.0–7.2	
			0.72		$e_1 = 1.8$ (fall)	$e_1 = 3.0$				
99]							2.8 ha	4.3–6.0 / 3.2–4.7		
00]			0.68–0.85			$e_1 = 2.8$–6.0		2.7–5.9	1.2–7.2	
01]							0.01–0.94 ha			
02]			0.95					c. 5–20	5–40	3
			0.89					c. 10–40		
97]	c. 0.18							c. 22.8		
02]						‡$e_a = 2.41$		22–50	<10–50	
						$e_a = 3.16$		22–42	<10–42	
03]							137–213 m			
04]										
7]										
5]							1.66–1.89 ha	7.2–9.1 max. / 1.31	0.74–29.9	None
06]	0.80	0.60						13.7–47.3 / 9.4–18.6 / 14.6–50 / 16.7–64.2 / 12.8		
06]	0.45–0.91									
07]			0.74							
7]										
7]	c. 0.15							6.4–25.4	up to 49.4 / c. 12.1	
7]										
08]										
09]							0.16–0.26 ha	0.94–5.7	0.32–5.7	
10]						$e_2 = 3.96$				
10]						$e_2 = 3.64$				
10]						$e_2 = 4.87$				

85

Table 4.2 (*continued*)

Species	Habitat Location [authority]	Seasonal or non-seasonal	Number of litters	By young of year	Litter size Embryo rate	Max % pregnant
Cricetine type (*continued*)						
Tatera brantsi	Highvelt S. Africa [111]	S	1			
Tatera indica	Desert India [89]					
Taterillus pygargus	Grassland Senegal [112					
Sciurid type						
Citellus leucurus	Desert Nevada [113]					
Citellus parryi	Verkhoyan'e USSR [114]				9.4	
Citellus pygmaeus	Semi-desert Caspian [115]					
	Desert Aral Karakumy, USSR [116]					
Citellus undulatus	Yakutia USSR [114]				8.4	
Cynomys ludovicianus	Grassland South Dakota [117]			No (only after 2 years)	4.9 (young)	
	Grassland Colorado [118]				4.2–5.0 3.3 (yearling)	0.98
Eutamias amoenus	Coniferous forest Washington [119] Minnesota [120]					
Eutamias minimus	Minnesota [120]					
Funambulus pennanti	Garden India [121], [122]					
Funisciurus anerythrus	[123]	Non				
Sciurus carolinensis	Woodland Britain [124]					
	Mixed forest North Carolina [125]					
Spermophilus franklinii	Meadows Manitoba [126]		1	No	7.5	
	Old field Manitoba [127]	S	1	No	9.4	
Spermophilus richardsonii	Shortgrass Wyoming [128]		1	No	6.6	
Spermophilus tridecemlineatus	Grassland Texas [129]	S	1–2		4.9–7	
Spermophilus undulatus	Tundra Alaska [130]					
Tamias striatus	New York [131] Deciduous and mixed forest Pennsylvania and Vermont [132]	S	2	Few	4	
Tamiasciurus hudsonicus	Coniferous forest British Columbia [133]				3.6–5.1	
	Coniferous forest Saskatchewan [134]				4	
	Mixed forest Alberta [135]	S			3.4 (1967) 4.3 (1968)	0.67 0.88
	Coniferous forest New York [136]	S	2		4.5	
	Coniferous forest British Columbia [137]		1–2	No		
Tamiscus emin	Forest Congo [123]	Non				
Zapodidae						
Zapus hudsonius	Wyoming [138]	Hibernates	1		5.1	
	Wyoming [139]		2		5.5	
Zapus princeps	Wyoming [140]	Hibernates	1			

	Survival (monthly)						Home range (area and/or maximum distance)	Density (ha^{-1})		
									Interseasonal	
	Adult	Young	Overwinter	Yearlong	Litters by season	e_0		Seasonal	Magnitude	Period (yrs)
[111]							♂ 0.49 ha ♀ 0.19 ha	14.8–27.1		
[89]			0.78			$e_1 = 2.0$	♂ 0.19 ha ♀ 0.19 ha			
[112]						$e_0 = 3.44$		2.5–9	1–9	
[113]							6 ha	0.06–0.35		
[114]										
[115]									Up to 14	
[116]									0.6–1.5	
[114]										
[117]		0.96						9.9–27.7	13.3–27.7	
[118]										
[119]							♂ 1.3–1.57 ha ♀ 0.84–1.0 ha			
[120]							♂ 1.29 ha ♀ 0.56 ha			
[120]							♂ 1.22 ha ♀ 0.66 ha			
[121], [122]							♂ 0.21 ha ♀ 0.15 ha			
[123]						$e_a = 5.68$			0.86–88.0 ha	
[124]							♂ 479 m ♀ 261 m		2.05	
[125]	0.95	0.80				$e_0 = 12$ $e_a = 21.8$		0.88–1.47	0.64–3.23	
[126]			0.79–0.93							
[127]	0.77–0.92									
[128]						$e_a = 2.1$	0.32 ha			
[129]	0.97	0.86					♂ 4.7 ha ♀ 1.4 ha		0.52–0.7	
[130]							0.3–3.5 ha			
[131] [132]	0.96	0.94		0.94	0.92 spring 0.94 summer					
[133]				0.96		$e_1 = 15.5$				
[134]										
[135]	0.66 annual	0.33 annual				$e_0 = 13.8$ $e_0 = 11.8$ (hunted)			0.51–0.61	2.6
[136]							♂ 2.4 ha ♀ 1.9 ha			
[137]										
[123]										
[138] [139] [140]	0.98	0.92					♂ 211 m ♀ 155 m	0.36–0.73	0.36–0.73	

Table 4.2 (*continued*)

Species	Habitat Location [authority]	Reproduction				
		Seasonal or non-seasonal	Number of litters	By young of year	Litter size Embyro rate	Max % pregnant
Heteromyid type						
Dipodomys merriami	Desert (enclosure) Nevada [141]					
	Desert (enclosure) Nevada [142]					
	Desert (enclosure) Nevada [143]	S	2		2.6	
	Desert California [144]					
	Desert Arizona [145]	S			2.02	
	(Captive) [146]				2.3	
	Desert New Mexico [147]	S	2	Yes		0.67
	Desert Arizona [96]					
Dipodomys microps	Desert (enclosure) Nevada [141]					
	Desert (enclosure) Nevada [142]	S				
Dipodomys nitratoides	(Captive) [146]				2.3	
Dipodomys ordii	Desert New Mexico [147]	S	4	Yes		0.94
Geocapromys ingrahami	Island Bahamas [148]				1	
Heteromys desmarestianus	(Review) [149]	Non				
Heteromys goldmani	(Review) [146]	Non			3–4	
Jaculus jaculus	Desert Sudan [150]			No?	3.8	
Liomys adspersus	Tropical forest Panama [151]	S	1.44		3.2	
Liomys pictus	(Captive) [146]	S			3.5	
Neotoma floridana	Savanna–woodland Oklahoma [152]	S	2+	No	3.2	
Neotoma fuscipes	Chaparral California [153]					
Neotoma lepida	Juniper–sagebrush Utah [154]	S				
Neotoma micropus	Prosopis–brushland Texas [155]				2.3	0.69
Perognathus californicus	(Captive) [146]				4.0	
Perognathus formosus	Desert (enclosure) Nevada [141]	S			5.6	0.90
	Desert (enclosure) Nevada [142]					
Perognathus longimembris	Desert (enclosure) Nevada [141]	S				
	Desert (enclosure) Nevada [142]					
	Desert California [144]	S				
Proechimys semispinosus	Tropical forest Panama [156]	Non				
Fossorial type						
Myospalax fontaniere	River Valley China [157]					
Spalacopus cyanus	Chile [158]				3.0	
Spalax microphthalmus	USSR [159]	S			2.83	0.38–0.50
Talpa europaea	Arable land N. USSR [160]		1		5.0	
	Arable land United Kingdom [161]	S	1			

	Survival (monthly)					e_0	Home range (area and/or maximum distance)	Density (ha^{-1})		
	Adult	Young	Overwinter	Yearlong	Litter by season			Seasonal	Interseasonal Magnitude	Period (yrs)
[141]				0.90		$e_1 = 6.6$		0.55–2.78	0.44–4.0	
[142]							♂ 1.76 ha ♀ 0.87 ha			
[143]										
[144]				0.85			1.92 ha	0.77–3.72	0.45–3.72	
[145]							♂ 0.19 ♀ 0.15			
[146] [147]							♂ 1.6 ha ♀ 1.6 ha		0.35	
[96]									7.8–15.6	
[141]				0.92		$e_1 = 8.4$		1.89–11.1	1.67–11.1	
[142]							♂ 0.93 ha ♀ 0.83 ha			
[146] [147]							♂ 1.4 ha ♀ 1.3 ha		0.15	
[148]								30		
[149] [146] [150]							0.56 ha		0.67	
[151]							♂ 0.57 ha ♀ 0.55 ha	5.5–10.0	5.4–11.0	
[146] [152]				0.65–0.73		$e_a = 2.2$	♂ 0.266 ha 62.5 m ♀ 0.16 ha	0–21.5	11.2–21.5	
[153]								1.2–6.1		
[154]							♂ 76.8 m ♀ 41.4 m		2.8, 7.6	
[155]				0.92–0.96		$e_1 = 3.12$	0.02 ha	9.9–30.9	9.9–30.9	
[146] [141]				0.94		$e_1 = 12.9$		12.7–27.8	2.78–27.8	
[142]							♂ 0.36 ha ♀ 0.24 ha			
[141]				0.96		$e_1 = 15.4$		0.78–3.3	0.78–5.0	
[142]							♂ 0.40 ha ♀ 0.51			
[144]				0.90					0.8–1.8	
[156]	0.86	0.75	0.88						8.1–10	
[157]							Flood plain 2.8 ha Terrace hills 1.87 ha Other hills 1.52 ha			
[158]							12 colonies in 4 ha, 15 individuals in each			
[159] [160]		0.70–0.88							1.7–7.0	
[161]										

Table 4.2 (*continued*)

Species	Habitat Location [authority]	Reproduction				
		Seasonal or non-seasonal	Number of litters	By young of year	Litter size Embryo rate	Max % pregnant
Fossorial type (*Continued*)						
Talpa europaea (*Continued*)	Arable land	S	1 (rarely 2)		2–6	
	United Kingdom [162]					
	Denmark [163]	S	1			
	Netherlands [164]				4.6 UK	
					5.0 Cent.	
					Europe	
	Arable land					
	United Kingdom [165]					
	Grassland					
	S. Poland [166]					
Thomomys talpoides	Farmland					
	Texas and Colorado [167]					
Thomomys bottae	Farmland	S		No	3–6	
	Texas and Colorado [168]					
	Grassland	S	1+		4.6	1.0
	California [169]					

† k_n, population cohort.   ‡ Mean life expectancy in months for adults.

1, Pearson (1967); 2, Martin (1971); 3, Taylor & Horner (1971); 4, Berry (1968); 5, Newsome (1971); 6, Newsome (1969a); 7, Newsome (1969b); 8, Kasatkin *et al.* (1969); 9, Negus *et al.* (1961); 10, Boshell & Rajagopalau (1968); 11, Tamarin & Malecha (1972); 12, French (1965); 13, Khamaganov (1968); 14, Brooks (1972); 15, Petryszyn & Fleharty (1972); 16, Dunaway & Kaye (1961); 17, Shvarts *et al.* (1969); 18, Yoshida (1970); 19, Bergstedt (1965); 20, Bobek (1971); 21, Bobek (1969); 22, Gebczyńska (1966); 23, Migula *et al.* (1974); 24, Radda (1969); 25, Hansson (1971a); 26, Zejda & Pelikan (1969); 27, Southern & Lowe (1968); 28, Hansson (1968); 29, Watts (1969); 30, Tanton (1969); 31, Smyth (1968); 32, Crawley (1970); 33, Ashby (1967); 34, Zejda (1972); 35, Morel & Meylan (1970); 36, Panteleev (1968); 37, Stoddart (1971); 38, Stoddart (1970); 39, Pelikan & Holisova (1969); 40, Panteleev (1967); 41, Kulik (1968); 42, Fuller *et al.* (1969); 43, Chitty & Phipps (1966); 44, Hansson (1969a); 45, Claude (1967); 46, Claude (1970); 47, Newson (1963); 48, Zejda (1971); 49, Semenov-Tyan-Shanskii (1970); 50, Ryszkowski (1971b); 51, Petrusewicz *et al.* (1971); 52, Gliwicz *et al.* (1968); 53, Bujalska *et al.* (1968); 54, Smyth (1966); 55, Naumov *et al.* (1969); 56, Okulova *et al.* (1971); 57, Pucek *et al.* (1968); 58, Ryszkowski & Truszkowski (1970); 59, Hsia (1966); 60, Koshkina (1965); 61, Koshkina (1969); 62, Schindler (1972); 63, Myllymäki *et al.* (1971); 64, Hansson (1968); 65, Evans (1973); 66, Wojciechowska (1969); 67, Spitz (1970); 68, Gromadzki & Trojan (1971); 69, Batzli & Pitelka (1971); 70, Pjastolova (1971); 71, Keller & Krebs (1970); 72, Buchalczyk *et al.* (1970); 73, Pjasto-lova (1971); 74, Gashwiler (1972); 75, Christian (1971); 76, van Vleck (1969); 77, Golley (1961); 78, Valentine & Kirkpatrick (1970); 79, Dapson (1968); 80, Borowski & Dehnel (1953); 81, Nosek *et al.* (1972); 82, Michielsen (1966); 83, Buckner (1969); 84, Tarkowski

	Survival (monthly)					Home range (area and/or maximum distance)	Density (ha^{-1})		
								Interseasonal	
Adult	Young	Overwinter	Yearlong	Litters by season	e_0		Seasonal	Magnitude	Period (yrs)
[162]									
[163]									
[164]						0.04 ha 46 m 123.6 m (tunnel)			
[165]						0.04 ha			
[166]	0.95								2.8–25.1
[167]									
[168]						0.02 ha	5–10	20–30	
[169]				♂ 0.93 ♀ 0.97	$e_6 = 7.6$	0.02 ha	26.7–82.1	20.7–84.1	

(1957); 85, Hansson (1968); 86, Pucek (1960); 87, Wise (1967); 88, Smirnov *et al.* (1971); 89, Prakash & Rana (1970); 90, Delaney (1971); 91, Prakash (1971); 92, Shubin & Bekenov (1971); 93, Linzey (1968); 94, Blus (1966); 95, Blair (1943*b*); 96, Chew & Chew (1970); 97, Terman (1968); 98, Howard (1949); 99, Metzger (1971); 100, Snyder (1956); 101, Sheppe (1966*b*); 102, Sadleir (1970); 103, Gashwiler (1971); 104, Dunmier (1960); 105, Sadleir (1965); 106, Petticrew & Sadleir (1970); 107, Beer & MacLeod (1966); 108, Delaney (1971); 109, Packard (1968); 110, Fisler (1971); 111, de Moor (1969); 112, Poulet (1972*a*); 113, Bradley (1967); 114, Solomonov (1970); 115, Shiranovich (1968); 116, Naumov *et al.* (1970); 117, King (1955); 118, Koford (1958); 119, Broadbrooks (1970); 120, Forbes (1966); 121, Prakash *et al.* (1968); 122, Prakash & Kametkar (1969); 123, Rahm (1970); 124, Taylor *et al.* (1971); 125, Barkalow *et al.* (1970); 126, Sowls (1948); 127, Iverson & Turner (1972); 128, Clark (1970); 129, McCarley (1966); 130, Carl (1971); 131, Yerger (1955); 132, Tryon & Snyder (1973); 133, Millar (1970*b*); 134, Davis & Sealander (1971); 135, Kemp & Keith (1970); 136, Layne (1954); 137, Millar (1970*a*); 138, Clark (1971); 139, Whitaker (1963); 140, Brown, L. N. (1970); 141, French *et al.* (1974); 142, Maza *et al.* (1973); 143, Bradley & Mauer (1971); 144, Chew & Butterworth (1964); 145, Reynolds (1960); 146, Eisenberg (1963); 147, Blair (1943*a*); 148, Clough (1972); 149, Fleming (1970); 150, Happold (1970); 151, Fleming (1971); 152, Goertz (1970); 153, Vogl (1967); 154, Stones & Hayward (1968); 155, Raun (1966); 156, Gliwicz (1973); 157, Chen *et al.* (1966); 158, Reig (1970); 159, Ovchinnikova (1969); 160, Ivanter (1969); 161, Deanesley & Allanson (1967); 162, Godfrey (1956); 163, Mohr (1933); 164, Raeck (1969); 165, Godfrey (1955); 166, Skoczen (1966); 167, Miller (1964); 168, Turner *et al.* (1973); 169, Howard & Childs (1959)

Table 4.3. *Summary of demographic and home range characteristics by ecological type*

	Survival rate low, reproductive rate high	
	Murid type	Microtine type
Reproduction		
Seasonal	Yes (few exceptions)	Yes
By young-of-the year	Yes	Yes
Number of litters	3.30	3.32
Litter size	6.14	5.86
% ♀♀ reproductive	66	81
Survival		
Life expectancy (months)	1.82	3.05
Yearlong (monthly prob.)	0.82 (1 observation)	0.64
Overwinter (monthly prob.)	0.25	0.74
Home range		
Area (ha)	0.50	0.89 (♂ 1.66, ♀ 0.58)
Maximum distance (m)	29.5 (♂ 24, ♀ 35)	177.3 (♂ 283, ♀ 71)
Density		
Seasonal	117.7	66.1
Interseasonal	15.7	67.3

	Survival rate moderate, reproductive rate moderate	
	Insectivora	Cricetine type
Reproduction		
Seasonal	Yes	Yes
By young-of-the-year	Some	Yes (some species)
Number of litters	3.00	3.06
Litter size	6.15	4.04
% ♀♀ reproductive	74.0	59.6
Survival		
Life expectancy (months)	7.4	3.65
Yearlong (monthly prob.)	0.91	0.73
Overwinter (monthly prob.)	—	0.81
Home range		
Area (ha)	0.52	1.09
Maximum distance (m)	152	97.3
Density		
Seasonal	10.13	15.4
Interseasonal	7.24	10.6

Table 4.3 (*continued*)

	Survival rate high, reproductive rate low	
	Sciuridae	Zapodidae
Reproduction		
Seasonal	Varies with latitude	Yes
By young-of-the-year	No	—
Number of litters	1.44	1.33
Litter size	5.53	5.30
% ♀♀ reproductive	84	55 (1 observation)
Survival		
Life expectancy (months)	12.53	—
Yearlong (monthly prob.)	0.91	0.98 (1 observation)
Overwinter (monthly prob.)	0.86	—
Home range		
Area (ha)	1.63 (♂ 1.81, ♀ 0.93)	—
Maximum distance (m)	370 (♂ 479, ♀ 261)	183 (♂ 211, ♀ 155)
Density		
Seasonal	6.73	0.54
Interseasonal	11.8	0.54

	Survival rate high, reproductive rate low	
	Heteromyid type	Fossorial
Reproduction		
Seasonal	Mostly seasonal	Yes
By young-of-the-year	Some	No
Number of litters	2.29	1.00
Litter size	3.05]	4.20
% ♀♀ reproductive	80	63
Survival		
Life expectancy (months)	9.28	7.6
Yearlong (monthly prob.)	0.87	0.95
Overwinter (monthly prob.)	—	—
Home range		
Area (ha)	0.77 (♂ 0.75, ♀ 0.62)	0.90
Maximum distance (m)	—	84.8
Density		
Seasonal	8.6	31.0
Interseasonal	8.5	23.9

The combination of high reproductive rate and high turnover rate of individuals often results in a population that is volatile, that is, characterized by strong fluctuations in density. Rodents of the microtine and murid types fall into this category.

The second major demographic grouping is characterized by species having a moderate reproductive rate, a medium survival rate, and a moderate population density. These populations are more stable than those of the previous group, and they are seldom found at high densities. The major taxonomic group of small mammals that compose this category is the subfamily Cricetinae. The Insectivora are close allies of this group, according to demographic characteristics, and are here included within it. It should be noted that all data for Insectivora refer to members of one subfamily, the Soricinae.

The third category of the demographic classification is the group characterized by a pattern of low reproductive rate, high survival, and rather low density populations. Seasonal dormancy is observed to occur in many species and appears to contribute to long survival. Populations are generally quite stable. Included in this category are the families Heteromyidae, Sciuridae, Zapodidae, and fossorial forms.

Group characteristics

In all of the major rodent families and subfamilies considered here, reproduction is limited to a particular season or part of the year. There are, however, some exceptions in each taxa, particularly where they have representatives in latitudes or habitats where climatic conditions are less strongly seasonal. For example, the Sciuridae, a strongly seasonal group in temperate latitudes in which most members hibernate during a large portion of the year, has representatives in tropical Africa (*Tamiscus* and *Funisciurus*) which are nonseasonal in reproduction. Representatives of the Microtinae and the Heteromyidae, under tropical conditions, may also show nonseasonality in reproduction. However, even within groups having both temperate and tropical representatives and having seasonal and non-seasonal reproduction, the total reproductive effort (*sensu* Williams, 1966) is very nearly the same. That is, the same number of young are produced per female, but spaced over a greater or lesser portion of the year (Fleming 1971).

In number of litters produced during a single season, the microtine type, murid type, Insectivora, and cricetine type may be categorized as high (>3.0), whereas the other groups may be considered low (<2.3).

Reproduction by young-of-the-year in the same season of their birth generally occurs in the groups with a high number of litters during the season and generally does not occur in those groups with a low number of litters per season. The murids, Insectivora, microtines, Sciuridae and Zapodidae are characterized by large litter size (>5.3). The cricetines, fossorial forms, and heteromyids are generally lower in litter size (<4.5). One subgroup of Cricetinae, including *Neotoma* and certain *Peromyscus* (*californicus* and *truei*, possibly *polionotus*), have unusually small litter sizes for the group. This may be a function of body size (large) or habitat (arid and warm). Ecologically, some of these forms better fit into the heteromyid group. In the Sciuridae, Heteromyidae, and Microtinae the prevalence of pregnancy or percentage of females reproductive in the population is very high (>80). It is intermediate in the Insectivora and relatively low (<70) in murids, cricetines, Zapodidae, and fossorial forms.

In the various methods of expression of longevity or survival in populations of small mammals, the microtines and murids are short-lived (e.g., life expectancy 1.8 months, survival 0.64/month) while the Sciuridae, heteromyids, Zapodidae, Insectivora, and fossorial forms are characterized by high survival or longer life-span (life expectancy 7·4 to 12.5 months, survival 0.87 to 0.98/month). These groups mentioned above are characterized by particularly high annual survival rates. The cricetine and microtine types are intermediate (life expectancy 3.1 to 3.6 months, survival 0.64 to 0.73/month.

Home range is large in the Cricetines and Sciuridae (1.1 to 1.6 ha). Other groups have home ranges of smaller size but there seems to be little consistency in published home range data.

Densities are generally lower in the stable, slow-breeding populations. Highest densities and greatest interseasonal range of density can be found in the murid and microtine types (66 to 118/ha). These may be considered as high density, unstable populations. The Sciuridae, heteromyids, cricetines, and insectivora have low density (7 to 15/ha), stable populations; fossorial types sometimes attaining intermediate density (31/ha). The Zapodidae evidently have very low population densities (0.54/ha).

Generalizations

By and large, the groups set up on a taxonomic basis also form natural groups on a demographic scheme. There is evidence in some groups, such as the heteromyids, zapodids, and the sciurids, that the habit of hibernation confers unusual longevity. The Heteromyidae also contain nonhibernators, but when the life histories of all are considered, they still evidently form a natural ecological group. Sciurids form a natural ecological group in spite of the fact that this group contains hibernators and nonhibernators, seasonal breeders and nonseasonal breeders, colonial and solitary species.

Evidence summarized here indicates that there is an inverse relationship between survival and production in small mammal groups (see Fig. 4.2 for illustration of these trends). Apparently small mammal populations have achieved success in maintaining numbers and in exploiting their habitats by two types of strategies. One of these is to achieve a rapid rate of increase by production in the form of high

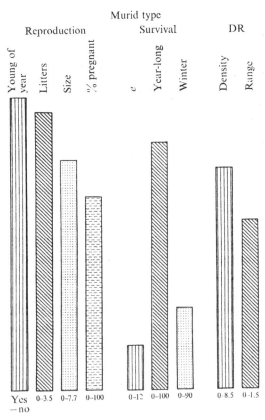

Fig. 4.2. Relative values of the demographic parameters. Each mean value for the group in Tables 4.3, 4.4, and 4.5 is plotted on a scale that represents the range of that value for all groups. Each bar, therefore, represents the percentage of the total range for that parameter.

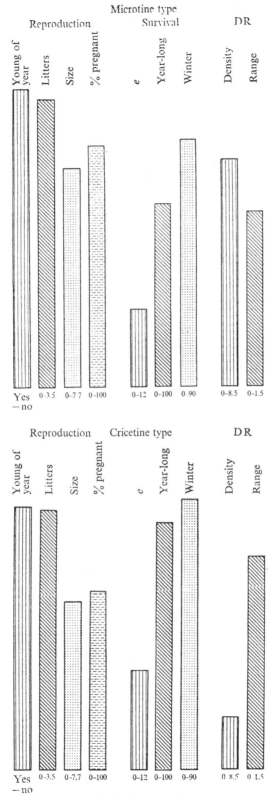

Fig. 4.2 (*continued*)

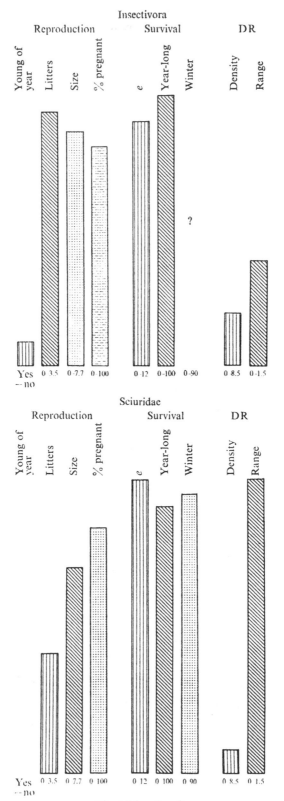

Fig. 4.2 (*continued*)

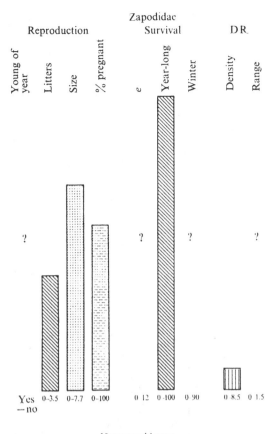

Reproduction — Zapodidae Survival — D R.

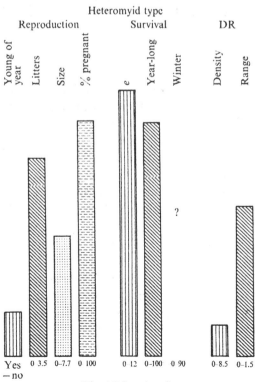

Heteromyid type

Reproduction — Survival — DR

Fig. 4.2 (*continued*)

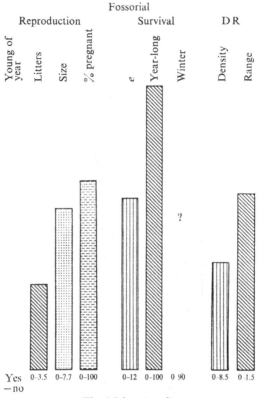

Fig. 4.2 (*continued*)

reproduction and rapid growth to maturity. The second means of achieving success is by effective survival of the reproductive members of the population. Small mammal populations adopt one or the other of these strategies (Fig. 4.3). Those species populations that are characterized by high rates of production are also characterized by low survival rates. Those that are characterized by good survival rates have comparatively small and infrequent litters, and development to maturity is slow. In the latter group, population density is generally controlled by environmental conditions acting on food supply of the population. The former type of population is sometimes controlled by intrinsic or density-dependent factors.

These strategies are defined as *r*-selection and *K*-selection by Mac-Arthur & Wilson (1967). The former group is characterized by a high rate of productivity and low investment of energy in individual off-spring, while the latter group is characterized by high energy input into individual offspring and consequent low relative productivity (Pianka, 1970). In unsaturated environments where density effects and competition are minimal, *r*-selection occurs; *K*-selection occurs where the environment is saturated with organisms and density effects are maximal.

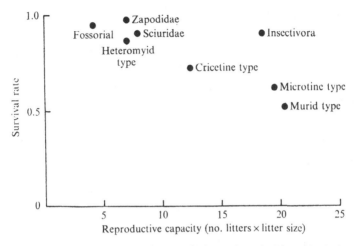

Fig. 4.3 Relationship between reproductive capacity and survival in ecological groups of small mammals. Survival is the mean monthly rate for the group, and reproductive capacity is the mean seasonal production of young.

Hence, offspring are developed for competition in the latter case and for exploitation in the former.

Figure 4.3 also indicates a separation of groups according to energetic relationships. Those groups to the left of the graph possess energy-saving behavioral characteristics, such as hibernation or heterothermy. Those on the right are characterized by high individual metabolic rates. The cricetine group contains some species that can lower body temperature to some degree. These species show fluctuations on a daily scale whereas the true hibernators fluctuate on a seasonal basis.

The relative importance of the various demographic parameters depends upon the strategy for success adopted by the species population. Those that depend upon high rate of production for success have optimized the parameters of litter size, growth to maturity, number of litters per season, and reproduction by young-of-the-year. Of lesser importance to these populations are the parameters affecting survival. These populations normally have high rates of turnover. Populations that have adopted the strategy of improved survival for their success have optimized the demographic parameters dealing with life-span or

101

longevity. They are characterized by high survival of young and adults and often conserve energy and adjust their period of active functioning to that period when environmental conditions are optimal by employing hibernation as a means of surviving adverse conditions.

It is important to note that species are not rigidly locked into a particular strategy of survival. Various populations of a given species living in different environments modify the relative amounts of energy devoted to longevity and production. This has been observed in a microtine in Europe and a cricetine in North America, where populations of both species have great altitudinal ranges. These cases demonstrate the existence of a balance between longevity and productivity in populations of small mammals.

This summary of demographic data from small mammal populations from most major taxonomic groups and from many parts of the world suggests that the evolutionary objective of these populations is to reproduce, but that the environment imposes limitations on the reproductive capacity; this is reflected in enhanced survival. Such a working hypothesis may serve to guide further surveys and research efforts.

These differing strategies of population success in small mammal populations have implications for artificial control of pest species. Most effective control can be achieved by designing the control program to attack those demographic parameters most critical to success of the population. For those dependent upon rapid rate of increase, control effort should be applied in a manner to affect productivity of the population, such as eliminating nest sites, modification of the sex ratio, or reducing fertility of the species. Effort devoted to altering the survival rate of the species, such as trapping or poisoning, would be expected to be less fruitful. For those species populations which have adopted the strategy of augmented longevity for their success, control measures designed to increase mortality would be expected to be more rewarding than control measures designed to further reduce the already limited reproduction capacity of the species.

This paper reports on work supported in part by National ScienceFoundation Grant GB–31862X2 to the Grassland Biome, US International Biological Programme, for 'Analysis of Structure, Function, and Utilization of Grassland Ecosystems'.

5. The role of dispersal in the demography of small mammals

W. Z. LIDICKER, JR

One feature of small mammal populations which is currently being given increasing attention is that of dispersal. The primary purpose of this chapter is to summarize the information currently available on dispersal in small mammals and to suggest some directions for future research. Commonly, dispersers have been either ignored or considered to be of little significance by population ecologists. In part this has been due to considerable difficulty in measuring dispersal, but it is also widely assumed that such movements are demographically unimportant. For example, dispersers are typically considered to be individuals that leave home only when conditions become intolerable, and then have a near-zero probability of surviving long enough to reproduce elsewhere. Widespread acceptance of the term 'gross mortality' to represent any losses from a population, whether from real mortality or from emigration, serves to reinforce the notion that emigrants are only briefly postponing their inevitably premature death. Moreover, population growth rates are all too frequently defined in terms of birth and death rates only. In fact, growth rates are a function of the rate of additions to a population, which are composed of births *and* immigration, minus the rate of losses, which are divided between deaths *and* emigration.

In contrast to these perfectly logical biases held by many ecologists, evolutionary theorists, including population geneticists, have generally considered 'migration' as a potentially important factor in their investigations. Differential movements of genes, for example, are acknowledged as one of the factors capable of disrupting Hardy–Weinberg equilibria. Similarly, gene flow can oppose selection acting to improve local adaptations; it can be one component in an adaptive strategy for a variable environment, oppose inbreeding, increase the probability of novel gene combinations, etc.

In my view it is now time for a re-appraisal of the role of dispersal in short-term population processes as well. Consequently, in this paper I will call attention to the accumulating evidence that strongly implicates dispersal as a generally important demographic parameter. In doing this, I wish to discourage the impression that I do not consider other factors important. Clearly such a view would be nonsense, and in fact I have advocated (Lidicker, 1973) a multi-factorial or community view of small mammal population dynamics. Of course, some ecologists have recognized the potential importance of dispersal, and I would like in particular to mention significant discussions by Andrewartha & Birch (1954; p. 124), Andrzejewski, Kajak & Pieczyńska (1963), Petrusewicz (1966a), and Krebs *et al.* (1973). In the present paper, dispersal will be shown to influence various population properties as well as, in some cases, the regulation of numbers. Clearly, if dispersal can affect the composition, size, and spacing of populations, it can influence the productivity of those populations as well. The focus will be on small mammal populations, but the conclusions drawn are, I believe, more generally applicable.

Additional objectives of this paper are: (1) to characterize two kinds of dispersal, each having different causes, demographic effects, and evolutionary bases; (2) to recommend the use of islands and enclosures to study the nature and influence of dispersal; and (3) to propose models for the possible role of dispersal in the regulation of numbers. Emphasis will thus be on the demographic aspects of dispersal. The extremely important evolutionary questions will not be dealt with adequately, although I have discussed some of the relevant issues in an earlier paper (Lidicker, 1962), and Gadgil (1971) has provided an extremely interesting discussion of the importance of environmental variations in space and time on the evolution of dispersal. Perhaps it will be sufficient to point out that models proposed here require only individual selection for their operation. However, they are also consistent with the possibility that group selection, in a manner such as proposed by Van Valen (1971), may have abetted individual selection in the evolution of dispersal behavior.

I should like to point out at the outset that I am using the term dispersal to refer to any movements of individual organisms or their propagules in which they leave their home area, sometimes establishing a new home area. This does not include short-term exploratory movements, or changes in the boundaries of a home range such that the new range includes at least part of the former. Dispersal thus produces homeless travelers (vagrants) who are in search of a new home.

When such movements result in individuals leaving or entering the population under study, this then becomes emigration and immigration respectively. These latter terms are thus defined in a demographic context, and are not the same as the geneticist's 'migration'. Migration usually implies that genetic material has moved (e.g., Parsons, 1963), whereas dispersal refers only to individuals. While dispersal is clearly necessary for gene flow, it is not sufficient.

Saturation and pre-saturation dispersal

Basic to the development of my propositions on the demographic role of dispersal is the notion that there are two qualitatively distinguishable kinds of dispersal, and these result in two corresponding kinds of emigration. One type I propose to call *saturation dispersal* and the other *pre-saturation dispersal* (both types come under the general definition of *emigration*). Previously (Lidicker, 1962) I had called attention to these two types, but used the term 'density responsive emigration' for the

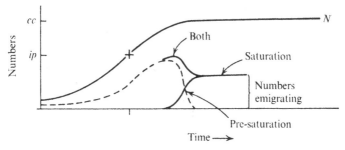

Fig. 5.1. Relation between changes in numbers of a population and both saturation and pre-saturation types of emigration; *cc* is carrying capacity and *ip* stands for inflection point.

latter. Fig. 5.1 illustrates the time course of these two kinds of emigration relative to changes in population numbers, and also suggests that both types may be exhibited by the same population.

Saturation emigration is the outward movement of surplus individuals from a population living at or near its carrying capacity. Such individuals (exhibiting saturation dispersal behavior) are faced with the immediate choice of staying in the population and almost certainly dying, or moving out and probably dying. They would represent, more often than not, social outcasts, juveniles, and very old individuals, those in poor condition, and in general those least able to cope with local conditions. For these reasons, they would be very susceptible to the various physical and biological hazards that they would surely encounter on their journeys, and would have only a very small chance of reaching a suitable location and reproducing successfully. Similarly, vagrants who do not become emigrants would not long survive under these conditions.

This is the traditional view of emigration, and contributes to the inclusion of such emigrants under the rubric 'gross mortality'. Wynne-Edwards (1962, p. 483) uses the term 'safety-valve' emigration for dispersal which relieves overpopulation, and Howard (1960) uses 'environmental dispersal'. These are similar but not identical notions to saturation emigration, as they are not tied so specifically to resource limitations. Many cases of mass movements of individuals (see Dymond, 1947, for examples) may be expressions of saturation dispersal.

Pre-saturation emigration is an exodus from a population before its habitat becomes saturated with that species. That is, it occurs during population growth, and may even begin very soon after growth starts. Moreover, such emigrants (exhibiting pre-saturation dispersal behavior) are not necessarily destitute either socially or economically, as there remains a surplus of resources in their home environment. Therefore, they will in general be in relatively good condition, and may include any sex and age group including pregnant females. Pre-saturation emigrants may be characterized by possessing a particular sensitivity to increasing densities, or they may have discovered some better home location during an exploratory excursion. Compared to saturation emigrants, such individuals can be expected to have a much greater chance of surviving and re-establishing themselves elsewhere.

Thus, these kinds of emigration differ not only in their timing with respect to population growth, but also produce emigrants of very different quality. In general they can be operationally identified by whether emigration occurs before carrying capacity is reached or after. Realistically, however, there should be some overlap as carrying capacity is approached, because at this stage individual variation and micro-geographic variation in conditions will cause some individuals to leave because resources have run out for them, while others will still be leaving 'voluntarily'.

Saturation emigration does not seem to raise any difficulties with respect to its origin by natural selection. Such dispersal behavior is an act of desperation which prolongs life and very slightly increases the chances of longer-term survival. On the other hand, the evolution of pre-saturation dispersal behavior is not so easily explained. It involves, after all, the assumption of all the risks of leaving home when it is not really necessary. In an earlier discussion (Lidicker, 1962), I offered quantitative (increased opportunity for mating), qualitative (heterosis and increased chance for producing favorable genetic recombinants), and diplomatic (avoidance of population crashes and predator buildups) reasons why dispersal behavior may be favored by natural selection acting on individuals. To this we can add an economic reason. As populations become denser, intra-specific interference may result in inefficient utilization of remaining resources. For example, resources may be scattered where previously they were clumped, the most desired food or shelter may be scarce, more time and energy may be spent in social interactions, or activity periods may be shifted to less optimal times. Consequently, individual fitness may be improved by leaving rather than continuing to utilize available resources inefficiently. Of course, if circumstances permitted the operation of group selection, this too could lead to selection for dispersal behavior (cf. Van Valen, 1971).

In general, it can be claimed that where conditions favor the evolution of dispersal behavior, it will be the pre-saturation type that will appear. This is simply because it is this type that leads to a class of wanderers that has a relatively high probability of passing on its genetic make-up to future generations. If these arguments are valid, one would expect to find evidence that in many species dispersal behavior has a genetic component. This is not the place to document in detail the rapidly expanding evidence in support of this prediction. However, one can point to the existence among organisms of numerous adaptations for long-range movement, for example life history stages particularly suited for dispersal, as one line of evidence. Another kind of evidence is the nearly ubiquitous presence of sex, age and/or developmental stage differences in a tendency to move. This sort of evidence has been emphasized by Howard (1965) in distinguishing 'innate' from 'environmental' dispersals. Particularly elegant examples of this are provided by cases among migratory birds in which adults depart early from the breeding areas, and leave the remaining resources to juveniles (e.g., Pitelka, 1959). Strain differences in dispersal tendencies have also been described for a number of insects as well as house mice, *Mus musculus* (Incerti & Pasquali, 1967). Finally, there is the evidence for dispersal polymorphisms within a single population. This last includes a variety of invertebrates as well as vertebrates. Earlier reviews by Johnston (1961) on birds, and Howard (1960) and Anderson (1970) on mammals have been supplemented by recent evidence for several species of small mammals, for example, *Microtus pennsylvanicus* and *M. ochrogaster* (Myers & Krebs, 1971), and possibly *Peromyscus polionotus* (M. H. Smith *et al.*, 1972).

A further consequence of this view of dispersal is that pre-saturation emigration will in fact be common. Until recently, few investigators have thought to search for this kind of movement so that good evidence remains sparse. Nonetheless, Andrewartha & Birch (1954, p. 124) considered emigration from sparse populations to be ubiquitous, and Andrzejewski, Kajak & Pieczyńska (1963) seem to have recognized it as widespread.

Again, there are few examples from the invertebrates, and even humans (Birdsell, 1957), but I will focus my attention on evidence from species of small mammals. Perhaps the best known species in this regard is *Mus musculus*. It is now well-established that social pressures motivate dispersal long before economic saturation of the habitat is reached (Naumov, 1940; Strecker, 1954; Andrzejewski & Wrocławek, 1962; DeLong, 1967; Newsome, 1969a; Anderson, 1970; Lidicker, unpublished results). In addition, at least four species of microtines seem to show this behavior. These are: *Clethrionomys glareolus* (Evans, 1942; Smythe, 1968; Crawley, 1969), *Microtus californicus* (Lidicker & Anderson, 1962; Krebs, 1966; Fig. 5.3), *M. pennsylvanicus* (Van Vleck, 1968; Grant, 1971; Myers & Krebs, 1971; Ambrose, 1973), and *M. ochrogaster* (Myers & Krebs, 1971). Four other species of rodents with similar behavior are *Rattus villosissimus* (Newsome, 1975), *Spermophilus beecheyi* (Evans & Holdenreid, 1943), *Sigmodon hispidus* (Joule & Cameron, unpublished results), and *Myocaster coypus* (Ryszkowski, 1966). Of these last, *R. villosissimus* shows particularly spectacular dispersal often covering hundreds of miles across the deserts of central Australia. One insectivore, *Sorex cinereus* (Buckner, 1966), shows evidence of possibly exhibiting this kind of mobility.

Although this list is brief, consisting of only ten species, it should be greatly expanded when investigators are interested in searching for pre-saturation dispersal. As it is, it suggests a preliminary generalization about the circumstances that may be conducive to the evolution of this kind of behavior. In general, the rodents on this list fluctuate strongly in numbers and live in small demes when their numbers are low. It may be that this life history style leads to selection for pre-saturation dispersal. I suspect that it will also be found to be common amongst rare species.

Effects on population properties

Gross mortality

As already mentioned, the classic view of emigration is to lump it with real mortality under the heading of gross mortality, which then accounts for all losses to the population. Clearly emigration can contribute importantly to the total losses suffered by a population, and, to the extent that emigration is of the saturation type, it may not matter much how these losses are constituted. Emigration would then be merely a substitute for deaths as it only briefly delays mortality.

It is well known that vagrants carry a high risk of death from lack of shelter, starvation, predation (e.g., Errington, 1946; Pielowski, 1962; Metzgar, 1967; Varshavski, 1937, cited by Stoddart, 1970; Ambrose, 1972), etc. If they are further compromised by being in poor condition when they leave home, their chances are even less. Andrzejewski & Wrocławek (1961) have shown that dispersing *Clethrionomys glareolus* and *Apodemus flavicollis* die in live traps more often than do residents, and Janion (1961) has found that dispersing female house mice carry more fleas than do settled individuals.

Pre-saturation emigration, of course, also contributes to population losses, but is not so appropriately grouped with mortality. Not only may it occur when death rates are very low, but it may involve a component of the population which is very different from that most subject to mortality.

Age structure and sex ratio

Whenever emigration is differential with respect to age or sex, changes in these properties of the resident population are to be expected. The literature on small mammals is rich with references to dispersal dominated by juveniles. Certainly this is a common mammalian pattern, and may produce breeding populations with fewer young than would be expected on the basis of recruitment rates. Correspondingly, new colonies or low density groups may be more youthful than expected. This is not always the case, however. For example, Puček & Olszewski (1971) report that movements of *Clethrionomys glareolus* and *Apodemus flavicollis* into trapped-out areas were random with respect to age, at least in autumn, and Pearson *et al.* (1968) found a wide range of ages among *Ctenomys talarum* moving onto the frontiers of colonies.

A second common claim is that males disperse more often or to a greater distance than do females. In fact this seems to be characteristic of the entire family Sciuridae, and may be true of many other groups as well. Such sexual bias in dispersal may affect age specific mortality schedules for the two sexes, since males would be more often at greater risk. It would also mean that established colonies, or demes, would have more females relative to new ones. Such sex ratio biases could secondarily have profound effects on social structure or reproduction.

Small mammals

Of particular interest is the discovery that in at least five species of microtine rodents, adult or even pregnant females can play a major dispersal role along with young. These five are *Arvicola terrestris* (Stoddart, 1970), *Clethrionomys glareolus* (Kikkawa, 1964), *Microtus pennsylvanicus* (Grant, 1971; Myers & Krebs, 1971), *M. ochrogaster* (Myers & Krebs, 1971), and *M. oeconomus* (Tast, 1966). In the last species, it is even claimed that adult females normally disperse just prior to parturition. In the black-tailed prairie dog (*Cynomys ludovicianus*) adult females also frequently emigrate, and in this case they leave their recently weaned young in possession of the old territory (King, 1955). Adult females of *Perognathus formosus* are also known to disperse (French *et al.*, 1968; N. R. French, personal communication).

It is surprising to find significant dispersal among adult females, and hence future confirmation of the above reports will be important. Where such a pattern is established, we can predict that it will be among species with pre-saturation dispersal, simply because such dispersers have a higher probability of successful re-establishment. There is of course also the interesting possibility that adult females may disperse to provide ecological space for their offspring.

Growth rates

Perhaps it is not surprising that emigration has been given little credit as a suppressant on population growth rates since it requires the prior recognition of pre-saturation emigration. Only this type can be effective in this regard, because it is the only type which occurs during rapid growth. Saturation emigration occurs only at or near equilibrium densities so can exert no effect on the rate at which a population approaches that level. Moreover, even pre-saturation dispersal, if occurring only within the boundaries of the population being studied, will exert only local effects on growth rates.

However, if pre-saturation emigration is occuring, then dispersal may exert a profound effect on population growth rates. It follows logically that, if losses due to emigration are occurring during periods of growth, the resulting growth rate will be correspondingly less. Such an effect could be critical in a strongly seasonal environment where the period of time available for reproduction may be severely limited. This should be especially true for *r*-strategists (being short-lived and emphasizing high fecundities) who are attempting to make maximal use of temporarily very favorable conditions.

110

The best documented case for the importance of emigration in influencing growth rates is that of *Microtus californicus* (California vole). Relatively accurate density estimates are available for a number of rapidly growing populations of the species, and some of these are compared in Fig. 5.2. Four of these populations represent cases where emigration was absent or at least restricted. The fastest growth rates are shown by the population living on Brooks Island (curves *E* and *F*) in San Francisco Bay (Lidicker, 1973), and by a population (curve *H*) in an experimental enclosure (Houlihan, 1963). The short plateau shown by this last population occurred when the population temporarily ran out of food and shelter. Rapid growth continued when this limitation was artificially relieved. Only slightly less rapid growth is shown by the Brooks Island east transect (Lidicker & Anderson, 1962). These data (curve *A*) represent the growth pattern during the spring of 1959 on the side of Brooks Island distal to the point of colonization. Although the island was still being colonized during this period, the growth pattern shown was relatively little influenced by dispersal, as colonization was nearly completed by the time the east side was well populated. The transect, moreover, was in an area of very favourable habitat.

The four curves with lower slopes in Fig. 5.2, represent unenclosed populations. One (curve *G*) is from the mainland adjacent to Brooks Island (Batzli & Pitelka, 1971), and another (curve *D*) is from the Berkeley Hills (Krebs, 1966). The last two (curves *B* and *C*) are additional transects from Brooks Island. In this case the transects represent the pattern of growth while colonization of the island was in progress, the west transect being very close to the point of initial colonization. Both show the presumed effect of extensive emigration slowing population growth rate. When enclosed and unenclosed populations are compared with respect to monthly growth rates (increment of growth relative to density at start of month), differences are also apparent. The unenclosed populations had generally lower monthly rates which were less variable and showed a gradual downward trend as peak numbers were approached. Populations with limited emigration, on the other hand, had higher monthly rates which were more variable and declined more abruptly to zero as the peak was achieved.

The potential effect of emigration on growth rates is further emphasized by a comparison of growth rates on Brooks Island, not only between different parts of the island, but also between spring periods, when colonization was taking place, and afterwards (Fig. 5.3). Fig. 5.2 has already drawn attention to the dramatic difference in growth rates

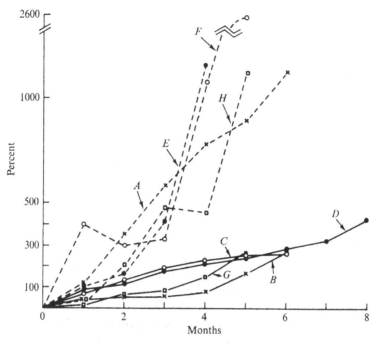

Fig. 5.2. Percentage increases (monthly intervals) from starting densities during periods of rapid increase to peaks in *Microtus californicus*. Solid lines refer to unenclosed populations and dashed lines to enclosed ones. *A*, Brooks Island, east transect (Lidicker & Anderson, 1962; Lidicker, 1973); *B*, Brooks Island, west transect (Lidicker & Anderson, 1962); *C*, Brooks Island, central transect (Lidicker & Anderson, 1962); *D*, Berkeley Hills, Tilden Park control (Krebs, 1966); *E*, Brooks Island, spring 1961 (Lidicker, 1973); *F*, Brooks Island, spring 1963 (Lidicker, 1973); *G*, Richmond Field Station (Batzli & Pitelka, 1971); *H*, enclosure (Houlihan, 1963).

on the east and west transects in 1959. This period of growth followed the arrival of the voles on the island in the summer of 1958. By the spring of 1960, all parts of the island showed similar rapid growth rates, the island then being completely colonized and net emigration presumably being nearly absent. The documentation presented here that populations of *Microtus californicus* in which emigration can occur grow at a much slower rate than similar populations in which emigration is inhibited, strongly supports the existence of a substantial amount of pre-saturation dispersal in this species.

Emigration has also been hypothesized to be one of six key factors explaining the three to four year population cycles exhibited by mainland populations of *M. californicus* (Lidicker, 1973). Its influence is manifest during the two years of the cycle when numbers are low. During this period, the species persists mainly in survival pockets or refuges of particularly favorable micro-habitat. Because of pre-saturation emigration, most of the reproduction which occurs during this time is channelled into colonization of empty habitat surrounding the refuges. Consequently, densities increase very little in the survival areas. Usually after two breeding seasons, the available habitat becomes colonized and the stage is set for an increase to peak densities. Clearly, there are other important factors involved, but without pre-saturation emigration the hypothesis proposes that the pattern of density changes in this species would be very different.

It seems possible that emigration may also be important in the demography of other microtines which show a three to four year cycle in numbers. By using enclosures Krebs *et al.* (1969) have shown that dispersal may be an important element in the demography of *Microtus pennsylvanicus* and *M. ochrogaster*, and in fact their enclosed populations generally had higher growth rates. Gentry (1968) has reported on a similar experiment with *M. pinetorum*, but does not compare growth

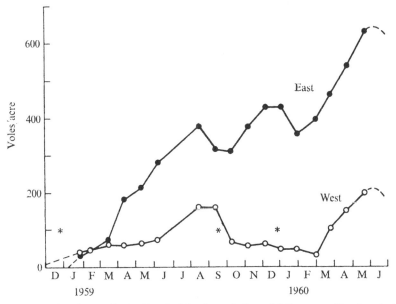

Fig. 5.3. Density changes in the Brooks Island population of *Microtus californicus* during the first two growing seasons following their arrival on the island. The west side of the island is where the original invaders established, and the east side is distal to that point. The east side also has a larger carrying capacity for voles. Asterisks represent times when heavy rains caused sprouting of new vegetation and thereby ended dry periods. Data from Lidicker & Anderson (1962) and Lidicker (1973).

rates between enclosed and unenclosed populations. The negative effect of emigration on population growth rates has also been demonstrated for *Clethrionomys glareolus* by Mazurkiewicz (1972). She has found that a population of this species living on Wild Apple Island (Wyspa Dzikiej Jabłoni) in Lake Bełdany grew to peak densities more rapidly, and reached higher numbers, than did an unrestricted population in the Kampinos National Park (Poland). Although other explanations for this difference are possible, the notion that it is due to pre-saturation emigration being frustrated on the island is appealing.

This last study also illustrates still another kind of effect that differential population growth rates can have. Mazurkiewicz (1972) has given evidence that in addition to the differences in population growth rates noted, individual growth rates were also different between her island and mainland populations. Individuals living in the unrestricted population grew more rapidly than did those from the island. A similar relationship may be present in other species as well. In such cases, emigration affects individual growth rates indirectly through its influence on population growth rates. Additionally, it would not be surprising if resident individuals were subject to modifications in fecundity and viability as a result of differing population growth rates.

Finally, we should consider that dispersal may produce immigrants as well as emigrants, and that an influx of immigrants into our subject population will produce enhancing effects on population growth rates. Clearly, if a population accepts immigrants they add directly to the growth rate by increasing the additions per unit of time. This is true whether or not the immigrants reproduce successfully in their new home.

Seasonal shifts of individuals between different habitats will clearly produce situations where immigration contributes massively to population growth. Many examples are known (see Brown, 1966, p. 131, for a good sample), but only a few will be mentioned here in order to indicate the nature and variety of the phenomenon. Regular seasonal movements are well known in *Mus musculus*, both in North America where winters are severe, and wheat fields in South Australia (Newsome, 1969*a*). A summer influx of *Apodemus sylvaticus* into grain fields has been described by several authors (see Crawley, 1970). Periodic re-invasions of seasonally flooded areas are characteristic of *Microtus oeconomus* (Tast, 1966), as well as of species associated with river plains (e.g., Sheppe, 1972; I. Linn, personal communication). Massive re-invasions can also be expected following burns (M. Delaney, personal communication). Finally, *Lemmus lemmus* seems to engage in regular spring and autumn

114

dispersal between summer and winter habitats (Kalela, 1961; Kalela *et al.*, 1971).

Various indirect effects of emigration on growth rate may also occur through influences on the social structure, reproductive activities, and/or genetic composition of the host population. Some of these will be considered below (see also Andrzejewski, Kajak & Pieczyńska, 1963).

Social structure

Dispersal movements can influence the social structure of both the residents remaining in the group from which dispersers have left and the host group into which the dispersers attempt, successfully or unsuccessfully, to immigrate. Little is known regarding the effects of emigration on residents. However, we can surmise that if emigration were combined with a resistance to immigration, social stability would be generally enhanced. This is because socially dominant individuals would be less likely to leave, those with a reduced inclination to move (physiologically or genetically) would tend to accumulate, and the addition of outsiders would be resisted. DeLong (1967) has suggested that this model applies to feral house mouse populations. He further points out that, if social stagnation becomes chronic, it can lead to reduced fecundity in the social group. Of course, the complete absence of emigration, as occurs in population cages, also leads in house mice to social stagnation and reduced reproduction (Southwick, 1955*a*; Crowcroft & Rowe, 1957; Petrusewicz, 1957, 1963; Lidicker, 1965). Additional effects on resident social structure might be expected to develop indirectly from emigration through influences on growth rate, sex ratio, age structure, individual growth and condition, etc.

More information is available on the influence of immigrants on host population social structure. We know, for example, that newcomers are generally added to the bottom of an existing social hierarchy so that there is a minimum of disruption (e.g., Andrzejewski, Petrusewicz & Walkova, 1963). An ingenious demonstration of the social avoidance of strangers by residents has been provided by Kołodziej *et al.* (1972) for *Clethrionomys glareolus*. Such avoidance or delegation to low rank may not always be the case, however, as immigrants may displace social dominants as well. One variable which can influence the success of immigrants in social integration is the number of immigrants relative to residents. If the number of immigrants is large, success is more likely as has been shown for *Mus musculus* by Andrzejewski, Petrusewicz &

Walkova (1963) and for *Peromyscus maniculatus* by Packer & Lidicker (unpublished results). In at least two studies, large numbers of colonizers have even produced a partial exodus of the residents (*Peromyscus maniculatus*, Healey, 1967; *Sigmodon hispidus*, Ramsey & Briese, 1971). Negative population growth rates were also reported to sometimes follow large introductions of aliens into established populations of *Rattus norvegicus* (Davis & Christian, 1956).

In addition to these direct effects on social structure, immigrants may become involved in social interactions which affect reproduction in important ways. One possibility is the disruption of pregnancies by strange males through olfactory contact with females in particularly sensitive stages of pregnancy (Bruce effect). This has been described for laboratory *Mus* (Bruce, 1959; Bruce & Parrott, 1960; Chipman & Fox, 1966*b*; Chipman *et al.*, 1966), feral *Mus* (Chipman & Fox, 1966*a*), and *Peromyscus maniculatus* (Eleftheriou *et al.*, 1962). A more common role appears to be that of stimulating reproduction. Such a role has already been mentioned for *Mus musculus*, and may be widespread among mammals. One such mechanism is the stimulation and/or synchronization of estrus by males (Whitten effect; see Whitten, 1966, for review), although it is not clear regarding the extent to which such stimulating males must be strangers. Among small mammals this effect is best known in house mice (e.g., Lamond, 1959; Marsden & Bronson, 1964; Whitten *et al.*, 1968). It has also been described for *Peromyscus maniculatus* (Bronson & Marsden, 1964), and the males of some microtines also stimulate behavioral estrous (e.g., Chitty & Austin, 1957, for *Microtus agrestis*; Richmond & Conaway, 1969, for *M. ochrogaster*).

Still another way in which immigrants may influence reproduction is through the release of reproductive inhibition in females which seems sometimes to accompany the act of dispersal. That is, when reproductively inhibited females (often subadults) leave home, they often rapidly become reproductively mature. This clearly is the case for *Mus musculus* (Crowcroft & Rowe, 1958; Lidicker, 1965), and may also be true for some species of *Microtus* (Richmond & Conaway, 1969; Myers & Krebs, 1971). It has also been found in *Peromyscus leucopus* (Sheppe, 1965), as well as various ungulates. It has been suggested by Richmond & Conaway (1969) that in these species any minor stress brings about an increase in FSH (follicle-stimulating hormone) release followed by estrous. Since dispersal is undoubtedly stressful, such a mechanism could explain the fact that dispersing females tend to rapidly become reproductively competent. In this way they not only increase their chances of social acceptance in a new home, but they are also immediately ready to contribute reproductively. A particularly extreme example of this strategy of reproductive readiness is the dispersal of pregnant females such as in *Microtus oeconomus* (Tast, 1966) and *M. pennsylvanicus* (Myers & Krebs, 1971). This kind of dispersal is probably an adaptation for rapid colonization and exploitation of temporarily favorable habitats.

The importance of frustrated dispersal

In order for dispersal to occur, individuals must not only be motivated to leave home but they must physically be able to do so. Thus physical barriers surrounding a population may prevent or severely reduce dispersal, even if strong motivation exists. Moreover, even in the absence of physical barriers, there generally needs to be some place that the potential disperser is willing to go to. These refuge areas I would like to refer to as dispersal sinks, modelling the term after the 'behavioral sink' proposed by Calhoun (1962). Such a sink will generally be some empty or unfilled suitable habitat, or perhaps marginal or even unsuitable habitat in which at least short-term survival is possible. In the absence of such a sink, a potential disperser will probably return home after making an exploratory excursion, although possibly a few individuals could persist for a limited time as vagrants while in search of a refuge.

Whenever motivation to disperse exists, but consummation is prevented by barriers or inhibited by the absence of an unfilled sink, a condition of frustrated dispersal exists. The level of this frustration will depend on the extent to which the stimulating forces providing the motivation exceed realized net dispersal. Dispersal does not, therefore, have to be absent to produce frustration, but only be at a rate less than would be required to relieve the pressure for dispersal that develops. Populations surrounded by severe physical barriers or filled dispersal sinks will be subject to a low emigration rate. Under these conditions, the appearance of almost any degree of motivation for dispersal will produce frustrated dispersal. Those few individuals who manage to survive as vagrants without even a temporary refuge or home would move randomly and not produce any reduction in the level of frustration. In species exhibiting pre-saturation dispersal, motivation to disperse will appear even at early stages in population growth, and at relatively low densities. With saturation dispersal, however, motivation to leave will only begin when the current carrying capacity is nearly reached. The inhibition of dispersal may, therefore, have quite different demographic effects in these two kinds of species.

It has become apparent to me that circumstances which serve to frustrate dispersal can be powerful, analytical tools for studying the normal role of dispersal, even when a population is ordinarily not subjected to such frustrations. In a manner analogous to the experimental removal of some organ in order to study its function, the population biologist can contrive to frustrate dispersal in an attempt to understand its normal role. An additional purpose for such manipulations is to permit the discovery of alternative density regulatory mechanisms which may be available to various species, but which are utilized only infrequently. We may thus be led to a better understanding of the ecology, behavior, and even physiology of our subject species. Experimental enclosures and natural islands readily serve these functions, and are to be recommended for further exploitation by ecologists studying small mammals.

Of course, the use of enclosures and islands by ecologists is not a new idea, and it will be of interest to consider some examples of such studies in order to illustrate what is known of the ways in which species of small mammals cope with frustrated dispersal. One tactic that is widely exploited in these circumstances is the increased usage of marginal habitats (especially where the absence of competitors makes such areas more suitable). This has been documented for *Microtus ochrogaster* on islands by McPherson & Krull (1972), for *M. pennsylvanicus* on islands by Webb (1965), for this species and *Peromyscus maniculatus* in enclosures by Grant (e.g., 1971), and for *P. leucopus* on islands by Sheppe (1965). If even this opportunity for dispersal is unavailable, however, there remain only two fundamental mechanisms for stopping growth; to decrease natality or to increase mortality rates. Both have been widely observed. Often, though, an isolated population is not able to achieve a long-term balance with its resources; it becomes too large, and then suffers a crash which may even result in its extinction. Direct evidence for this last point has been provided by Crowell (1973) involving *Microtus pennsylvanicus* living on islands in the Gulf of Maine, by Lidicker & Anderson (1962, p. 504) for *M. californicus* on a small islet in San Francisco Bay, and by Sheppe (1965) for *Peromyscus leucopus* on an island in Ontario.

The achievement of unusually high densities along with some form of reduction in birth rates has been found in several microtines and in *Mus musculus. Mus* has been most widely studied in this regard, and its ability to stop breeding under high density, socially stabilized conditions has already been mentioned. This is primarily accomplished by progressively increasing inhibition of reproductive maturation in females (see previous references and Lidicker, unpublished results). An increased frequency of pseudo-pregnancies may also be a factor to some extent (van der Lee & Boot, 1955). Lund (1970) placed *Arvicola terrestris* in enclosures and produced high densities and a drastic fall in reproductive rates. A particularly interesting case is that of *Clethrionomys glareolus* living on Wild Apple Island in the Mazurian Lake District, studied extensively by a group at the Polish Institute of Ecology. Bujalska (1970) has proposed that reproduction is controlled in this population by the territorial behavior of adult females, and there is also a strongly marked inhibition of growth in young born late in the breeding season (Bujalska & Gliwicz, 1968; Petrusewicz *et al.*, 1971). A long-term adaptation to island life is illustrated by the Skomer vole (*Clethrionomys glareolus skomerensis*) long isolated on a small Welsh island. This population lives at chronically high densities, and has an exceptionally short breeding season and slow maturation rate (Fullagar *et al.*, 1963; Jewell, 1966).

There are other examples in which increased mortality rates were emphasized. Louch (1956) found high mortality rates accompanying high densities of *Microtus pennsylvanicus* in small pens. Krebs *et al.* (1969) describe the fate of populations of both *M. pennsylvanicus* and *M. ochrogaster* in large enclosures (8372 m^2) in southern Indiana. These populations reached very high densities, ate out their food supply, and dropped back to control levels or below. No adjustments in reproduction were made by these populations, leading at first to excessive numbers and then to high mortality rates. In small population cages, increase in numbers of *Oryzomys palustris* and (to a major extent) *Peromyscus truei* was stopped by increasing mortality through aggression (Lidicker, 1965). Most of these deaths were of neonates within three days of birth. Occasionally a young animal survived and compensated for deaths among adults who became badly wounded. Also, house mice living in large complex enclosures characteristically cannibalize or desert their young under high densities (Brown, 1953; Southwick, 1955b; Anderson, 1961; Lidicker, unpublished results). Complete reproductive inhibition may not be possible under these conditions because of the continual stimulation stemming from interactions among the multiple social groups present. *Peromyscus maniculatus* is apparently another species which uses both reproductive inhibition and mortality of nestlings to cope with frustrated dispersal (Lidicker, 1965).

In spite of the variety of these examples, it still seems premature to draw generalizations from them regarding possible relationships between the kind of dispersal normally exhibited by each species and its mode of coping with dispersal inhibitions.

Emigration as a regulating factor

Although in theory emigration can act as a major factor regulating population density, there are few data directly bearing on this question. The potential for emigration to act in this way stems from its role in the population growth equation, as has already been pointed out. In considering this question, it is important to remember that the question of the manner of density regulation should be applied to the same defined population as is the measure of emigration. The term carrying capacity is used in the following discussion to specify the maximum numbers of individuals that an area can support based on the availability of essential resources such as space, food, water, and shelter.

In considering the role of emigration in the regulation of numbers we must, for convenience and clarity, subdivide our treatment into three levels of importance. The simplest role that emigration can play in this regard is as one of several factors which account for the total losses suffered by a population when its growth rate becomes zero at carrying capacity (i.e., a contributing factor). Under these conditions, if emigration should become frustrated, some other form of loss would soon increase to compensate for it. Thus, emigration is contributing to regulation of numbers, but is not an essential element in the process. This may be a common pattern among small mammal species, and a good example is provided by the Arctic ground squirrel, *Spermophilus undulatus* (Carl, 1971).

The second level of involvement is where emigration acts as a key factor in stopping population growth at carrying capacity. This is similar to the situation above, but now emigration makes up a very large fraction of the losses. It thus requires a large dispersal sink in order to accommodate long-term, large-scale emigration. If emigration becomes inhibited, under these conditions, population density would probably shoot up above carrying capacity, and then crash. Some long-term damage to the habitat may also be a consequence. This pattern would be enhanced under conditions in which demes were asynchronous in their fluctuations in number, because the presence of low populations in the vicinity of high ones would ensure the existence of unfilled habitat to accommodate the dispersers. Both these levels of involvement could in principle occur in species showing only saturation emigration or some combination of saturation and pre-saturation types.

An example of emigration playing a key role in regulation may have been provided by Pearson (1963) in his study of two outbreaks of feral house mice. He concluded that dissipation of the peak populations was primarily the result of emigration, although mortality from predation and disease played secondary roles. In these cases, the outbreaks were local leaving large areas of suitable habitat available as dispersal sinks.

The third potential role that emigration may play in density regulation is actually to prevent numbers from reaching carrying capacity (so that the equilibrium density, K, is then below carrying capacity). This possibility was suggested in an earlier paper (Lidicker, 1962), and is clearly the most controversial of the three proposed roles. Its existence, however, would seem to be predictable solely from the presumed reality of pre-saturation dispersal. Some of the difficulties inherent in understanding how this kind of dispersal behavior might evolve have already been discussed. However, if pre-saturation dispersal is real, and there seems to be good evidence that it is, then the potential exists for emigration to play this kind of role, at least occasionally. Several models for how this role can be fulfilled by emigration will be proposed; all require the involvement of pre-saturation dispersal.

The first model (I) combines the action of frustrated dispersal and a temporarily very favorable carrying capacity. Given these conditions, it is hypothesized that there will be a hyper-development of various behavioral and/or physiological traits which have evolved to stimulate pre-saturation dispersal. But, since emigration is inhibited and potential emigrants do not immediately become mortality statistics (the population not having yet reached its carrying capacity), the stage is set for over-stimulation of all members of the group. These hyper-developed traits, such as aggressive behavior, are hypothesized to increase mortality or decrease natality sufficiently to stop population growth. The higher the carrying capacity, the greater would be the opportunity for this syndrome to develop. It can be thought of as a case where the existence of proximate regulating factors forces the population to be temporarily out of phase with its ultimate regulating forces.

Model I is illustrated in Fig. 5.4, which also contrasts the situation prevailing if saturation emigration only were to be present. In the model, emigration is shown at first to perform only a key factor role, density increasing to the carrying capacity of the habitat. The carrying capacity is then imagined to increase abruptly to a new higher level. When only saturation emigration is present, numbers increase promptly to the new

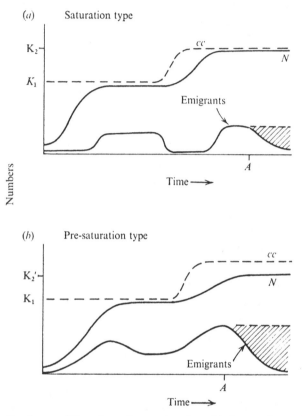

Fig. 5.4. Density changes (*N*) and numbers emigrating with increased carrying capacity (*cc*) when (*a*) only saturation dispersal is present and (*b*) when pre-saturation dispersal occurs along with the hyper-development of traits which stimulate such dispersal (Model I). Hatched areas indicate the extent of frustrated dispersal which is initiated at time *A*; K_i refer to equilibrium densities.

level. Even when emigration becomes frustrated (time *A* in the figure), no demographic effect is evident since individuals which cannot emigrate simply become mortality victims. If emigration were playing a key factor role, numbers might increase abruptly after *A*, and then crash. With pre-saturation dispersal, however, the population is shown to level off at a new *K*-level, this time below the carrying capacity, at a time shortly after emigration becomes frustrated.

The temporary nature of the proposed increase in carrying capacity is emphasized, because the situation in which numbers were being regulated below resource levels would not likely be stable in evolutionary time. This is simply because selection would favor individuals who were more tolerant of the conditions generated by frustrated dispersal and hence could more fully utilize available resources. Equilibrium densities would in time then approach the new carrying capacity. This notion is illustrated in Fig. 5.5. In this example it is imagined that a species displaying pre-saturation dispersal colonizes an island which has a very favorable carrying capacity for that species. Being an island, dispersal becomes frustrated as soon as colonization of all suitable parts of the

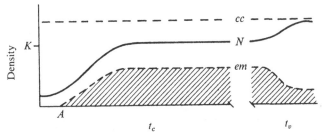

Fig. 5.5. Hypothetical regulation of numbers below carrying capacity in an island population of a species showing pre-saturation dispersal. Hatched area indicates extent of frustrated dispersal; t_c is ecological time, t_v evolutionary time, *em* number of potential emigrants, and other symbols as in Fig. 5.4.

island is complete. Because of this, and the extremely favourable level of resources, the population becomes regulated below carrying capacity. This condition would prevail through ecological time (t_c) as shown in the figure. After many years, however, selection would reduce the perceived level of frustrated dispersal, allowing the density to increase to carrying capacity.

A second model (II) contains the same basic elements as the first, (pre-saturation dispersal and its occasional frustration), but in this case large dispersal sinks along with large-scale pre-saturation dispersal keep low density populations from growing toward capacity. If and when the sinks become filled, dispersal is inhibited, and densities build up rapidly.

This model is illustrated in Fig. 5.6., and may apply to several species of *Microtus*. The enclosure experiments by Krebs and his students (Krebs *et al.*, 1969) mentioned earlier are also relevant here. Enclosed populations of *M. pennsylvanicus* and *M. ochrogaster* reached much higher densities than did unenclosed controls. Moreover, in one

123

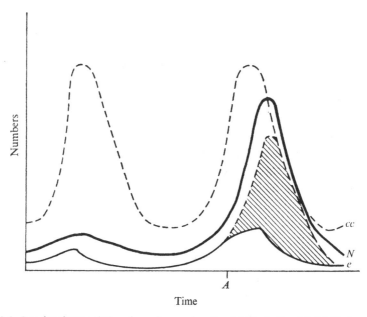

Fig. 5.6. Density changes (*N*) and numbers emigrating (*e*) illustrating Model II. A strongly fluctuating carrying capacity (*cc*) is coupled with strong pre-saturation dispersal. At time *A* dispersal becomes frustrated and a large increase in numbers follows. The extent of frustrated dispersal is indicated by hatching.

enclosure *ochrogaster* reached high numbers three years running. Hatt (1930) also reported that *M. pennsylvanicus* in enclosures reached very high densities and damaged vegetation. Similarly, Gentry (1968) enclosed two populations of *M. pinetorum* (in pens of two acres each) and found that they reached unusually high densities. These results suggest that when emigration is not prevented these species of *Microtus* are keeping their numbers significantly below the peaks in carrying capacity. This idea is supported by Myers & Krebs (1971) who report that 59 to 69 per cent of the *M. pennsylvanicus* in their study populations dispersed. In addition, I have suggested (Lidicker, 1973) that, in *M. californicus*, emigration is one important element in keeping densities from increasing to carrying capacities for two or more years at a time. Eventually, the dispersal sinks become filled, and numbers increase to peak densities, only to be stopped by seasonally declining carrying capacities. Perhaps in these species of *Microtus*, long-term evolution away from this pattern is prevented by the fact that the carrying capacity of their habitat varies greatly with season, making it only temporarily very favorable.

124

There is at least one example of a species which appears to show almost continuous regulation of population size below carrying capacity, that is, even temporary outbreaks do not occur. Moreover, the process is stable in evolutionary time. This is the North American beaver (*Castor canadensis*), for which we must provide still another model. Model III can be considered only a modification of II. In the beaver a population is equivalent to the family group. It is well-known that the numbers of beavers in any family group are very strictly regulated (e.g., Bradt, 1938; Taylor, 1970). There is typically one pair of adults, a litter less than one year old, and a litter of yearlings. When the young reach two years of age, they are firmly driven from the colony and disperse. Immigration into a population is impossible unless one of the adults should die. When a new colony is established, it is living far below its carrying capacity. As the area in the vicinity of the house and dam become exploited over the years, the carrying capacity of the area gradually diminishes. Eventually the site is abandoned, whereupon it gradually recovers. Thus, the beaver model requires only pre-saturation dispersal and a permanent dispersal sink, so that dispersal does not become frustrated permitting temporary high densities.

Since beavers exploit a resource which is only slowly renewable (they 'eat into the capital'), it would be a poor strategy to permit each population to build up in numbers until short-term carrying capacity was reached. If this were to happen, the carrying capacity of the site would soon drop to near zero, and would then recover only very slowly. *Microtus* also exploit the capital of their food supply, but since this is herbaceous rather than woody, they are able to recover more quickly following peak densities. Wiegert & Owen (1971) also argue that herbivores will evolve feeding styles that avoid long-term oscillations in their food populations.

Model III may also be applicable to prairie dogs (*Cynomys*). According to King (1955) these social rodents produce two kinds of emigrants (yearlings which are mostly males and older adults which are mostly females), and do not ordinarily destroy their food resources. A very large dispersal sink is assured by the widely spaced arrangement of their colonies.

Discussion and summary

Emphasis in this chapter has focused on the potential importance of dispersal movements as demographic agents. This represents a significant shift in perspective from the widespread view that such movements can be largely ignored, either because they merely substitute for other forms of mortality or because they cause internal perturbations more likely to have social than demographic effects. In part this change has been a consequence of the accumulating evidence that many species of small mammals are organized in a demic dispersion pattern, even in the centers of distribution (cf. Anderson, 1970). With this has come the realization that knowledge of dispersal may be critical to an understanding of demographic phenomena. It must be cautioned, however, that the relative importance and nature of dispersal will undoubtedly be found to vary greatly among different species (see also Chapter 4). I expect further that not infrequently variation within species will also be found. Particularly is this to be sought in comparisons of populations on the edges of a species' range with those from the center. A demic structure, for example, is likely to be found on the periphery of ranges even where it is not characteristic elsewhere.

A second reason for giving increased attention to dispersal is the recognition that there are two kinds of dispersal (pre-saturation and saturation), each having a different timing with respect to phase of population growth and producing qualitatively quite different kinds of emigrants. Moreover, pre-saturation dispersal can have greatly different demographic effects compared to the saturation type. The available evidence for pre-saturation dispersal in small mammal species has been outlined, but evidence is also available for numerous other organisms as well. An intriguing aspect of the existence of pre-saturation dispersal is the probability that it has a genetic basis, and that a population may even be polymorphic for dispersal tendencies. The evidence for this has been summarized. However, it is of further interest to consider whether or not the phenomenon of cyclomorphosis exhibited by some species which have multiple generations per year may incorporate such a polymorphism. Reviews of the phenomenon have been provided by Shvarts *et al.* (1964) and Anderson (1970), and the latter suggests that a colonizing and survival cohort can often be recognized. Of course we do not know if the differences observed among such seasonal cohorts are phenotypic, genotypic, or both. The possibility that such shifts in the average properties of individuals in a population are at least in part genetic must be seriously considered in view of the recently available evidence that seasonal shifts in gene frequencies can occur in small mammals (Semeonoff & Robertson, 1968, for *Microtus agrestis*; Tamarin & Krebs, 1969, for *M. pennsylvanicus* and *M. ochrogaster*;

Berry & Murphy, 1971, for *Mus musculus*; and McCollum, 1975, for *Microtus californicus*).

One generalization about pre-saturation dispersal may tentatively be offered even at this very early stage in our knowledge, namely, that this type of dispersal may often be associated with (1) colonizing species such as *Mus musculus*, and (2) those species with feeding styles such that they can affect the future supply of their food. In the latter case, I am referring to species which can, and do, consume more than the surplus production of their food supply. These would include herbivores which can damage their food plants to the extent that the plant's reproductive abilities are reduced, or carnivores who kill prey individuals with high reproductive values. In such species, high densities imperil future carrying capacities and natural selection may then favor individuals who disperse before crashes occur (this is in part the 'diplomatic advantage' of Lidicker, 1962, and is also suggested by Evans, 1942, and Andrzejewski, Kajak & Pieczyńska, 1963). The fact that the meager evidence for pre-saturation dispersal so far available turns up in genera such as *Microtus* and *Castor*, and not in mast-feeders such as *Apodemus* and *Peromyscus* lends support to this generalization (see also discussion of herbivory by Wiegert & Owen, 1971).

Among the population properties which can be influenced by dispersal are 'gross mortality', age structure, sex ratio, growth rates, and social structure. In considering these effects, it is important to distinguish the role of emigrants from that of immigrants. This is especially true when social effects and secondary influences on reproduction are being studied.

Two new concepts are considered essential for discussion of the role of dispersal in the regulation of numbers. One is the notion of 'frustrated dispersal' which is thought to exist whenever dispersal is inadequate relative to its motivation. Space which provides an outlet for wandering impulses is termed a 'dispersal sink'. If inadequate sinks exist or if physical barriers prevent access to them, dispersal becomes frustrated. Such circumstances can have interesting demographic consequences, and where pre-saturation emigration occurs, can even lead to equilibrium numbers which are below current carrying capacity. The study of populations exhibiting frustrated dispersal is recommended as one means of understanding the normal demographic role of dispersal.

Dispersal is thought to act at three levels of involvement in the regulation of densities: as a contributing factor, as a key factor, and in the establishment of equilibrium densities below carrying capacity. Only the last is considered controversial, and three models are proposed suggesting how emigration may act in this manner. All require the presence of pre-saturation emigration, and only one involves a situation in which the separation between numbers and resources is both chronic and evolutionarily stable.

The initial impetus for the preparation of the theoretical portion of this paper came from Drs G. N. Cameron and M. H. Smith when they invited me to participate in a symposium on population regulation for the 1973 annual meeting of the American Society of Mammalogists. I am grateful for their confidence and contagious enthusiasm.

The further development of this paper, and my participation in this IV Symposium of the Working Group on Small Mammals, was only possible by the encouragement and assistance of Drs K. Petrusewicz and K. Andrzejewska. I am extremely appreciative of their efforts on my behalf. Dr Petrusewicz further stimulated my thinking on dispersal and provided many helpful suggestions in the course of several intensive discussions.

I am particularly grateful to my colleagues Drs O. P. Pearson and F. A. Pitelka for carefully reading an early draft of this paper and making many helpful suggestions. Mr R. Glenn Ford gave me the idea for Fig. 5.2, and Mr S. F. Smith not only critically read the manuscript but also provided some help with the literature. Warm thanks are due to many of the participants in the Symposium at Dziekanów, and especially to Dr G. O. Batzli, for stimulating discussion of this chapter.

6. Morpho-physiological characteristics as indices of population processes

S. S. SHVARTS

The use of morpho-physiological indices can be of great importance in solving such ecological problems as determination of the condition of the population and prediction of its future adaptation to environmental conditions, intraspecific variability and speciation. This chapter describes the general features of the method; more details are presented by Shvarts *et al.* (1968).

Population state and its fate

When the method of morpho-physiological indices is used to predict the fate of a population, the following parameters are of particular importance: body weight, amount and chemical characteristic of body fat, proportions of the body and skull, relative weights of the liver and kidney, development of the adrenals, hypophysis and thymus, hematological indices and the content of vitamin A and glycogen in liver.

The essence of the morpho-physiological approach consists of obtaining information on the condition of the population and, therefore, on its probable fate, from a series of indices estimated for a representative group of animals. Indices that are easy to establish under natural field conditions should be used, as an experienced technical assistant can then determine a set of such indices during one working day and obtain in this way an idea of the range of population 'welfare'. Some general recommendations are presented below concerning the application of morpho-physiological indices to determination of the condition of rodent populations and, therefore, to prediction of their dynamics of numbers. These recommendations are fully applicable only to areas where the species do not breed in winter (or where their breeding is considerably reduced at that period). Our general theoretical considerations are based on the assumption that the range of population variability is finally determined by the natality and mortality rates in the animals.

Indices characterizing energy balance are used as the criteria of physiological condition. Thus, when the population is disturbed the body weight of animals is reduced, the expected curve of the seasonal changes in body weight is altered, fat reserves are low, the relation of liver weight to body weight is below average, glycogen content in the liver is low, adrenals are hypertrophic and relative brain weight is high. In some cases, when food resources are scarce or high energy losses are not replenished, a decrease in hemoglobin and protein content in blood also occurs; when there is inadequate moisture in food, a decrease in the hemoglobin content and in the number of erythrocytes is accompanied by relatively high content of protein in plasma, although the effect of anaemia caused by infectious factors must be excluded by measurement of the number of leucocytes and the rate of sedimentation of erythrocytes.

Before analyzing some examples of the method, one important point should be emphasized. High variability of morpho-physiological indices and their responses to the environmental conditions may produce the false impression that differences in indices between populations or among intrapopulation groups result not from ecological conditions but from random factors.

For example, in the far northern valley of the Ob River the majority of the large islands, with very similar environmental conditions, are inhabited by *Microtus oeconomus* (Pyastolova, 1964). There is no regular exchange of individuals among the various populations, although there are occasional migrations of individuals from island to island. When these island populations are compared we would expect that there should not be significant morpho-physiological differences between them. Indices showing the development of some internal characters of *M. oeconomus* living on two adjacent islands are presented in Table 6.1. These data indicate that our expectation is correct: there

Table 6.1. *Morpho-physiological indices of 4 body organs (in ‰; adrenals in mg/kg) in* Microtus oeconomus *inhabiting 'Southern and 'Northern' Islands of the Ob Valley, based on 22 animals on Southern Is. and 31 from Northern Is. M, mean value of indices; CV, coefficient of variation*

	Heart		Liver		Kidney		Adrenals	
Index	Southern	Northern	Southern	Northern	Southern	Northern	Southern	Northern
M	6.75	6.68	48.7	49.4	7.48	7.55	195	197
CV	0.204	0.103	1.8	0.9	0.27	0.13	10.1	10.7

are no differences between the two populations. This result is understandable since similar environments make uniform those features of the population that are most variable and sensitive to even the slightest changes in environmental conditions.

Perhaps the most important conclusion obtained by use of morpho-physiological indices is that the average values obtained for populations or intrapopulation groups of animals are never random but always biologically determined. They are not random because being changeable they sensibly respond to environmental changes. When populations develop under identical environmental conditions, it is seldom possible to demonstrate differences in their indices even if complex investigations are made.

This conclusion permits us to determine a 'morpho-physiological standard' for particular populations. The deviations from this standard provide a basis for predicting the probable dynamics of population numbers. Certainly, this standard should not be static. For example, it should reflect regular seasonal changes in the particular indices. The examples presented below indicate that such 'dynamic standards' can be quite clearly expressed (Figs. 6.1–6.5). For many years similar results have been obtained for dozens of species of mammals under different environmental conditions. These significantly enrich and supplement the life-tables for these species and can be used in solving many theoretical and practical problems.

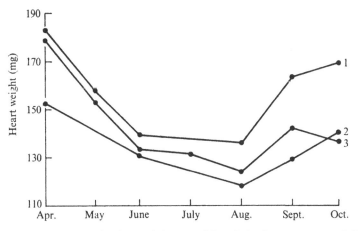

Fig. 6.1. Seasonal changes in the total heart weight of *Apodemus agrarius* of 15–20 g. 1, breeding females; 2, non-breeding females; 3, males.

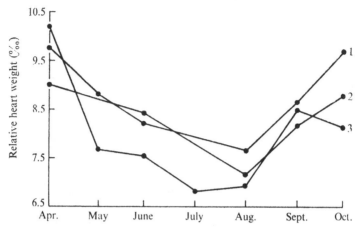

Fig. 6.2. Seasonal changes in the relative heart weight (‰ i.e., mg/g) of *Apodemus agrarius* of 15–20 g. Explanation of indices as in Fig. 6.1.

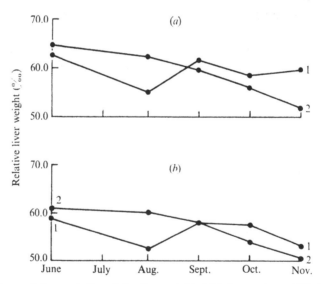

Fig. 6.3. Seasonal changes in the relative liver weight of (*a*) *A. agrarius* and (*b*) *A. sylvaticus* from two younger age groups. 1, body weight 15–20 g; 2, body weight <15 g.

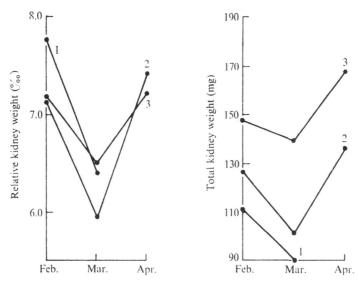

Fig. 6.4. Changes in the total and relative kidney weight (mg/g) for three weight classes of *Clethrionomys glareolus* in the early spring. 1, 10–15 g; 2, 15–20 g; 3, 20–25 g.

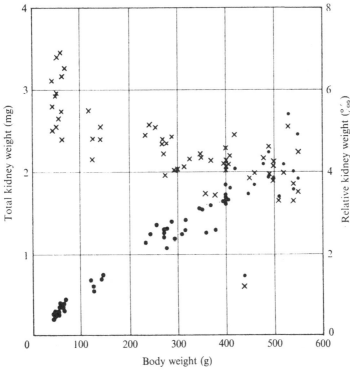

Fig. 6.5. Changes in the total and relative kidney weight of young growing *Ondatra zibethicus* (Salekhard region, July–September). ●, total kidney weight; ×, relative kidney weight.

It should also be noted that the 'dynamic standards' should take into account not only age or seasonal changes, but also include such important details as differences in the development of animals born at different seasons of the year (Fig. 6.6).

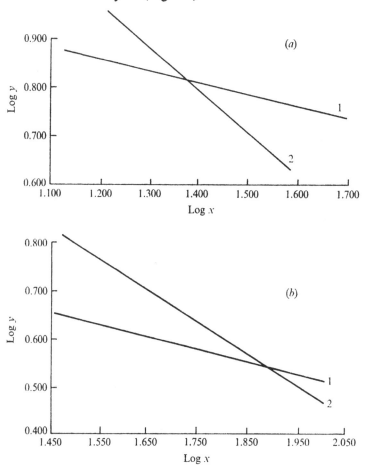

Fig. 6.6. Changes in relative kidney weight (y) with body size (x). (a) *Lagurus lagurus*. 1, $y = 14.6x^{-0.254}$ (in winter); 2, $y = 107.29x^{-0.89}$ (in summer). (b) *Microtus oeconomus*. 1, $y = 57.46x^{-0.644}$ (in May); 2, $y = 11.25x^{-0.268}$ (in January–April).

When the population cannot be investigated in detail and the standard of different indices is not known, it is sufficient to compare the indices from various intrapopulation groups of animals, as illustrated by the following example.

The liver weight in pregnant and lactating females is higher than in males. This is because a source of glycogen is necessary to provide food for the embryo. This relationship was shown for a subarctic population of *M. oeconomus* whose numerical dynamics have been studied in the Yamal Peninsula since 1957. Data over five years were analyzed separately for a younger age class with body weights of 20 to 40 g, and an older age class, with body weights of 40 to 60 g. The results expressed in parts per thousand (‰) of liver to body weight areas follows:

	1959	1960	1961	1962	1968
Younger					
Female	68.4	56.3	62.9	59.3	63.8
Male	44.2	42.2	56.8	46.9	61.8
Older					
Female	62.5	53.9	63.6	63.8	63.8
Male	56.4	52.4	50.2	45.9	61.2

It can be seen that the liver weights of the females during breeding periods are always greater than those of the males. Under very favorable environmental conditions, however, the difference in this index between males and females is smaller as all members of the population can accumulate maximal amounts of glycogen (for example see 1968 above). It should be added that the glycogen content in the livers of the voles under study usually ranges from 0.7 to 4.5 per cent. When the liver weight is about 60‰, the glycogen content averages 9 per cent.

When the relative liver weight in males is considerably lower than in females, we can predict that the energy balance of the populations is strained and that the older animals must soon die. This condition is also indicated by hypertrophy of the adrenals accompanied by a significant decrease in the liver weight.

It is well known that the adrenals are much larger in females than in males. We should, however, recall the 'morpho-physiological standard' once again, and note that this relationship has different quantitative values for different species (Fig. 6.7). During the breeding period the increase in the adrenals of females is very rapid (Fig. 6.8), and this can be regarded as a standard morpho-physiological condition. But, when hypertrophy of the adrenals exceeds the range of this standard (Fig. 6.9), it can be regarded as a sign of possible collapse of the population.

The initial stages of the stress situation for the population can be detected by comparing the degree of hypertrophy of the adrenals in pregnant females at different ages. The degree of adrenal hypertrophy is higher in the old and very young females. It is also indicated by an early involution of the thymus in young animals of both sexes, while significant development of thymus in breeding animals indicates that the population is in good condition.

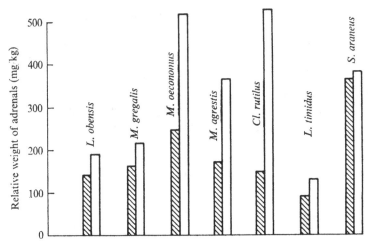

Fig. 6.7 Average relative weights of adrenals in males and females of different species of mammals in the breeding period. Animals caught in one region at the same time were selected for the comparison. Hatched columns, males; open columns, females.

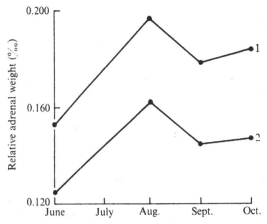

Fig. 6.8 Relative adrenal weight (mg/g) in 1, breeding females and 2, non-breeding females of *Clethrionomys glareolus*. Weight class 15–20 g.

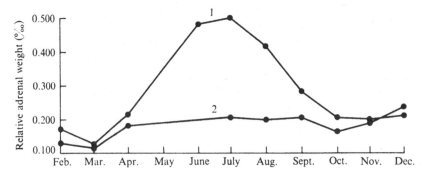

Fig. 6.9. Relative adrenal weight (mg/g) in 1, breeding females and 2, males of *Clethrionomys glareolus*. Weight class 15–20 g.

The normal changes in the weight of thymus are shown for *Clethrionomys glareolus* (Fig. 6.10). In cases where the increase in the weight of the gland is not high in the spring, it can be assumed that breeding will be inhibited.

The changes in the weight of the above mentioned organs also can be related to the dynamics of the content of vitamin A in the liver. Vitamin

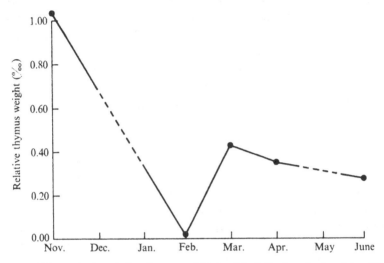

Fig. 6.10. Changes in the relative thymus weight (mg/g) of *C. glareolus* in winter. Weight class 15–20 g.

A is used continuously and can be stored only when its input, or the input of carotene from food, exceeds the normal output for any period of life. If the output is increased and the amount of vitamin A supplied from food is too low, then the stored resources are utilized. For example, Smirnov & Shvarts (1957) have shown that vitamin A is regularly stored in the liver of the muskrat, *Ondatra zibethicus*, during autumn and winter. As the parts of the plants without chlorophyll and carotene are more often the food of the muskrat in autumn and winter than in summer, resources of the vitamin in the liver would be expected to decline. However, during the summer period, when the breeding of adults occurs and young are growing, vitamin A is intensively used and there is no possibility to store larger deposits, while in winter (the non-breeding period), vitamin output is lower and it can be stored in an amount exceeding 8 to 10 times the summer level. When the bodies of water become free from ice the mating period begins and these stored resources are used during a short period of 1 to 2 weeks, even though the aquatic vegetation grows rapidly and contains more carotene than that below the ice in winter.

A significant difference has been recorded in the level of stored resources between males and females. Males use their deposits more rapidly, but at the end of the mating period can restore them, while females cannot restore their deposits during the whole period of pregnancy and lactation. It is interesting that the muskrats experimentally supplied at the mating period with additional food containing vitamins (carrots), even in a quite limited quantity, had a greater percentage of their offspring in autumn than control animals fed on potatoes. The addition of the vitamin resulted not only in the improved condition of particular animals, but also in a rise in the productivity of the whole population. This means that, through the analysis of the content of vitamins in the animals, it is possible to find a key for the control of their reproductive capacity and dynamics of numbers.

Information on the typical differences in the vitamin A reserves between males and females is useful when the functioning of the population is concerned. The content of vitamin A in the muskrat males during the summer period ranges from 30 to 40 mg/100 g and in females it is lower than 10 mg/100 g. Significant deviation from these values may be used as a good basis for the prediction of number dynamics, especially when it is compared with the other morpho-physiological indices mentioned above. For example, the lack of differences in the liver weights between pregnant and non-pregnant females indicates that food conditions are very favorable for the population. This conclusion may be supported by the analysis of the vitamin A content in the animals' livers: if there are no significant differences in its content not only between pregnant females and males but also between nursing females and males, it is an index of the excellent condition of the population.

There are differences in the mortality rates of animals at different periods of the year, and it is during transitional periods (early spring and late autumn in the far north), that the mortality rate usually increases rapidly. In many cases, the rate of this process is of great importance when predicting the number of rodents. The following set of indices may be used, together with an analysis of meteorological conditions and population structure, as an indicator of a possible rapid increase in mortality at the period of autumnal cold: low increase in the liver weight in autumn particularly for breeding animals; significant differences between older and younger generations; early hypertrophy of adrenals being pronounced during a long period (attention should be paid to the differences between age classes); no significant increase in the content of vitamin A in the liver (in this case it is useful to compare breeding females and males); thymus involuted even in the animals of the youngest age classes. The opposite combination of indices indicates that the animals are in good physiological condition at the beginning of winter.

The study of rodents in the spring provides information on the condition of the population surviving the winter and the probability of increased mortality resulting from an increase in physiological stress when food resources decline and unfavorable environmental conditions prevail before plant growth begins. It also permits us to predict the breeding success of the population.

Rodents in good physiological condition at this time of year are characterized by the following set of indices: high percentage of body fat, high content of vitamin A in liver until the beginning of major vegetation growth and a rapid increase in vitamin A, in males and females, with the development of green food plants. In younger age classes that survive the winter, there is a regular increase in the weight of the thymus, relative and absolute increases in the weight of the kidney, no significant decrease in the weight of the liver, no significant hypertrophy of the adrenals in males and no considerable difference in adrenal hypertrophy among females of different ages.

The beginning of the breeding period can be assessed in advance due to an increase in hypophysis weight (Table 6.2), and, in some cases, a

Table 6.2. *Seasonal changes in relative weight of the hypophysis of the muskrat from a forest–steppe of Transural (in mg/kg)*

Sex, age	Month								
	Oct.	Nov.	Dec.	Jan.	Feb.	Mar.	Apr.	May	June
Young males	10.8	11.6	10.9	11.5	13.7	–	16.8	18.2	16.2
Young females	11.3	8.6	9.8	11.8	9.7	–	19.2	24.5	31.0
Adult males	16.5	11.5	13.3	14.9	16.5	–	–	19.9	–
Adult females	20.8	15.4	14.0	17.5	22.4	–	1.92	33.6	–

significant increase in the weight of the animals (Figs. 6.11, 6.12) and a pronounced increase in kidney weight (Fig. 6.13). The earlier the morpho-physiological changes occur in the spring, including the differences among animals of different sex and age, the earlier breeding begins.

139

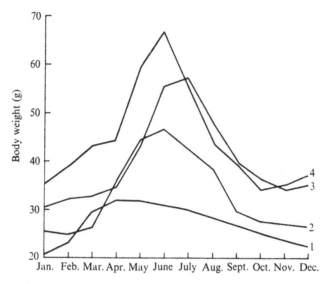

Fig. 6.11. Seasonal changes in the average body weight of male voles in a vivarium. 1, *Lagurus lagurus*; 2, *Microtus g. gregalis*; 3, *M. g. major*; 4, *M. middendorffi*.

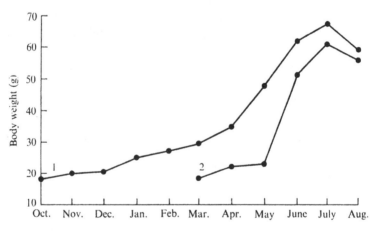

Fig. 6.12. Seasonal changes in the average body weight of *Microtus gregalis* males born at different seasons. 1, August–September; 2, January–February.

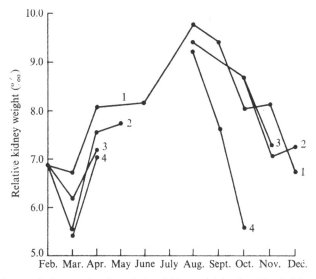

Fig. 6.13. Seasonal changes in the relative kidney weight (mg/g) of *Microtus arvalis*. Weight classes: 1, 10–15 g; 2, 15–20 g; 3, 20–25 g; 4, >20 g.

As already noted, it is useful to pay attention to the liver weight of breeding females. If the average relative liver weight in females is lower than in males, it is almost a certain signal of a disastrous population condition. Also, the content of vitamins in the liver, size of adrenals and, in young females, the weight of thymus should be measured.

When young animals are studied, the proportions of the body and skull may provide important information. For example, significant increase in the relative length of the skull, as compared with a standard, indicates that the growth rate of the animals is delayed. This point needs special emphasis; if the regularities of the formation of the proportions of the body and skull are known, it is possible to reconstruct the fate of the population using simpler indices. The practical and theoretical significance of this is evident.

The body proportions in animals exhibit general rules, since there is a relationship between the constitution of the animals and their rate of growth at different periods of ontogenesis. These rules were found long ago for domestic animals and recently their importance for wild animals has been described in studies at the Institute of Animal and Plant Ecology, Academy of Science, USSR, on laboratory colonies of *Lagurus lagurus*, two subspecies of *Microtus gregalis*, *Microtus middendorffi* and also hybrids between the subspecies of *M. gregalis*. In Table 6.3, cranial characteristics are presented for two size groups of *L.*

141

Table 6.3. *Cranial characters of two size groups of Lagurus lagurus in relation to their growth rate (experimental populations). Measurements are in millimetres, plus or minus one standard deviation*

Indices	Length of body	Condylobasal length of skull	Breadth of skull	Breadth of interorbital (inter-eye) distance	Length of line of teeth	Length of visceral part of skull	Height of skull
Group of slowly growing animals†	84.0 ± 0.523	21.9 ± 0.226	12.9 ± 0.160	2.57 ± 0.049	5.37 ± 0.086	7.98 ± 0.139	7.1 ± 0.105
Group of rapidly growing animals†	82.8 ± 0.688	20.5 ± 0.192	12.5 ± 0.108	2.67 ± 0.037	5.06 ± 0.049	7.15 ± 0.102	7.4 ± 0.219
Group of slowly growing animals‡	95.0 ± 0.617	22.5 ± 0.244	13.7 ± 0.152	2.60 ± 0.052	5.50 ± 0.048	8.20 ± 0.098	7.3 ± 0.071
Group of rapidly growing animals‡	96.2 ± 1.17	22.0 ± 0.141	13.4 ± 0.124	2.80 ± 0.046	5.30 ± 0.049	7.83 ± 0.078	7.6 ± 0.066

† Body length 80–85 mm.   ‡ Body length 92–105 mm.

lagurus, each group being divided into slowly and rapidly growing individuals. These data indicate that the proportions of body and skull for animals of the same size with different growth rates are significantly different (for detailed description of the experiments see Shvarts, 1962). In slowly growing animals, the condylobasal length of the skull, the breadth of the skull, the length of the line of tenth and visceral part are greater, while the height of the skull and the breadth of the inter-orbital distance are smaller than in rapidly growing animals. These differences do not depend on the age of the animals compared; however, the differences among animals of the same body size growing at different rates are comparable with the differences that exist among animals of different sizes. For example, there are larger differences in the relative length of the skull between rapidly growing *L. lagurus* of the first group (body length 80 to 85 mm) and those growing slowly than between 'average' animals of 86 to 90 mm and 70 to 80 mm. These are very significant differences.

Analysis of the available material shows that the rate of growth during the first period of an animal's life leaves an indelible trace on the constitutional characters of adult and old individuals. For example, the relation between the rate of growth of rodents in the first month of life and the relative length of their skull was established using the method of multiple correlation. The following coefficients were calculated: r, coefficient of correlation; p, level of significance of the correlation. The results were as follows:

Lagurus lagurus	$r = -0.97 \pm 0.06, p = 0.01$
Microtus gregalis (north subsp.)	$r = -0.370 \pm 0.16, p = 0.05$
Microtus middendorffi	$r = -0.67 \pm 0.09, p = 0.01$

Such correlations were also obtained in experiments conducted using other methods, and the conclusions were verified in the field for species whose age could be determined with sufficient precision. Thus, the muskrat was selected because of its age can be readily estimated using the method described by Smirnov & Shvarts (1959). It was shown that the width of the skull was smaller in rapidly growing than in slowly growing individuals; also, the visceral part of the skull was shorter and the relative length of the skull was smaller when growth was rapid. The laboratory and field data were in full accordance.

One further remark should be made here: it is certain that changes in the rate of growth are followed by changes in the proportions of body and skull and also probable that this relationship for different species is essentially similar. For example, in all the species under study, an increase in the growth rate results in a decrease in the relative length of the skull. The reason for this relationship is that the skull grows at a more constant rate than other body parts and when the rate of body growth increases, a difference in the proportions must be produced in relation to the linear increase in body size. This pattern has also been observed for some other measurements, but so far there is not sufficient evidence to show that a definite relation between the growth rate of the animals and development of particular body character is always realized. There are probably exceptions to this rule.

The results indicate that the characteristic features of body proportions in particular animals reflect characteristic features of development of particular populations. In a monograph by Shvarts *et al.* (1968), there are some examples indicating that on the basis of skull proportions and other constitutional features of the animal body the whole history of the population and intrapopulation groups can be reconstructed. Further, through the method of morpho-physiological indices the response of various intrapopulation groups of animals to changes in environmental conditions can be estimated and, therefore, a deeper insight into the interrelationships between population structure and dynamics of numbers can be reached.

Some of the methods presented for estimating the condition of rodent populations are still not quite certain, but such criteria as the relative weight of liver and adrenals are available and should be taken into account in the prediction of numbers of rodents. They should also be introduced into the practices for predicting mass occurrences of rodents.

Differences between populations

To estimate the biological characteristics of the population, it is necessary to extend the set of indices used. Allometric indices, which characterize the relative rate of various parts of the body and organs, are of particular importance as are the content of myoglobin in different groups of muscles, the electrophoretic analysis of proteins in blood plasma, etc. The analysis of morpho-physiological indices for arctic and mountain species of mammals indicates that any isolated or even partially isolated population is morpho-physiologically specific. Not only different species, but also various populations of the same species adapt to environmental conditions in different ways. This problem has

144

been discussed thoroughly in a monograph on subarctic mammals (Shvarts, 1969) and, therefore, only two examples will be presented here.

Adaptations of subarctic populations of *Microtus oeconomus* to life in the far north consist of a considerable increase in natality, an increase in the growth rate and early sexual maturity. Lemmings, which are typical subarctic rodents, adapt differently. Their breeding period is prolonged due to breeding in winter, their fertility is lower than *M. oeconomus*, the amount of food consumed is larger and their resting metabolism is lowered.

Certainly, identical changes in the morpho-physiological indices of Siberian lemmings and *M. oeconomus* must have different ecological meanings. An increase in the indices of metabolic rate during the breeding period represents an ecological standard for *M. oeconomus* while for lemmings it represents a deviation from the standard.

The importance of the comparative analysis of allometric relationships in the development of various anatomical structures may be illustrated by the following example. The growth of the skull in southern forms of *Microtus gregalis* is accompanied by a decrease in the interorbital breadth which in very large voles can affect the normal growth of teeth and can also affect the development of the eyes. The northern populations of this species, *Microtus gregalis major*, are larger than southern ones and it has been shown experimentally that the differences between the subspecies of *M. gregalis* are hereditary (Fig. 6.14). The contradiction in skull development is solved in a rather particular way – negative

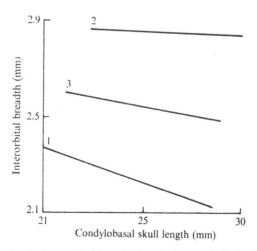

Fig. 6.14. Changes in the interorbital breadth in relation to skull size for 1, *M. gregalis gregalis*; 2, *M. gregalis major*; 3, hybrids.

145

allometry is replaced by isometry (Ishchenko, 1966). Certainly, when the skull proportions in various forms of the same species are studied, different systems of estimates should be used.

Large amounts of material concerned with these problems have been collected by Toktosunov (1973), who studied morpho-physiological indices in 16 species of mammals occuring in Tyan-Shan. He found that in mountain forms of these animals the following indices of their ecology and physiology were changed: (1) type of population dynamics, (2) relative length of the intestine and appendix, (3) size of the liver and regularities of its growth, (4) content of glycogen in the liver and iodine in fat, (5) rate of accumulation of vitamin A in the liver, (6) myoglobin content in different organs and (7) metabolic rate and hematological indices. Also, each population of each species was morpho-physiologically specific The specific character of the population is determined, on the one hand, by the life conditions (even at the same altitude the effectiveness and intensity of natural selection is different) and, on the other hand, by the genetic composition of the original group of animals. The following example will illustrate these interesting findings.

In the comparative study on the changes of the relative length of the intestine with age in different mammal populations, there was no significant difference in the length of the intestine between young *Citellus relictus* from the Toktogul population and young ground squirrels from the Ak-Ulak population; but, the relative length of the intestine in individuals from the Ak-Ulak population increased with age,

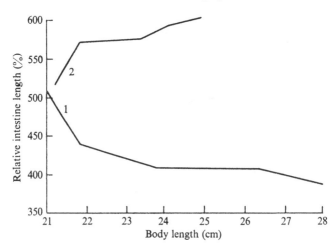

Fig. 6.15. Age changes in the relative intestine length of *Citellus relictus* in spring. 1, Toktogul population; 2, Ak-ulak population.

while that of individuals from the Toktogul population decreased. As a result, the length of the intestine of grown animals in the Toktogul population is 1·5 times smaller than that in Ak-Ulak. These data readily show the biological distinction of various populations. Some examples of this regularity are presented in Figs. 6.15, and 6.16.

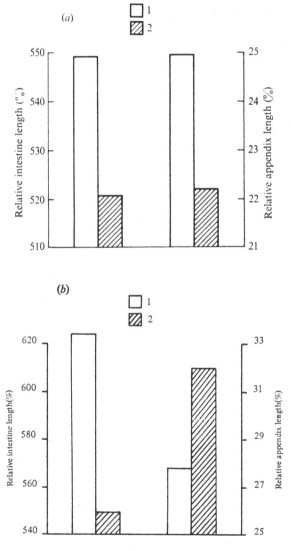

Fig. 6.16. Relative intestine and appendix length in (*a*) males and (*b*) females of 2 populations of *Citellus fulvus* in spring. 1, Sla-arshinskaya population; 2, Norusovskaya population.

It is known that changes in the liver weight depend on the condition of an animal and the season of the year. Toktosunov (1973) has shown that the seasonal dynamics of relative liver weight in animals of various populations of the same species is different. Thus the liver index (relative liver weight) in adult *Meriones erythrourus* of the Ak-Ulak population increases from spring (28.7‰) to winter (42.0‰), while for individuals of the same species from the Issik-Kul population it is about 33.8‰ in spring, reaches a maximum of 42.8‰ in summer and is lower in winter than in spring. These differences are certainly due to the differences of life conditions of the populations, but within the scope of this study it is more important to emphasize the interpopulation differences characterized not only by absolute characteristics but also by the type of their seasonal variations.

The analysis of the changes in the relative weight of liver with age also reveals significant differences among populations. For instance, the

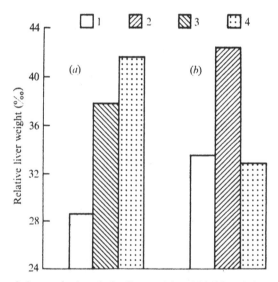

Fig. 6.17. Seasonal changes in the relative liver weight of (*a*) Aksuekskaya population and (*b*) Issyk-kulskaya population of *Meriones erythrourus*. 1, spring; 2, summer; 3, autumn; 4, winter.

liver index in *M. gregalis* of the Arpinsk population continuously increased with age while the weight of liver in individuals of the Son-Kul population reached the maximum in individuals with a body weight of about 32 g after which the value of the liver index dropped. Similar data can be reported for almost all species studied (Fig. 6.17).

An investigation by Toktosunov has shown that the content of myoglobin in individuals from the majority of high-mountain populations is increased. *Ellobius talpinus* may serve as an example, as all mountain populations of this species have an increased myoglobin content. The changes in the myoglobin content of some muscles are

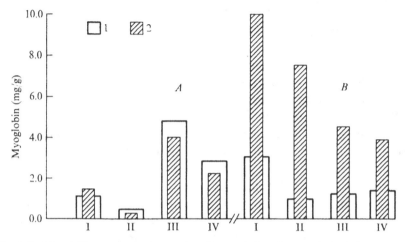

Fig. 6.18. Myoglobin content in the muscles of 2 populations (*A* and *B*) of *Ellobius talpinus* at different seasons (1 and 2). Muscles: I, myocardium; II, pectoral; III, foreleg; IV, hindleg.

clearly pronounced (Fig. 6.18); in other muscles they are insignificant or the myoglobin content drops with altitude. However, in animals of different populations, significant differences in the myoglobin content occurs in different muscles (Fig. 6.18). We consider these data to be of special importance as they show that the same physiological effect in various populations is reached in different ways, and, therefore, that various populations of the same species are morpho-physiologically specific. The investigation of this distinction is of great theoretical and practical interest.

Intraspecific variation and speciation

To study the ecological regularities of intraspecific variability and speciation, the ecological techniques should be supplemented by immunological and karyological methods and by study of the genome.

For illustration, we will describe the results obtained by the method of morpho-physiological indices when two interesting problems are analyzed: the origin of sibling species and the genetics of intraspecific variability and speciation.

In our laboratory the immunological divergence in subfamily Microtinae was studied by Syuzyumova (1973) and Zhukov (1973). These investigations indicate that the immunological distance reflecting biochemical distinction of organisms at both intraspecific and interspecific levels does not correspond to the range of morpho-physiological differences among the forms compared. This means that biochemical characters of animals are determined not only by adaptation of various forms to different environmental conditions but also by genetic differences which are not directly related to the adaptive characters of populations, subspecies and species.

If two populations occur in environmental conditions so similar that stabilizing selection is more advantageous than creative selection, then selection will favor and stabilize the optimum phenotype. But, as the genetic structure of the population can differ originally and this differentiation gradually increases due to random mutations, then the stabilizing selection of optimum phenotype must inevitably lead to a reorganization of genotypes to neutralize the effect of new mutations. As a result, genetic differences are being developed among populations with very similar morpho-physiological characters and the range of genetic differences does not coincide with the range of morpho-physiological differences.

Any isolated or even partially isolated population is genetically distinct. To explain this distinction there is no need to refer to the doubtful hypothesis of neutral mutations; it can be satisfactorily explained by the theory of stabilizing selection developed by Shmal'-gauzen (1946), and extending this theory by the concept of optimum phenotype (Shvarts, 1968). It is supported also by the results of studies indicating that the degree of immunological divergence does not coincide with the range of morpho-physiological divergence. It is theoretically possible that the differences among populations resulting from stabilizing selection will have such an effect that the formation of a different genotype, as compared with genotypes of other populations of the species, will result in genetic isolation. Certainly, this process is slow and is delayed by inevitable interpopulation hybridization. Therefore, real sibling species which show no greater differences between each other than the common interpopulation differences occur very rarely, but, in the cases when the ecological–physiological characters of a differentiating population slightly exceed their average values, the process of interpopulation hybridization is weakened and increasing genetic isolation rapidly accelerates speciation processes.

Thus, the formation of sibling species can be explained on the basis of stabilizing selection and should be included in modern evolution theory. These data also provide materials for the analysis of another interesting problem, i.e., the transplantational immunity in a number of vole species, represented by a large number of populations, the study of which brought Syuzyumova (1973) to an important concept of immunological unity of the species.

Any subspecies or any population of the species can distinguish 'its own' species from a 'strange' one. The unity of the species at the tissue level and specific distinction of tissues refer to all levels of intraspecific differentiation. This conclusion is supported by the character of tissue incompatibility of heterotransplants without the dependence on intraspecific differences in donors. In addition, it has been shown that the 'mean' immunological differences among populations do not exceed the differences between extreme variants of one population. Taking into account that the immunological distance reflects the range of biochemical differences among the organisms, which, in turn, reflects the range of genetic differences (hereditary information is realized through the processes of protein synthesis), it is suggested that the immunological unity of the species excludes the possibility of the formation of a hiatus among any intraspecific forms. Thus, the formal criterion of a species, hiatus, which was noticed by taxonomists long ago, reflects its biological essence, i.e., genetic unity, which appears in the form of intrapopulation variation in biochemical characters.

The results indicated that more detailed investigations on the nature of interspecific and intraspecific differences are needed. Therefore, a comparative study on the structure of the genome of two significantly ecologically diversified subspecies and a different species of the same genus was carried out. In this study arctic (Yamal) and a forest–steppe (Southern Transurals) subspecies of *Microtus oeconomus* were compared with *Microtus arvalis*.

DNA from liver preserved in ethanol of *M. o. oeconomus*, *M. o. chahlovi* and *M. arvalis* was analyzed in the Laboratory of Bioorganic Chemistry of Moscow University. Nucleotide composition of DNA was determined by G. P. Micrshnichenko, K. M. Val'sko-Roman and N. B. Petrov using the methods of Antonov & Belozerskii (1972).

It was found that the nucleotide composition of all three forms of voles was practically identical (42–43 per cent cytosine + guanine),

151

which is not surprising as the composition of DNA in mammals fluctuates within narrow limits (Antonov, 1973). The analysis of the degree of unity of pyrimidine nucleotides in DNA (β) indicates that the DNA of *M. o. oeconomus* and *M. arvalis* cannot be distinguished using this index ($\Delta\beta = 0.03$), but both of these forms differ significantly from *M. o. chahlovi* ($\Delta\beta = 0.23$ and 0.26 respectively). Significant differences were observed in the fraction of monopyrimidine fragments. Identical results were obtained when 'genospectra' of DNA were compared. Statistically significant differences (using Fisher's criterion) were found when the genospectra of *M. oeconomus* and *M. arvalis* were compared (7.5) and also when both of these forms were compared with *M. o. chahlovi* (19.7 and 27.0). These are preliminary results as only one portion of the DNA of each form of voles was used in the analyses.

Nevertheless, the results suggest that the genetic material of *M. o. oeconomus* is more similar to DNA of *M. arvalis* than to that of *M. o. chahlovi*. The greatest differences were found when the original structures of DNA of *M. arvalis* and *M. o. chahlovi* were compared. The data indicate that the genetic differences between the two subspecies are greater than those between species. Morpho-physiological characters of arctic *M. oeconomus* coded in DNA required greater changes at the genome level than the differences between ancestral (forest–steppe) forms of *M. oeconomus* and *M. arvalis*; the genetic cost of subspecific differences was higher than that of specific differences. Immunological investigations indicate that this result could be theoretically predicted and it was used as a basis for appropriate experiments. The specific characteristics are determined by the quality and not by the quantity of genetically coded characters of the animals.

We hope that the examples presented indicate that a complex of morpho-physiological and biochemical methods can be successfully used to solve both theoretical and practical problems in mammals.

7. Biological production in small mammal populations

K. PETRUSEWICZ & L. HANSSON

In Chapter 1.2, production (P) is defined as the total amount of biomass or energy produced by a population through growth and reproduction over a period of time less that lost to maintenance. This definition can be expressed by the formula

$$P = A - R, \tag{7.1}$$

where A is assimilation (energy flow), and R is respiration or maintenance.

The above formula can seldom be applied to production estimates in practice. We usually estimate P and R in other ways and then calculate assimilation from the formula

$$A = P + R,$$

(also see Chapter 8).

The definition of production as the biomass produced over a time period indicates that production comprises not only the increase in biomass in that time (T), i.e., $\Delta B_{(T)}$, but also the biomass produced but not present at the end of the study period, i.e., the biomass that has been eliminated through emigration or death as well as losses through moulting and other processes. This concept can be expressed by the formula

$$P_{(T)} = (B_T - B_0) + E_{(T)} = \Delta B_{(T)} + E_{(T)}, \tag{7.2}$$

where $P_{(T)}$ stands for production within the time period T, B_T and B_0 are biomass at times T and 0 respectively, and $E_{(T)}$ is elimination within the time period T.

Production of new biomass may be due to body growth (P_g) and reproduction (P_r):

$$P = P_r + P_g. \tag{7.3}$$

A question arises as to what should be represented as production due to reproduction in mammals: the biomass of newborn or the weight at birth and growth until weaning? Many authors (Davis & Golley, 1963; Petrusewicz, 1967a; Walkowa, 1967; Petrusewicz & Walkowa, 1968)

153

suggest that the growth until weaning should be attributed to reproduction P_r. Here a distinction will be made between production due to biomass increase of unweaned animals (P_g) and that due to the number of young born and growth of suckling animals (γ_r, Petrusewicz, 1968, 1969). For some investigations, it can be useful to distinguish between production of newborn (P_b), production due to reproduction (P_r) (i.e., production of newborn and production of growth until weaning) and production of growth (P_g).

These definitions have been generally accepted in small mammal studies. In somewhat older literature the word 'productivity' is often used synonymously with 'reproduction' as is still the case in game management (Beasom, 1970; Roseberry & Klimistra, 1970).

Production can be expressed in various units such as live weight, dry weight, ash-free dry weight and energy. Interrelations between these various units and the advantages of using various measures to express the productivity concept are discussed in Chapter 8.

An accurate knowledge of reproduction is of great importance for correct production estimates because it accounts for approximately 50 per cent and can amount to 80 per cent of the total production (Petrusewicz *et al.*, 1968). For this reason we will discuss the advantages of various methods of making reproduction estimates.

Finally, it is extremely important to be conscious of the accuracy of the results obtained. This subject is also discussed in later sections.

Determination of reproduction

For the calculation of the number of individuals born (γ_r) during a certain time period (T) it is necessary to have the following empirical data: (1) litter size, L; (2) time period of pregnancy, t_p; and (3) numbers (standing crop) of pregnant females over the period, N_p, from which we can calculate the average numbers of pregnant females $\bar{N}_p$. The number of pregnant females can be found empirically, or can be calculated from the numbers (N), sex ratio ($s = N_\female : N$), and the pregnancy ratio ($f = N_p : N_\female$) using the formula

$$N_p = N \cdot s \cdot f. \tag{7.4}$$

In addition, mortality of pregnant females must be known for precise calculation of γ_r.

Based upon these data, two formulae have been used most often for estimating the number of newborn in a small mammal population:

(1) The first is based on a constant daily birth rate (Elster, 1953; Petrusewicz, 1968; Bujalska *et al.*, 1968):

$$\gamma_r = \bar{N} \cdot s \cdot f \cdot T \cdot L/t_p = \bar{N}_p \cdot T \cdot L/\bar{t}_p. \tag{7.5}$$

(2) The latter is based on an instantaneous rate of population growth applied to small mammals (Golley, 1960), with the assumption that the sex ratio is 1:1, and $S = N_{\female}/N = \frac{1}{2}$:

$$\gamma_r = \bar{N} \cdot T \cdot f/t_p \cdot \ln\left(\frac{L}{2} + 1\right). \tag{7.6}$$

Petrusewicz (1968, 1969) found that both formulae give only an approximation of the number of newborn and their accuracy depends on various demographic and population factors. In particular, each of these methods gives different results depending on whether the number of pregnant females is increasing or decreasing. This is due to the fact that the dynamics of numbers of pregnant females depends on mortality which reduces N_p, and also on the number of impregnations (an increase in N_p) as well as on the number of births (a decrease in N_p).

Formula (7.6), based on the instantaneous birth rate always under-estimates the number of newborn. The underestimate ranges from 45 to 70 per cent and usually approximates 50 per cent of the real value for animals with a fecundity similar to that of *Clethrionomys glareolus* or *Microtus*, i.e., animals with a litter size of five (Petrusewicz, 1968).

Formula (7.5) usually overestimates the results; at the same time the overestimation for the period of decrease in the number of pregnant females is different from that for the period of increase. During the period of decrease, the overestimate is 2 to 15 per cent and usually does not exceed 5 per cent for animals with a fecundity of about five. During the period of the increase, the overestimate is larger, and amounts to 5 to 20 per cent for animals which have the vole reproduction type.

Analyzing simulated populations with $L = 5$ and $t_p = 20$ days, Petrusewicz (1968) suggested the formula

$$\gamma_r = \left(\frac{\bar{N}_p T}{\bar{t}_p} - \frac{E_{p(T)} + \Delta N_{p(T)}}{2}\right) \cdot L, \tag{7.7}$$

where $E_{p(T)}$ is the number of pregnant females eliminated during the time (T) and $\Delta N_{p(T)} = N_{p(T)} - N_{p(T_0)}$. This formula gives a correct estimate of animals born assuming that half of the pregnant females disappearing during T gave birth before dying.

155

But (7.7) is of little use in field investigations because it is very difficult to estimate the elimination of pregnant females. Approximate estimates may be obtained for the period of increase in the number of pregnant females by using

$$\gamma_r = [(\bar{N}_p \cdot T/t_p) - \Delta N_{p(T)}/2] \cdot L, \qquad (7.8)$$

and for the periods of decrease by (7.5)

$$\gamma_r = \bar{N}_p T \cdot L/t_p. \qquad (7.5)$$

As mentioned earlier, the estimate of the number of newborn is based mainly on N_p, L and t_p. The litter size (L) is the easiest reproductive parameter to determine. It is a physiological index, pertaining to an individual of a species and, therefore, is more stable than an ecological parameter (Table 7.1). For example, although, one litter of *Clethrionomys* may contain 1 to 10 newborn, the average for 30 litters or 30 pregnant females is a reliable value which does not differ significantly from an average value for 50 or 100 litters (Zejda, 1966; Bujalska *et al.*, 1968). Of course, the litter size changes with the age of an animal, with living conditions, and with season of the year, but these differences are usually rather small. For *Clethrionomys glareolus*, one of the best known small mammals, under field conditions litter size ranges from 4.0 to 6.1 (Table 7.1). Litter size has been estimated for many other species and even its seasonal variability is known.

The duration of pregnancy (t_p) is also a physiological parameter, and is, therefore, relatively stable despite the fact that during lactation pregnancy may be prolonged (Wrangel, 1940; Naumov, 1948).However, this parameter has been studied less than litter size. Besides, it is very difficult to estimate t_p in the field since it requires intensive trapping by the catch–mark–release method to find the beginning and the end of the pregnancy period. For that reason, the gestation period is often estimated under laboratory conditions. However, extrapolation of laboratory data to the field always contains the danger of an error. For example, pregnancy under field conditions for *C. glareolus* was estimated as 22 days, while in the laboratory, it usually equals 17 to 18 days (Bujalska & Ryszkowski, 1966).

The number of pregnant females (N_p) is the most difficult reproductive parameter to estimate and can be the main source of error. As we have already said, the sex ratio (s) in (7.4) may vary, but the real difficulty is in the estimation of N and f. Pregnancy ratio (f) has a full range of variability, from 0 to 100 per cent, since none of the females or all of

Table 7.1. *Examples of physiological parameters*: *mean values of litter size* (L) *and time of pregnancy* (t_p) *in days. Litter size obtained from autopsy of field animals and counts of laboratory-born litters. Pregnancy time from vaginal smears of field animals.*

Species	Place	Time	L	t_p	Author
C. glareolus	Great Britain France	Breeding season	3.8–3.9		Zejda (1966)
C. glareolus	North Germany	Breeding season	5.0		Zejda (1966)
C. glareolus	CSSR	Summer	5.0		Zejda (1966)
C. glareolus	CSSR	Winter	3.9		Zejda (1966)
C. glareolus	USSR	Breeding season	5.3–6.1		Zejda (1966)
C. glareolus	Poland	Breeding season	4.9–5.6	22(17–28)	Bujalska & Ryszkow-ski (1966)
C. glareolus	North Sweden	Spring	5.9		Hansson (1969b)
C. glareolus	Laboratory	—	3.6		Buchalczyk (1970)
C. glareolus	Laboratory	—		14–30	Drożdż (1965)
C. glareolus	Poland	Breeding season	4.8	—	Ryszkowski & Trusz-kowski (1970)
M. agrestis	Great Britain	Breeding season	3.0–5.3		Chitty (1952)
M. agrestis	North Sweden	Spring	6.8		Hansson (1969b)
M. agrestis	Laboratory	Spring	3.9	21	Ranson (1934, 1941)
Peromyscus maniculatus	North America	Breeding season	3.5–6.8		Smith & McGinnis (1968)
Peromyscus leucopus	North America	Breeding season	3.5–5.8		Smith & McGinnis (1968)
Mus musculus L.	Great Britain 'natural habitats'	Breeding season	5.6		Laurie (1946)
Mus musculus L.	Cold stores	Breeding season	6.4		Laurie (1946)
Mus musculus L. ('Swiss')	Laboratory	—	7.5	19.4	Myrcha & Walkowa (1968) Walkowa (unpub-lished results)
Mus musculus L.	North Norway	August	5.7		Andersson & Hans-son (1966)

them may be pregnant. Further, the number of animals in the population may range from zero to several thousand per hectare. For that reason, the number of pregnant females ($\bar{N}_p$) must always be estimated for a specific area and time period.

Various measurements of reproduction rates were discussed by Davis & Golley (1963) and Bobek (1969). Published field estimates of the number of small mammals born per hectare and year are listed in

157

Table 7.2. *Field estimates of reproduction. Figures given per plot were recalculated on a per hectare basis. SM, standard-minimum snap trapping; CMR, catch-mark–release live trapping; P, computations according to Petrusewicz (1970); G, according to Golley (1969); B, according to Bobek (1969); Mod, modified method.*

Ecosystem	Place, years	Species	Sampling method	Estimation of γ_r	γ_r/ha · yr	Author
Forest type ecosystems						
Conifer (pine) forest	Poland 1967–69	*Clethrionomys glareolus* (Schr.)	SM	P	0–22	Ryszkowski (1971b)
Conifer (spruce) forest	Bulgaria 1967–68	*Clethrionomys glareolus* (Schr.)	SM	P	53–59	Petrusewicz et al. (1972)
Pine oak forest	Poland 1967–69	*Clethrionomys glareolus* (Schr.)	SM	P	6–69	Ryszkowski (1971b)
Deciduous (beech) forest	Poland 1965–66	*Clethrionomys glareolus* (Schr.)	Mod. SM	B	91	Bobek (1969)
Deciduous (mixed) forest	Poland 1967–69	*Clethrionomys glareolus* (Schr.)	SM	P	52–132	Ryszkowski (1971b)
Deciduous (mixed) forest	Poland 1966–69	*Clethrionomys glareolus* (Schr.)	CMR	P	253–273	Petrusewicz et al. (1968, 1971)
Deciduous (mixed) forest	Poland 1967–69	*Clethrionomys glareolus* (Schr.)	Mod. SM	B	55–179	Bobek (1971)
Deciduous (mixed) forest	Poland 1967–69	*Apodemus flavicollis* (Melch.)	Mod. SM	B	18–159	Bobek (1971)
Tropical rain forest	Panama 1968–69	*Proechimys semispinosus*	CMR	Mod. P	20	Gliwicz (1933)
Grassland type ecosystems						
Agrocoenose	Poland 1964–68	*Lepus europaeus* (Pall.)	Mod. SM	P	0.81	Petrusewicz (1970)
Agrocoenose	Poland 1963–67	*Microtus arvalis* (Pall.)	CMR	Mod. P	1083	Trojan (1969)
Abandoned fields	Finland 1961	*Microtus agrestis* (L.)	CMR	Mod. P	2150	Myllymäki (1969)
Abandoned fields	Michigan, USA 1956–57	*Microtus pennsylvanicus* (Ord.)	CMR	G	385	Golley (1960)

Table 7.2. The estimates vary greatly; however, for the same type of small mammals there is usually greater reproduction in grassland than in forest ecosystems.

Methods for determining production

We are not able to determine the production of a population empirically under field conditions. In order to estimate production, we must find a number of ecological parameters, such as numbers (N), the number of new-born individuals (γ_r), mortality, the average lifetime ($\bar{t}$), or the time of presence ($\bar{t}'$), i.e., the length of life within a defined time period (T) (see Chapter 1.2). We must also determine a number of physiological parameters, such as weight increase at a definite age (or at a given weight), or, better, the individual growth curve, the calorific value of tissue, etc. Based on these two sets of parameters, the production of the population can be calculated. A detailed list of the principles for calculating production is given by Petrusewicz & Macfadyen (1970), so only the most frequently used methods in small mammal studies are reported below.

The growth–survival curve method

A convenient graphic method of calculating production for a cohort (Figs. 7.1, 7.2) was introduced by Allen (1950) and then described and discussed in detail by Ness & Dugdale (1959). Nowdays it is largely known as a growth–survivorship curve method (Petrusewicz & Macfadyen, 1970).

Using this method, one must plot survival and individual growth curves (Fig. 7.1) first, and then straighten the individual growth curve along the abscissa (*x*-axis). In this way we convert the abscissa from a time axis into the individual weight axis (Fig. 7.2).

The consecutive numbers of survivors in a population (cohort) are plotted against the corresponding average weights. The area between the curve and the coordinate axes represents the production in the weight units. A simple mathematical treatment of this relationship was demonstrated by Petrusewicz (1970) and Petrusewicz *et al.* (1971). Normally, different cohorts, born at various seasons, must be treated separately as they have different growth and survival rates.

This method of calculation is easy to use and has great advantages. It permits calculation of production for any period of the cohort life.

159

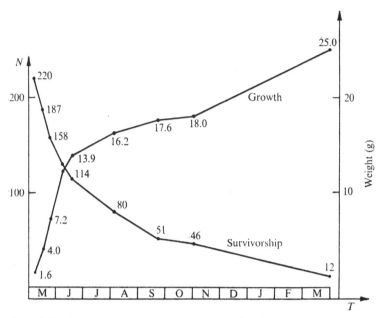

Fig. 7.1. Individual growth and survivorship curves of spring cohort of *C. glareolus* as basis for growth–survivorship curve. On the survivorship curve numbers (*N*) is equal to number of discrete individuals (*γ*) since it represents the dynamics of one cohort. Data from Gliwicz *et al.* (1968) and Bujalska & Gliwicz (1968).

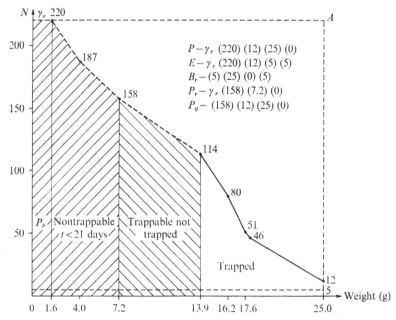

Fig. 7.2. Growth–survivorship curve (data from Fig. 7.1). The growth curve is straightened on the abscissa and the number of individuals plotted against weight; the area below the growth–survivorship curve represents production.

160

For example, biomass (production) of newborn individuals is a rectangle $\gamma_r(220)$ (1.6) (0). The value of production of trappable animals, production of overwintered animals, and others, can be estimated with this method (see Fig. 7.2). The difficulty is in finding data for individual growth and survival curves. Data, especially for the latter, is not easy to find on small mammals which reproduce throughout the growing season or even during all of the year at varying rates.

The starting point in the calculation is the total number of animals born in the population or cohort, or the total number of animals present in the population at the beginning of the growing season when the calculation of the production of overwintered animals is concerned. In species that do not breed in winter, the overwintered animals may be considered as one cohort, because, despite the differences in age, there are very small differences in their mortality and growth rates (Petrusewicz *et al.*, 1971; Bujalska & Gliwicz, 1968).

For further estimation of growth and survival, either marked animals or a reliable age indicator (see Chapter 3) is necessary. Data for laboratory-bred animals are sometimes used to construct the growth curve. It is possible to find very accurately the relationship between the age and growth under laboratory conditions, but the extrapolation of these data to the field may result in large error (Bujalska *et al.*, 1968; Petrusewicz, 1970). A way around this problem is to find the shape of the growth curve under laboratory conditions, and then, having some reliable data from the field, to extrapolate a field curve based on the shape of the laboratory growth curve. In this calculation growth curves have usually been smoothed between successive weight measurements (Bujalska & Gliwicz, 1968; Bobek, 1969; Hansson, 1971a). While care must be taken not to include pregnant females in the calculation of the growth curve, it was found that double inclusion of young, both as embryos and newborn, resulted in an error of only 4 per cent in one examination of a laboratory population (Walkowa, 1967). At some periods, though, when 100 per cent of the females are pregnant (Bujalska *et al.*, 1968), the error resulting from the double inclusion of embryos and newborn may be significant.

To construct a survival curve for animals trapped in the field whose ages are determined by means of age indices, abundant material should be collected from as large an area as possible. It may be observed that the number in a particular age class increases instead of decreases in consecutive censuses when animals are trapped in an area of 4 to 6 ha using the standard-minimum method (Petrusewicz, unpublished results).

Small mammals

This is the result of immigration, and it may be observed rather frequently at the peak of population numbers. In order to calculate the change in the number of survivors between censuses, it can be assumed that the daily mortality rate is either constant (straight-lined) or is instantaneous (there is a constant proportion disappearing per time unit). Gliwicz *et al.* (1968) demonstrated that measurements of survival in the trappable population are best fitted by a straight line.

The turnover method

As we have stated previously (see Chapter 1.2) one of the formulae describing production is

$$P_{(T)} = \bar{B}_{(T)} \cdot \theta_B \qquad (7.9)$$

where $\bar{B}_{(T)}$ is the average biomass for the time period, and θ_B is the turnover of biomass.

It is usually impossible to determine the turnover of biomass in small mammals directly in the field, and this essential index, characterizing important ecological features of the population, can only be found when production and average biomass have been found in other ways. Nevertheless, various modifications of (7.9) have been used to estimate production.

The production of an individual born during time interval T is its maximum weight in that time; for the individuals dying in the time interval that will be the weight at the moment of death and for individuals that outlive T it will be the weight at the end of that time interval. Now, the production of γ individuals will equal the product of the number of individuals and their average maximum weight W^+ (the weight at the moment of death or at the end of the study period):

$$P_t = \gamma \cdot W^+. \qquad (7.10)$$

To calculate the total production, the production of the adults (usually overwintered) that were present in the population at the moment the study began should be added.

Further calculations are based on a formula by Petrusewicz (1966*b*):

$$\gamma = NT/\bar{i}' = N \cdot \theta_N \qquad (7.11)$$

where $\bar{i}'$ is the average length of residency within the study period T, $\bar{N}$ is the average number and $\theta_N = T/\bar{i}'$, is the turnover of individuals.

Formulae 7.10 and 7.11 result in:

$$P_{(T)} = \bar{N} \cdot T \cdot W^+/\bar{i}' = \bar{N} \cdot W^+ \cdot \theta_N. \qquad (7.12)$$

In order to compute θ_N the mean length of life should be known. This statistic can be obtained from life tables or from survival curves, but it is often deduced from a supposed constant mortality rate. By fitting successive changes in a cohort to the equation

$$N\bar{t} = N_0\,e^{-\mu t} \tag{7.13}$$

estimates of μ the instantaneous mortality rate can be obtained from

$$\mu = \frac{\ln N_0 - \ln N_T}{t}. \tag{7.14}$$

The mean length of life $\bar{t}' = 1/\mu$ (Petrusewicz, 1966; Ryszkowski, 1967). $\theta_N = T/\bar{t}$, or if the year is taken as a unity $\theta_N = 1/\bar{t}$.

Note that in (7.11) $\gamma \cdot \bar{t}' = \bar{N}T$, is completely correct only when the average length of life within time period T, is taken into account and not the average lifetime ($\bar{t}$). In the case of a complete survival curve which accounts for the death of all the animals in a cohort, the time of presence and the length of life are the same. However, for a specific time period such as one year or one growing season, it is the average time of presence during the period that should be taken into account.

Formula 7.12 was applied to four experimental mouse populations (Walkowa & Petrusewicz, 1967), the production of which was known since every individual was weighed frequently. The results calculated by (7.12) represented 101 to 108 per cent of the real value.

It is, however, impossible to know the weight at the moment of death in the field, so Petrusewicz (1966b) and Grodziński (1968) have proposed replacing W^+ (the average weight at death) in (7.12) by the weight reached at the age of the average lifetime ($\bar{W}_t$) of a statistical individual:

$$P = \gamma_{(T)} \cdot \bar{W}_t = \bar{N}T/\bar{t}' \cdot W_t = \bar{N} \cdot \bar{W}_t \cdot \theta_N. \tag{7.15}$$

The accuracy of the estimate obtained by (7.15) was checked for the confined mouse populations (Walkowa & Petrusewicz, 1967). (As the animals in these populations were weighed often, their weight at death was known.) Production values obtained by means of (7.15) were 150 to 188 per cent of that found empirically and, therefore, the following modification has been introduced. Instead of the total average lifetime of an individual in the population being used, the average lifetime is calculated separately for three intervals; 0 to 3 weeks, 4 to 19 weeks, and over 20 weeks. The weights corresponding to the average length of life within each of these three time periods are derived from the growth

163

curve. Then, the production of the population is calculated for each of the three time periods using (7.15). When summed, these values give a result varying from 102 to 110 per cent (an average of 107 per cent) of the real production value; a result that is fully satisfactory for an ecologist.

Other methods

Field methods for estimating production have been diverse and a variety of computations have been suggested, many of which have been used only by a single author. Examples include the following:

For determining P_g of trappable animals Trojan (1969) used the formula

$$P_g = \bar{N} \cdot \Delta W \cdot T.$$

In this weight-gain method, ΔW must be measured at different seasons and also in different weight classes (Golley, 1960). To obtain the total production, P_r must be measured separately and added.

During studies of predation on small mammals, Pearson (1966) combined the method of catch–mark–release and scat analysis. From scat analysis, the number of animals eaten was determined giving a rough estimate of elimination. With such data, it is possible to estimate production from (7.2).

A very simplified estimate may be obtained from the average ratio of production to respiration (P/R) evaluated separately for different groups of small mammals. This ratio is quite stable for a given species. If population respiration is known, population production can be calculated from the formula

$$P = R \cdot P/R.$$

Productivity estimates based on the standard–minimim method

The standard-minimum method (Grodziński Pucek & Ryszkowski, 1966) was recommended (Golley *et al.*, 1968; Gentry *et al.*, 1968; Ryszkowski, 1969; Hansson, 1970; Pelikan, 1970) for estimating productivity of small mammals during the IBP. The animals should be sampled on a large plot three times; at the beginning of reproduction and the time of peak numbers in one year, and the beginning of reproduction the next year. The technique is appropriate for environments where there is strong seasonality in reproduction and population numbers.

This method can be quite accurate. For example, the estimates of production from simulated standard-minimum censuses during a long-term intensive live-trapping program (Bujalska *et al.*, 1968) differed by only 3 per cent from those computed from data from the total program in which all animals present were enumerated. However, the time periods of the simulated standard-minimum censuses were chosen very appropriately, with an early-summer census at the time of the peak number of pregnant females. Other studies (Myllymäki, 1969; Hansson, 1971*a*; Petrusewicz *et al.*, 1972) have shown the peak in pregnancy to come later in summer, at least in *M. agrestis* and *C. glareolus* and if the census misses this peak then it might represent the real population much less accurately.

By supplementing spring and autumn standard-minimum trapping with traplines during high summer, Bobek (1969, 1971) was able to construct life tables and determine the mean length of life. Hence, the production could be estimated by the growth–survival curve method. From this experience Bobek (1969) recommended that standard–minimum trapping be performed in spring, summer and late autumn. In contrast, Bujalska *et al.* (1968) and Petrusewicz *et al.* (1968) calculated production from only two censuses a year, in June and autumn.

Further, from the standard–minimum samples alone, Hansson (1971*a*) calculated production by assuming a constant mortality rate and applying the turnover method, as was also done by Grodziński, Pucek & Ryszkowski, (1966) and Grodziński (1971*a*).

A prerequisite for accurate calculations are good age indicators. For species in which it is difficult to determine age and which have a great annual variation in reproduction, only rough estimates of production are possible.

Accuracy

Physiological parameters

The accuracy of productivity values depends upon the accuracy of the empirically found parameters, and the correctness of the formula which is used for calculating these parameters. The physiological parameters, such as time of pregnancy, litter size and growth, certainly vary under different conditions, but the range of amplitude of their changes is restricted by their specific properties. They also show less dependence on environmental conditions than ecological parameters such as numbers (N) and mortality (M). The physiological parameters

165

may be tabulated for most sampling conditions and need not be estimated during each field census. After estimations have been made on a sufficient amount of material, they might be used for a whole climatic zone (Tables 7.1, 7.2). For example Zejda (1966) obtained estimates of litter size in *C. glareolus* in Czechoslovakia from a large data set for each month. We have shown that these averages also apply to *C. glareolus* in Poland.

While litter sizes found in laboratories are sometimes smaller than those in the field (Frank, 1957; Negus & Pinter, 1965), Ryszkowski & Truszkowski (1970) found only small differences in litter size directly observed in nest boxes under field conditions and estimates made by section of pregnant females.

Growth rates and weight at given ages vary with many conditions. Geographical changes in weight are known within a species (Shvarts, 1969). Further, growth rate depends upon seasonal variability (Shvarts *et al.*, 1964), the phase of the population cycle (Krebs, 1966), density (Grodziński, Pucek & Ryszkowski, 1966; Bobek, 1971) and food conditions (Ashby, 1967; Hansson, 1971*b*). Nevertheless, growth rate and weight at a given age have a rather limited variability. For example, adult *C. glareolus* can weigh from 18 to 27 grams; adult hares weigh from 3.7 to 5.5 kilograms.

Growth curves can be obtained easily under laboratory conditions, but these data can seldom be extrapolated from laboratory to field conditions. For example, Bujalska & Gliwicz (1968) have shown that the growth rate of *C. glareolus* in the field is significantly smaller than in the laboratory as investigated by Drożdż (1965). Petrusewicz (1970) found that the rate of growth of the European hare is significantly higher in the field than shown by Pilarska (1969) in the laboratory.

Ecological parameters

The natural variability in ecological parameters is larger than that of the physiological parameters considered above. Parameters such as numbers, mortality and pregnancy ratio may change greatly, e.g., from 0 to 100 per cent pregnancy in females and between zero and some thousand animals per hectare in density. These parameters have to be estimated at each field census.

Density estimation has been discussed in Chapter 2. Here we will only mention that density is a basic parameter in productivity investigations, and correct density estimation is often a difficult task. It seems possible to get rather good estimates for animals such as *Microtus arvalis* (Andrezejewski & Gliwicz, 1969) and the snow-shoe hare (Adams,1959), and for free living *Mus musculus* (Petrusewicz & Andrezejewski, 1962), and *Peromyscus polionotus* (Smith & McGinnis, 1968). Good estimates have also been obtained for isolated populations on islands (Petrusewicz *et al.*, 1968; Gliwicz, 1973). Further, during IBP, remarkable progress was achieved in estimating densities of forest type small mammals (Ryszkowski *et al.*, 1971; M. H. Smith *et al.*, 1971), (also see Chapter 2). We are now able to obtain comparable data.

Present methods of density estimation center mainly around the trappable part of the population. However, the nontrappable part, comprised of nestlings and small juveniles which die between trapping periods, can be responsible for as much as 81 per cent of the production (Petrusewicz *et al.*, 1968). The nontrappable part of a population is usually estimated by rather crude methods based on reproduction data. However, Ryszkowski & Truszkowski (1970) have shown that the nest box technique is very promising in direct evaluation of natality and survival of unweaned bank voles under field conditions. For the determination of density by the standard–minimum method, Ryszkowski *et al.* (1971) have given formulae for computing the dispersion around the estimate for one plot. Such estimates are often supposed to be representative for whole ecosystems, but replicates of standard–minimum plots are rare. In two series of duplicates, Ryszkowski (1971*a*) found considerable differences in the estimates.

Approximations

At sparse censuses, and in species with unreliable age indicators and irregular breeding periods, approximations of the suggested formulae are necessary in order that reproduction and production can be estimated.

With the turnover method, turnover of biomass cannot be found from field data. It has been replaced by turnover of numbers (θ_N) by Grodziński (1968), but the latter ratio may be higher. For example, Bobek (1969) compared estimates obtained by various combinations of turnover of biomass (θ_B) and of numbers with growth–survival computations. The use of turnover of numbers instead of biomass caused an increase of 130 per cent in the estimate.

A constant mortality rate is often assumed, and Petrusewicz *et al.* (1968), did find a constant number of trappable animals dying per unit

167

time. For computations of mortality for short periods (about two months) and excluding times of heavy mortality, the difference between instantaneous and straight-lined mortality of trappable animals may be negligible. Nestlings and juveniles show, however, a much higher mortality than adults. At present, we know very little about survival rate in the nest and, therefore, the exponential rate of mortality should be used for whole population estimates or for estimates over long time periods.

Composite estimates

Productivity estimates are products of two or more stochastically variable factors. Thus, the final results will be burdened with rather large errors. For this review, a calculation was performed of the final coefficient of variation when this variation was 5, 20, and 50 per cent respectively in the separate factors (Table 7.3). It is obvious such

Table 7.3. *Total coefficient of variation of composite estimates at various levels of the individual coefficients of variations*

| | Coefficient of variation of individual factors | | |
Estimate	5%	20%	50%
$\gamma_r = \dfrac{\bar{N} \cdot s \cdot f \cdot T \cdot L}{t_p}$	11	45	112
$P = \bar{N} \cdot \bar{W}_t \cdot \theta_N$	9	35	87
$P = \bar{N} \cdot W^+ \cdot \theta_N$	10	40	100

composite formulae as that for the number born will increase the error of estimates considerably and very large samples are needed to obtain narrow confidence limits. The common use of one or several decimals in final productivity estimates is, therefore, improper.

Field estimates: agreement and generalizations

What generalizations about reproduction and production are possible from the methods described above? Tables 7.2 and 7.4 have been used for examination of this question. In these compilations, the estimates of production and reproduction from grassland type ecosystems are about one magnitude higher than those of forest type ecosystems. While the grassland studies have been performed with a great variety of methods and we cannot exclude the possibility that some of the observed differences are due to the methods, we believe that the greatest differences between ecosystems are real. Presumably these differences in production in forest and grassland ecosystems are related to the pressures on the small mammals in these systems and their strategies for survival. Later chapters (Chapters 10, 11, 12 and 13) will focus on these phenomena.

Estimates from a series of forest ecosystems, ranging from poor pine to luxuriant deciduous forest (Tables 7.2, 7.4) showed an increasing gradient in reproduction and production. The methods of estimation agreed rather well.

Most studies have shown that reproduction contributes more than 50 per cent of the total production of the population. Petrusewicz *et al.* (1971) found that the main part of production in *C. glareolus*, living in a temperate zone forest, is due to spring and summer cohorts.

Petrusewicz and his coworkers (1971), generalized that such parameters of reproduction and production as number born and the basic stock in early spring show an annual stability. Comparing four pine forest ecosystems, Ryszkowski (1971*b*) found the highest stability of reproducing females during the period with the best shelter and food conditions for the bank voles. However, Bobek (1971), who studied a deciduous forest containing oak with a peak acorn production, showed the same parameters to be extremely variable. Grodziński (1971*b*) also indicated very great annual variations in these parameters in taiga ecosystems.

At increasing rates of exploitation of laboratory mouse populations, Walkowa (1970, 1971) found a peak production at 32 to 34 per cent exploitation. Larger percentages led to a decrease in production. The production increase with exploitation was mainly due to increasing survival of nestlings and these results were confirmed on a free-living *Mus musculus* population (Adamczyk & Walkowa, 1971). Petrusewicz (1970) found that in average years 40 per cent exploitation during the hunting season (autumn–beginning of winter) of the standing crop of hares (*Lepus europaeus*) did not damage the next year's production.

Maximum production of most small mammals in balanced populalations (that is without mass occurrence) seems to be around 3500 g/ha·yr or 5000 kcal/ha·yr (Table 7.3). This production rate is on the

Table 7.4. *Field estimates of production. Figures given per plot were recalculated on a per hectare basis. SM = standard–minimum snap trapping; CMR, catch–mark–release live trapping; GS, computations according to growth–survival method; T, according to turnover method; W, according to weight–gain method; Mod., modified method.*

Ecosystem	Place, Years	Species	Sampling method	Estimation of P	P_r	% total production	Author
Forest type ecosystems							
Tajga (spruce) forest	Alaska, USA 1963	*Clithrionomys rutilus* (Pall.)	CMR	T	730		Grodziński (1971a)
		Tamiasciorus hudsonicus	CMR	T	530		Gradziński (1971a)
		Microtus oeconomus (Pall.)	CMR	T	270		Grodziński (1971a)
		5 small mammals spp.	CMR	T	1670		Grodziński (1971a)
Mountain spruce forest	Bulgaria 1967–1968	*C. glareolus*	SM	GS	738–918	65–70	Petrusewicz et al. (1972)
Temperate spruce forest	Sweden 1971–1973	5 small mammal spp.	SM	T	530–1130		Hansson (unpublished results)
Deciduous beech forest	Poland 1965–1968	*A. flavicollis*	Mod. SM	GS	210–941	48	Gradziński et al. (1970)
Deciduous beech forest	Poland 1965–1968	*C. glareolus*	Mod. SM	GS	244–1142	56	Grodziński et al. (1970)
Deciduous beech forest	Poland 1965–1966	*C. glareolus*	Mod. SM	GS	1142	58	Bobek (1969)

Habitat	Location / Years	Species	Method	Type	Number	%	Reference
Deciduous mixed forest	Poland 1966–1969	C. glareolus	CMR	GS	2857–3265	51–58	Petrusewicz et al. (1968, 1971)
Deciduous mixed forest	Poland 1967–1969	C. glareolus	Mod. SM	GS	771–1977	58	Bobek (1971)
Deciduous mixed forest	Poland 1967–1969	A. flavicollis	Mod. SM	GS	370–589	46–51	Bobek (1971)
Tropical rain forest	Panama 1968–1969	P. semispinosus	CMR	GS	4000	43	Gliwicz (1973)
Grassland type ecosystems							
Agrocoenose	Poland 1964–1968	L. europaeus	Mod. SM	GS	1800	95	Petrusewicz (1970)
Agrocoenose	Poland 1963–1967	M. arvalis	CMR	W	33427	92	Trojan (1969)
Mountain grazing meadows	Poland 1966	2 small mammal spp.	Mod. SM	T	651		Grodziński, Pucek & Ryszkowski, (1966)
Abandoned fields	Michigan, USA 1953–1957	M. pennsylvanius	CMR	W	5170		Golley (1960)
Abandoned fields	S. Carolina, USA 1953–1957	Peromyscus polionotus (Wagn)	CMR	W	600–1500		Odum et al. (1962)
Shrub desert	Arizona, USA 1958	Dipodomys merriami	CMR	GS	350	90	Chew & Chew (1970)
Shrub desert	Arizona, USA 1958	7 small mammal spp.	CMR	GS	610		Chew & Chew (1970)
Reforestation	Sweden 1968–1973	M. agrestis	SM	T	1290–3470		Hansson (unpublished results)
Reforestation	Sweden 1968–1973	7 small mammal spp.	SM	T	2490–5150		Hansson (unpublished results)

same general level as that for large herbivorous mammals (Wiegert & Evans, 1967; Medwecka-Kornaś *et al.*, 1973), but is less than in some herbivorous insects. However, the production of small mammals is inconsiderable in some ecosystems such as deserts.

The authors of this chapter wish to thank many of our colleagues who, during the IV Small Mammals Working Group Meeting, have made many remarks, additions and criticism which helped to improve our chapter; especially Drs Golley, Grodziński and Ryszkowski.

8. Ecological energetics of small mammals

W. GRODZIŃSKI & B. A. WUNDER

Small mammals have been a favorite subject for physiological studies for many years, but most studies have dealt with classical questions of mechanism under laboratory conditions and only in the past twenty to thirty years have physiologists concentrated on small wild mammals, attempting to investigate comparative adaptive strategies for coping with environmental circumstances. This new approach has developed into the field of physiological ecology (Dill *et al.*, 1964; Slonim, 1961). Many of the earlier studies dealt with temperature regulation and metabolism but did not necessarily place them into a field context.

Small mammal ecologists began to approach ecosystem and/or population function from the point of view of energy flow by the early 1960s (Golley, 1960; Odum *et al.*, 1962). With the advent of the IBP studies it became evident that relatively little information was available concerning small mammal productivity and energy flow. Thus the study of ecological energetics has developed and been greatly expanded in many countries. Such studies have drawn upon physiology and physiological techniques but have attempted to frame them into ecologically realistic conditions. Consequently, many papers on mammal energetics have been published during this past decade by IBP investigators and others. It is our desire to attempt a review and synthesis of these studies. The concepts of energy flow through animal populations have their origins in the physics of thermodynamics. The same thermodynamic concepts have been used successfully in developing general theories of growth (Lotka, 1956).

One of the most important concepts developed by ecosystem theory relates to energy flow through ecosystems. Indeed, much of the effort expended by IBP in synthesizing models has been based on the central theme of energy flow and analysis of the component parts through which such energy may cascade. Small mammals constitute one such component and thus analysis of energy flow through their populations has been one goal of the IBP. Although most studies indicate that a relatively small percentage of the total energy flux through a particular

173

ecosystem passes through the small mammal populations, they may function as control gates rather than major processors of energy.

Energy flow, or assimilation (A), can be described for an animal population by two well-known equations (Petrusewicz, 1967a): $A = P + R$ and $A = C - FU$, where P represents the amount of energy incorporated into animal tissue, R is the amount of energy used for maintenance or respiration, C is the energy of food intake and FU the amount of energy lost through feces and urine. These two general equations have been modified to define more precisely the balance of energy flow through small mammal populations (Golley, 1962; Grodziński *et al.*, 1966):

$$A_T = K_b(\bar{N}B \cdot \theta_B) + M(\bar{N}B)_T \tag{8.1}$$

$$A_T = K_c(CW)_T - K_e(FU \cdot W)_T \tag{8.2}$$

where:

$\bar{N}$ = average numbers, or animal density,
B = mean biomass of an animal,
θ_B = turnover of biomass in the population,
M = metabolic rate, usually measured by gas exchange,
K_b = caloric value of the mammalian body,
K_c and K_e = caloric value of food and rejecta, respectively,
C = consumption (food intake) during a given time (units in W),
FU = feces and urine produce in the same time as C,
W = dry weight, or biomass, of food and excrement,
T = period of time for which energy flow balance is computed.

In order to determine the flow of energy through a small mammal population the parameters listed in either equation must be determined. Population numbers ($\bar{N}$), biomass (B) and turnover rate (θ_B) are determined from field studies. Within the restriction of present techniques most of the bioenergetic parameters are determined in the laboratory. Investigation of the bioenergetic parameters involves the study of respiration (metabolic rate; M), calorimetry (caloric values of mammalian production, food and excrements; K_b, K_c, K_e) and feeding balance (food consumption, feces and urine production; C, FU).

Our goal in this chapter is to cover three basic points concerning small mammal energetics. First, in order to relate small mammals to primary production, we shall discuss food utilization or consumption (how much energy small mammals consume and from what sources). Second, the energy which goes into small mammal production is important; this

174

relates to the amount of energy which small mammals provide for higher trophic levels. Thirdly, we will consider respiration since much of the energy processed by small mammals is used for maintenance. Cost of maintenance can also be determined from food assimilation.

Data and methods

Material and sources of data

Data for this chapter were gathered from a great variety of papers published primarily during the last decade in both American and European ecological and physiological journals. We have given special attention to information available in previous IBP publications. The IBP Handbook Series (Golley & Buechner, 1969; Petrusewicz & Macfadyen, 1970; Grodziński, Klekowski & Duncan, 1975) and previous proceedings of the IBP Small Mammal Working Group (Petrusewicz, 1967c; Petrusewicz & Ryszkowski, 1970; Palmén, 1971) have provided much relevant information. In addition, many recent texts and reviews on general animal bioenergetics have been useful in the preparation of our synthesis (e.g. Kleiber, 1961; Folk, 1966; Kalabukhov, 1969; Hart, 1971; Gessaman, 1973).

For the purposes of this paper it is important to define what we mean by a 'small mammal'. For our discussion we are restricting ourselves primarily to animals ranging in size from 3 to 300 grams. Furthermore, due to these size restrictions and the availability of sufficient data, we will consider only animals falling in the following taxonomic groups: Insectivora, Rodentia, Lagomorpha, and Carnivora. Only a few examples from Chiroptera were included.

Within these taxa, mammals gain food energy in a variety of ways each of which may modify the particular patterns and amounts of energy turnover or energy flow shown within populations of each species. For example, carnivores and insectivores feed from a different trophic level than do herbivores. Furthermore, their efficiencies of assimilation and the amounts of energy spent searching for food are different from herbivores. Within the herbivores, different feeding strategies necessitate different expenditures of energy. For example, seed eaters must, in a sense, hunt for their food, have larger home ranges (Brown, 1966), and thus conceivably spend more energy searching for food than do foliage-feeding herbivores such as microtine rodents (McNab, 1963).

Small mammals

Within the groups of mammals listed above, we find a variety of strategies for temperature regulation each of which may affect energy balance and energy flow somewhat differently. Mammals which remain homeothermic will expend more energy on maintenance than will those which demonstrate some degree of heterothermy. Many mammals show daily fluctuations in level of body temperature regulation, thus reducing the energy needed for maintenance during certain periods (Morhardt, 1970) while others show daily bouts of torpor (Bartholomew & Hudson, 1964); and of course, hibernators and estivators show reduced energy expenditure during part of the year.

Laboratory and field methods

A variety of techniques have been used for investigating the bio-energetics of small mammals (Golley, 1967). One of the concerns of the IBP has been to unify at least some basic techniques. This unification was attempted through an IBP training course in bioenergetics dealing primarily with the laboratory methods of calorimetry, respirometry and feeding ecology (Grodziński & Klekowski, 1968, Grodziński et al., 1975). Independent from the IBP some new field techniques have been introduced recently for studying homeotherm energetics under field conditions. These procedures include such techniques as radioisotopes, heavy water, and radiotelemetry (see Gessaman, 1973 for review).

Direct calorimetry. Most measurements of energy exchange between animals and their environment and, thus, their energy flow have been made using indirect measures of such things as oxygen consumption or carbon dioxide production. Such measurements give information about heat production and heat loss only if certain assumptions hold true (i.e., caloric equivalents of oxygen consumption, stability of body temperature and body heat content, etc.). Direct measures of such exchange are obviously desirable; however, simultaneous measures of heat production and loss have been few (Caldwell et al., 1966; Hammel et al., 1968). Investigators employing direct calorimetry used a simple animal calorimeter which consisted of a metal container lined with a net of thermocouples to measure direct heat loss. Due to the intricacy of construction and operation of direct systems, indirect calorimetry is probably a much better system to use in order to answer ecologically related questions about energy flow in animals.

Indirect calorimetry (respirometry). The indirect systems described here are much easier to operate than are direct measurement systems and allow more space in which the animals can operate during measurements (e.g., ADMR). Within the sorts of variation which we find in other parameters needed for calculating population energy flows, indirect systems do not provide unduly large sources of error even if RQ-values vary for oxygen–caloric conversions.

Closed circuit system respirometers. The simplest systems for measuring respiration (oxygen consumption) indirectly are closed system respirometers. The application of this technique depends simply upon physical gas laws. Animals are placed in closed containers with chemical agents to absorb CO_2 and water. As the animals consume oxygen, the CO_2 and water produced are absorbed and thus any change in volume (after appropriate temperature and pressure corrections) is due to the oxygen consumed. As oxygen is used, it may be replaced by adding specific aliquots, or through a tube attached to an oxygen reservoir, or one may simply monitor volume decreases.

Two of the most widely used and effective systems are the Morrison respirometer (described in Morrison, 1951; and Morrison & Grodziński, 1968) and the Kalabukhov–Skvortzov respirometer (described in Kalabukhov, 1962).

The advantages of closed systems are that they are relatively inexpensive to develop and maintain and are fairly simple to operate. The major disadantage is that in some instances gas concentrations may reach low levels and affect metabolism of the animals being studied.

Open circuit systems. In an open flow system the animals are usually placed in a relatively small volume respirometer, ambient air is passed at a known rate of flow through this chamber and the concentrations of oxygen in the inflowing and outflowing gases are determined. The advantages of this system are its ease of operation, speed of response and accuracy. Among the best oxygen analyzers for use in such systems are those which utilize the paramagnetic properties of oxygen or thermoconductivity differences such as the Beckman analyzers or the Kipp and Zonen Diaferometer. It should be pointed out that there are a number of technical problems in using such systems. In addition to the problems of the logisitics and physics of setting up a flow system, one must consider problems of removing CO_2 from incoming gases and the use of appropriate equations for calculation of oxygen consumption (Depocas & Hart, 1957). Hill (1972) has recently re-evaluated problems of CO_2 in inlet and/or analyzed air streams and found that calculation errors of up to 38 per cent can be made depending upon whether CO_2 is removed from inlet air or analyzed air. From Hill's analysis we would suggest that air going to the oxygen analyzer have CO_2 removed or flow rates be adjusted so that the volume fraction of oxygen in inlet air minus outlet air is greater than 0.00303. Such conditions will introduce an error of no more than 2 per cent which, considering the usual variation in oxygen consumption of a quietly resting mammal, is often of little significance.

Feeding trials. Studies of food utilization in mammals may be carried out on the intact animals *in vivo*, or outside them, i.e., *in vitro*. Digestibility *in vitro* is often employed when studying ruminants but is seldom used for studying non-ruminant small mammals.

In the nutrition studies of small wild mammals *in vivo* two methods are generally applied: the balance method and the tracer technique. The classical balance method is carried out in a metabolic cage, where all food consumed and all feces and urine produced are measured (Drożdż, 1966, 1968*b*). With such a method both digestibility and assimilation of energy, organic matter, or any nutrients can be determined. The tracer methods require addition of a marker (indicator substance) to food with subsequent analysis of its content in feces. Some colored markers can be used, e.g., chromic oxide, but recently radioisotopes such as ^{51}Cr (Petrides & Stewart, 1970) were successfully employed for small mammals. The tracer method may also utilize the natural ash content in food and excrement as a tracer. The analysis of ash content compares food and feces, and in the case of snap-trapped mammals it is limited to their stomach and colon contents (Johnson & Maxell, 1966; Johnson & Groepper, 1970). The latter technique has several limitations, but allows an estimation of digestibility of natural food by wild animals in their natural environment. All tracer methods will determine only digestibility (coefficient of digestibility) while the balance methods give both digestibility and assimilation.

Analysis of gross body composition and energy content. Some rather standard techniques have been applied by ecologists for determining the caloric value of production and for analysing body composition in animals. For total caloric values adiabatic bomb calorimetry has been utilized successfully (Gorecki, 1965*a, b*). Paine (1971) has reviewed the techniques and pitfalls associated with bomb determination, total ash or fat content of an animal. Fat can be determined by the Soxhlet ether extraction technique (Startin, 1969; Sawicka-Kapusta, 1970; Dawson, 1970), total ash by oxidation in a muffle furnace (Startin, 1969; Sawicka-Kapusta, 1970; Paine, 1971), and protein by the Kjeldahl method (as in Startin, 1969), using protein = 6.25 × nitrogen.

Because of individual and seasonal variation, ecologists have tried to develop techniques to gather average sample data from populations rather than simply measuring individual animals or parts of animals (Odum *et al.*, 1962).

Techniques for estimating metabolism in the field. For verifying models of energy flow through small mammal components of ecosystems and for getting some idea of what it costs small mammals to exist in the field, several methods have recently been devised for measuring metabolic rate of animals in the field. Such methods include: (1) radioisotope excretion rates, (2) $D_2^{18}O$ turnover rates and (3) bio-telemetry of heart rate. The strengths and weaknesses of most of these techniques have recently been reviewed (Gessaman, 1973). Most techniques are not completely satisfactory at present. To date no radioisotope excretion rates have been found to correlate well enough with metabolism to be used for field estimations of metabolic rate (Golley *et al.*, 1965; Wagner, 1970; Chew, 1971). Gessaman feels that the most promising technique is the $D_2^{18}O$ method; however, it is quite expensive, complicated and gives only a long-term average for integrated metabolic rate; and Mullen (1973) indicates that cost for isotopes becomes prohibitive for animals weighing more than 1 kg. Perhaps the technique that is next most promising is heart rate biotelemetry: however, we need much more data correlating heart rate and metabolic rate, especially during activity. Although involving expensive equipment, it is less expensive and involved than the $D_2^{18}O$ method but is limited to animals large enough to carry telemetry capsules. More work needs to be done in validating all of these techniques.

Bioenergetics parameters for energy budgets

Energy for maintenance (respiration)

In any discussion of maintenance energy turnover for mammals, three measures of metabolism are usually mentioned. These are basal metabolism (BMR), resting metabolism (RMR) and average daily metabolic rate (ADMR). For purposes of ecological investigations, basal metabolic rates are of little use due to the strict requirements for their determination (Benedict, 1938). For example, it is very difficult to determine when a small mammal is post-absorptive; in fact, in the field they may never reach the state of being completely inactive and post-absorptive. Thus, for our purposes RMR and ADMR are much more useful measures of energy turnover. RMR refers to the metabolic rate of an animal at rest under a prescribed set of environmental circumstances. It is usually measured as an acute situation (i.e., responses over a period of 1 or 2 hours). ADMR necessitates at least a 24-hour measurement period, and is mean metabolic rate measured over that time under a prescribed set of environmental circumstances. Thus, it includes not only energy expended at rest but also in voluntary activity during the 24 hour period (Grodziński & Górecki, 1967). The energy components which are considered in each of these various measures of metabolism are summarized in Table 8.1.

Table 8.1. *Components of three metabolic measures*†

	Components			
Metabolic measure	Basal metabolism	Thermo-regulatory metabolism	Specific dynamic effect	Activity metabolism (locomotion)
Basal metabolic rate (BMR)	+	−	−	−
Resting metabolic rate (RMR)	+	+	+?	−
Average daily metabolic rate (ADMR)	+	+	+	+

† Modified from Gessaman (1973).

In the following we shall consider one model for energy budgets based upon RMR studies and one model based upon ADMR studies.

Basal and resting metabolism. There are many data available concerning the metabolic rates of mammals at rest under various ambient conditions. Many of these are summarized in Hart (1971) and in Kalabukov (1969). The relationships between body size and 'basal' metabolic rate has long been established (reviewed in Kleiber, 1961) although there are some groups of mammals which show slight variations (i.e., heteromyid rodents, fossorial rodents and some desert species show characteristically low metabolic rates while those for microtine rodents are high). Although we frequently refer to such values as basal they are usually resting rates in thermoneutrality. It is not our purpose here to elaborate on the wealth of data available but merely to use the relationships correlating metabolic rate and body size generated from these data. 'Basal' metabolic rate can be estimated using the following formula which has been modified from the Brody–Proctor equation by Morrison *et al.* (1959).

$$\text{BMR} = 3.8 \; W^{-0.25} \qquad (8.3)$$

with MR in cc $O_2/g \cdot h$, W in g.

Metabolism below thermoneutrality can be estimated if we know the relationship between metabolism and ambient temperature. This relationship is called thermal conductance and can be estimated from the allometric equation of Herreid & Kessel (1967):

$$\text{TC} = 1.05 \; W^{-0.50} \qquad (8.4)$$

with TC in cc $O_2/g \cdot h \cdot {}^\circ C$, W in g.

Hart (1971, p. 77) has recently reviewed the data relating to this relationship and derived a similar relationship.

Average daily metabolism. ADMR measurements are made with the animals in large containers so that they can be active, have access to a nest and food and water (Grodziński & Górecki, 1967; Morrison & Grodziński, 1968). Consequently this measure contains basal metabolism, the metabolic equivalent of energy for thermoregulation and activity as well as the energy of SDA (specific dynamic action or the calorigenic effect of food).

Analyses of the relation of ADMR to body size in several species of voles, mice and squirrels have shown that it is intraspecifically allometric. However, the exponents for these intraspecific relationships (expressed as metabolism per whole animal) have been found to be close to 0.5 and not 0.75 which is the well-known exponent for BMR (Hansson & Grodziński, 1970; Grodziński, 1971*a, b*; Drożdż *et al.*, 1971; Górecki, 1971). Recently two general interspecific functions of ADMR against body weight were computed, one for small rodents and one for insectivores (French, Grant & Grodziński, unpublished data). These relationships were computed from 72 data points of ADMR for 36 species of rodents and 8 species of insectivores (Fig. 8.1). The regression line for rodents ranges from a 7 g pocket mouse (*Perognathus*) to a 370 g hamster (*Cricetus*) and the insectivore regression from a 3 g lesser shrew (*Sorex minutus*) to a 21 g short-tailed shrew (*Blarina*

181

Small mammals

brevicauda). The two equations have significantly different intercepts but the exponent is close to 0.50 in both cases (−0.57 and −0.46). By forcing the regressions into an average slope of 0.50 the following

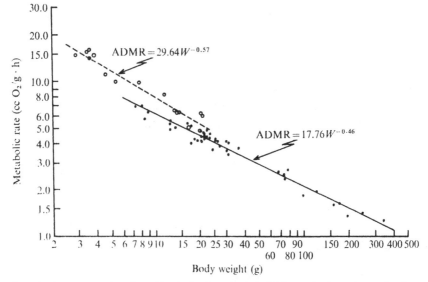

Fig. 8.1. ADMR as a function of body size in rodents (solid lines) and insectivores (broken lines). Data are taken from French ,Grant & Grodziński (see text).

equations were generated to predict ADMR for animals exposed to 20 °C:

$$\text{Rodent ADMR} = 19.94 \ W^{-0.50} \tag{8.5}$$

$$\text{Insectivore ADMR} = 26.80 \ W^{-0.50} \tag{8.6}$$

with ADMR in cc $O_2/g \cdot h$, W in g.

Physiologically ADMR is not a very well defined measurement; however, it is probably the most natural and ecological measure of metabolism possible under laboratory conditions.

Consumption and assimilation

Investigations of feeding ecology should provide answers to several questions important for studies of energy flow through small mammal populations; for example, what the food consumption by a small mammal is, what part of the consumed energy is assimilated and what fractions are passed on with rejecta, i.e., feces and urine. A more general question is that of how much food is available for small mammals in various ecosystems. This has been defined as 'that food which is

182

easy to find and is being chosen and eaten by these animals' (Grodziński, 1968). The estimation of available food is usually based on some knowledge of mammalian food habits. The food habits of wild small mammals may be studied with various approaches, but should preferably be determined both by analysis of stomach contents (or feces) and by applying food preference tests of choice ('cafeteria test'). The food available for rodents has already been estimated in several forest and grassland ecosystems. In various types of forests this constitutes only a few per cent (4 – 13 per cent) of the total primary plant production (Grodziński, 1968), but in the grassland ecosystems, including cultivated fields, a majority of the above-ground plant production can be considered as potential food for herbivorous grazing rodents (Golley, 1960; Grodziński *et al.*, 1966; Trojan, 1970; Batzli & Pitelka, 1971). Granivorous rodents have available only a small fraction of plant production (Odum *et al.*, 1962; Pearson, 1964).

The utilization of food energy is shown in a general scheme in Fig. 8.2. Note that the terminology employed by ecologists and nutritionists

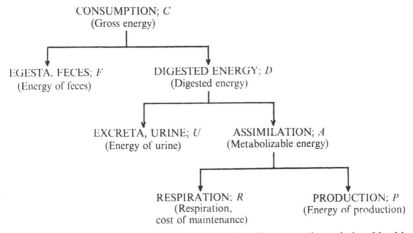

Fig. 8.2. A scheme of energy flow through a mammal and/or mammal population. Nutrition terminology is given in parentheses. Note that the terminology used by ecologists is slightly different (slightly modified from Drożdz, 1968).

is slightly different, but this is not of primary importance. It is clearly visible in this diagram that consumption (C) exceeds assimilation (A) by the amount of energy lost through feces and urine (FU). Thus, in order to calculate total consumption ecologists have recently studied assimilation and digestibility (or digested energy) in small mammals. We present in Table 8.2 a review of all the available data on digestibility and

Table 8.2. *Digestibility and assimilation of natural and laboratory foods in small mammals: as a percentage of gross energy, GE or organic matter, OM. B, balance method; T, tracer method; I, insectivore; C, carnivore; H, grazing herbivore; O, omnivore; G, granivore;*

No.	Species	Body wt (g)	Feeding type	Method	As % of GE or OM	Digestibility coefficient range	Digestibility coefficient av.	Assimilation coefficient range	Assimilation coefficient av.	Reference
	INSECTIVORA Soricidae									
1	*Sorex cinereus* Masked shrew	3.6	I	B	GE			93–95	94.0	Buckner, 1964
2	*Sorex arcticus* Arctic shrew	5.4	I	B	GE				88.0	Buckner, 1964
3	*Sorex araneus* Common shrew	8.5	I	B	GE		94.9			Hawkins & Jewell, 1962
4	*Microsorex hoyi* Pygmy shrew	3.5	I	B	GE				83.0	Buckner, 1964
5	*Cryptotis parva* Least shrew	3.6	I	B	GE		90.1		87.8	Barrett, 1969
6	*Neomys fodiens* European water-shrew	12.4	I	B	GE		92.5			Hawkins, & Jewell, 1962
7	*Blarina brevicauda* Short-tailed shrew	20.1	I	B	GE				80.0	Buckner, 1964
	CHIROPTERA Vespertilionidae									
8	*Lasiurus cinereus* Hoary bat	23.6	I	B	GE		81.0			Brisbin, 1966
	CARNIVORA Mustelidae									
9	*Mustela rixosa* Least weasel	60.0	C	B	GE		89.9			Golley, 1960

No.	Species / Common name	Weight	Diet	Habitat	Activity					Reference
10	*Mustela nivalis* Weasel		C	B	GE		82.6			Bobek & Grodziński (unpublished results)
	Felidae									
11	*Lynx rufus* Bobcat	6250.0	C	B	GE		90.6		82.6	Golley *et al.*, 1965
	LAGOMORPHA Ochotonidae									
12	*Ochotona princeps* Pika	171	H	T	OM	54–76	68.0			Johnson & Maxell, 1966
	Leporidae									
13	*Sylvilagus floridanus* Eastern cottontail	312.0	H	B	GE		60.0			Davis & Golley, 1963 (Golley & Amerson, unpublished results)
14	*Lepus europaeus* European hare	3800.0	H	B	GE	79.9–80.2	80.1	77.7–78.5	78.0	Myrcha, 1968
	RODENTIA Sciuridae									
15	*Eutamias minimus* Least chipmunk	34.0	G(O)	T	OM		75.2			Maxell, 1973
16	*Spermophilus richardsonii* Richardson's ground squirrel	303.0	O	B	OM	79.9–82.4	82.2			Johnson & Groepper, 1970
		250.0		T	OM		81.2			Maxell, 1973
				T	OM		46.1			Maxell, 1973
17	*Spermophilus tridecemlineatus* Thirteen-lined ground squirrel	132.0	O	B	OM		81.0			Johnson & Groepper, 1970
		67.0		T	OM		82.3			Maxell, 1973
18	*Spermophilus lateralis* Golden-mantled ground squirrel	270.0	O	T	OM		33.3			Maxell, 1973
19	*Cynomys ludovicianus* Black-tailed prairie dog	885.0	H	B	GE		85.8			Hansen & Cavender, 1973
				B	OM		85.9			Maxell, 1973
20	*Cynomys leucurus* White-tailed prairie dog	1050.0	H	T	OM		24.6			Maxell, 1973

Table 8.2 (continued)

No.	Species	Body wt (g)	Feeding type	Method	As % of GE or OM	Digestibility coefficient range	Digestibility coefficient av.	Assimilation coefficient range	Assimilation coefficient av.	Reference
	RODENTIA (continued)									
21	Sciurus carolinensis Gray squirrel	520.0	O	B	GE		73.8		71.2	Ludwick et al., 1968
	Gliridae									
22	Glis glis Fat dormouse	147.0	G	B	GE	90.9–91.3	91.1	87.8–88.1	88.0	Gebczynski et al., 1972
	Geomyidae									
23	Thomomys talpoides Northern pocket gopher	73.0	H	T	OM		57.5			Maxell, 1973
	Heteromyidae									
24	Perognathus fasciatus Olive-backed pocket mouse	11.0	G	B	OM	95.1–95.3	95.2			Johnson & Greopper, 1970
				T	OM	93.2–95.2	94.2			
25	Dipodomys ordii Ord's kangaroo rat	60.0	G	B	OM	95.1–97.7	96.4			Johnson & Greopper, 1970
	Cricetidae Cricetinae	67.0		T	OM	91.1–93.1	92.1			Maxell, 1973
							88.3			
26	Oryzomys palustris Marsh rice rat	37.0	O	B	GE	88–95	92.2			Sharp, 1967
27	Peromyscus maniculatus Deer mouse	20.0	O	B	OM		77.0			Johnson & Groepper, 1970
		22.0		T	OM	85.8–89.5	87.0			Maxell, 1973
				T	OM		71.3			
28	Peromyscus polionotus Old-field mouse	13.0	G	B	GE		93.9			Davenport, 1960 (after Davis & Golley, 1963)
				B	GE		87.0			Caldwell & Connell, 1968
29	Onychomys leucogaster Northern grasshopper mouse	30.0	I	T	OM		62.2			Maxell, 1973
30	Sigmodon hispidus Hispid cotton rat	100.0	H	B	GE		91.2		86.5	Golley, 1962

No.	Species / Common name	Weight (g)	FH	T	Comp.	Range	Mean	Value	Reference
31	*Neotoma cinerea* / Bushy-tailed wood rat	297.0	H	T	OM		52.3		Maxell, 1973
	Microtinae								
32	*Clethrionomys glareolus* / Red bank vole	23.0	O	B	GE	77.4–92.9	86.8	82.9	Drożdż, 1968a, 1970
				B	OM	75.8–90.7	84.9		Drożdż, 1968a
				B	GE	86.2–88.5	78.9		Kaczmarski, 1966
33	*Clethrionomys gapperi* / Southern red-backed vole	20.0	O	B	OM	81.8–93.1	87.4		Johnson & Groepper, 1970
		21.0		T	OM				
34	*Microtus arvalis* / Common vole	22.0	H	B	GE	70.4–92.3	81.3	77.5	Drożdż, 1968a, 1970
				B	OM	74.5–94.0	84.3		Drożdż, 1968a
				B	GE		91.0		Migula, 1969
				B				8.77	Hansson, 1971b, c
35	*Microtus agrestis* / Field vole	23.0	H	B	GE	33.0–60.0	50.7		
				B	OM	32.0–59.0	50.3		
36	*Microtus oeconomus* / Tundra vole	28.0	H	B	GE	68.7–73.9	71.3	69.3	Gebczynska, 1970
						67.2–71.4			
37	*Microtus pennsylvanicus* / Meadow vole	46.0	H	B	GE	82.2–89.8	86.0		Golley, 1960
		29.0		B	OM		81.1		Johnson & Groepper, 1970
				T	OM	72.0–76.3	74.2		Keys & Van Soest, 1970
				B	OM	43.0–62.9	51.7		
38	*Arvicola terrestris* / Water vole	74.0	H	B	GE	55.2–91.2	73.2	67.7	Drożdż et al., 1971
	Muridae								
39	*Mus musculus* / House mouse	16.0	O	B	OM		79.5		Johnson & Groepper, 1970
				T	OM	91.0–94.8	92.9		
40	*Apodemus flavicollis* / Yellow-necked field mouse	27.0	G(O)	B	GE	81.4–92.2	88.2	86.1	Drożdż, 1968a, 1970
				B	OM	82.8–89.5	86.3		Drożdż, 1968a
				B	GE	78.7–90.9	90.0		Turcek, 1956
41	*Apodemus agrarius* / Striped field mouse	21.0	G	B	GE	89.6–90.5	90.0	88.9	Drożdż, 1968a, 1970
				B	OM	86.8–92.1	89.4		Drożdż, 1968a
						88.9–89.0			
	Zapodidae								
42	*Zapus hudsonicus* / Meadow jumping mouse	16.0	O	B	OM		71.5		Johnson & Groepper, 1970
				T	OM		94.8		
	Castoridae								
43	*Castor canadensis* / American beaver	12998.0	H				69.0		Cowan et al., 1957 (after Davis & Golley, 1963)

assimilation in 43 species of non-ruminant small mammals. This list contains data on the utilization of both natural and laboratory food. The data have been determined in different trials by the balance method (B) and the tracer method (T). In the first case both the coefficient of digestibility and coefficient of assimilation as a percentage of gross energy (GE) are known. With the second method only the digestibility coefficient as a percentage of organic matter or dry matter (OM) was determined. The digestibility of energy and of organic matter is usually quite similar. We outlined data for 43 small mammals: 29 rodent species, 7 insectivores, 3 lagomorphs, 3 carnivores and 1 chiropteran. It is difficult to analyze such a long list of heterogeneous data and for this reason all species have been divided, according to feeding habits, into four general categories: (1) grazing herbivores, (2) omnivores, (3) granivores, and (4) insectivores and carnivores (Table 8.3, Fig. 8.3). It should be recognized that these are artificial divisions based upon the major feeding habits of the particular rodent, as most rodents are surely omnivorous. These main feeding types among wild mammals have been previously distinguished by Davis & Golley (1963). Animals fall into a particular category due in part to the anatomy of their digestive tract and in part to the availability of particular foods to them. All non-ruminant herbivores must have a large caecum.

Within these feeding categories the average digestibility and assimilation were computed using the original data. In some species, for which only the digestibility coefficient was available, the assimilation coefficient was estimated by subtracting from digestibility 2–3 per cent for the energy lost as urine. Grazing herbivore species, like many voles and lagomorphs, have the lowest level of digestibility and assimilation (av. 65–67 per cent). A higher level of utilization of food energy is represented by various omnivores such as mice, some voles, ground squirrels, etc. (av. 75–77 per cent). The highest digestibility and assimilation is reached by granivore rodents (dormice, pocket mice and field mice) as well as insectivores and also some carnivores. In these animals assimilation approaches 90 per cent, and in small shrews it may even be slightly higher.

Digestibility and assimilation depend on many factors, but primarily upon the quality of food itself. Bulky food is less digestible than concentrated food and hence the laboratory chow employed in many feeding trials has a very high percentage assimilation. Thus, data based on such experiments may be of limited use for ecologists. Digestibility also depends on the chemical composition of the food, mainly on the

Table 8.3. *Digestibility and assimilation of natural foods in small mammals, as percentage of energy or organic matter consumed*

Feeding type	Number of species (number of data used)	Digestibility coefficient (%)	Assimilation coefficient (%)
Grazing herbivore	14 (14)	67	65
Omnivore	10 (12)	77	75
Granivore	6 (8)	90	88
Insectivore	9 (9)	–	85
Small shrews	6 (6)	–	90
Carnivore	2 (2)	90	–

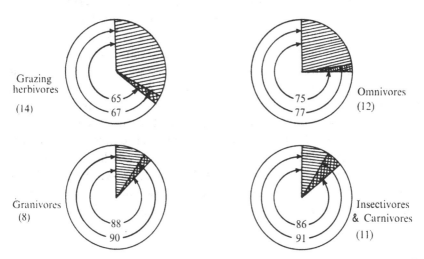

Fig. 8.3. Utilization of food energy by small mammals. The top number on each scheme is the coefficient of assimilation and the bottom number the coefficient of digestibility. Both numbers are given as percentages. The lined areas represent energy lost with feces and the cross-hatched areas energy lost through urine.

content of poorly digested fiber which is found in cell wall constituents, CWC (Van Soest, 1966). Out of this fraction only cellulose and hemi-cellulose are utilized while lignin is not digested, so digestibility is a function of the lignin content of the food. Low protein content, or high content of mineral constituents, e.g., silica or calcium, are other factors limiting digestibility. Another well-known relationship is represented by the effect of the amount of consumed food on its utilization. Finally, the age of a small mammal and its physiological state may to some extent be relevant.

Small mammals

Estimation of the consumption of the whole population may be based upon the assimilation coefficients discussed here. In such computations we usually start with assimilation determined as a sum of production and respiration. The simplest way of estimating total consumption is by adding energy of feces and urine to assimilation. In general the average coefficients can be employed for different categories of small mammals (i.e., by adding to the assimilation of herbivores 35 per cent, omnivores 25 per cent, granivores 12 per cent, insectivores and carnivores 10–15 per cent). This is a 'desperate' and very inaccurate estimation, and the real one should be based on more specific data (Table 8.2) There is a great need for analysis of food utilization in natural conditions since this parameter is critical for computations of the total energy flow. Estimation of food assimilation could also be used to determine the cost of maintenance (R) in small mammals. If an animal does not change body weight during a feeding trial (i.e., $P = 0$) and also does not change the composition of its body, we can consider assimilation as respiration (R). Such an approach has been used by several investigators (Odum *et al.*, 1962; Sharp, 1967; Drożdż, 1968*b*). This approach to the investigation of energy assimilation is basically the same as that outlined for nutrient assimilation by Gentry *et al.* in Chapter 9. Comparison of results from both feeding trials and respirometric studies show fairly close agreement. The feeding method usually gives higher estimates than those using respiration, especially in ADMR tests. Animals studied in metabolic cages do not have nests available and this may explain part of the reasons for higher estimates (Drożdż, 1968*b*).

Modifiers of energy expenditure

Size. The effect of body size on the metabolic rate of small mammals was previously discussed on pp. 181–2. In general, the smaller the mammal, the greater the metabolic rate. The power function of this relationship is different, however, for BMR or RMR and for ADMR.

Shape. Brown & Lasiewski (1972) have shown that in extreme cases the shape of mammals can affect their metabolic requirements. Long thin weasels (*Mustela frenata*) when placed under cold stress at 5 °C have metabolic rates 50–100 per cent greater than that of normal-shaped mammals of the same weights. Their metabolic coefficient or thermal conductance (TC) is about 50 per cent greater than that predicted by the equation of Herreid & Kessel (1967).

190

Reproduction conditions. A great many studies have shown that the energetic cost of lactation in cows is high. However, relatively little information is available concerning the energetic cost of reproduction in small mammals. To our knowledge, in small wild mammals, only three rodent species have been studied recently (Kaczmarski, 1966; Trouan & Wojciechowska, 1967; Migula, 1969; Myrcha *et al.*, 1969). (see Table 8.4). In both mice and voles a slight increase in energy

Table 8.4. *Additional energy requirements of reproducing females during gestation and lactation (values recalculated from references)*

Rodent species	Percentage increase†		Reference
	Assimilation	Respiration‡	
Clethrionomys glareolus Bank vole	57.8	49.4	Kaczmarski, 1966
Mus musculus House mouse	77.7	64.7	Myrcha *et al.*, 1969
Microtus arvalis	82.5	69.3	Migula, 1969
Common vole	80.5	69.4	Trojan & Wojceichowska, 1969
Average	72.3	61.2	
Correction factor (rounded off)	1.70	1.60	

† In comparison with non-reproducing females of the same body sizes.
‡ Respiration represents assimilation minus production of litters including placentae and fetal membranes.

requirements during pregnancy has been found and a very great increase in energy cost during lactation (Fig. 8.4). Reproduction can increase metabolic cost (respiration) by 50–70 per cent.

Having such an effect on metabolism, reproduction can be quite an important factor in the energy budgets of small mammal populations. However, the magnitude of the effect on a population basis depends upon the number of reproducing females and the length of the breeding season.

Season: acclimatization. The season of the year affects the metabolism of many small mammals. There has been considerable confusion in the literature regarding this topic. Part of this confusion arises from the manner in which data are gathered and the terminology associated with such studies (see Hart, 1971, pp. 41–5). Acclimation refers to changes resulting from exposure to controlled variables (usually one at a time and, for our purposes, usually temperature) in the laboratory. Acclimatization refers to modifications in the animal stimulated by seasonal changes in the field. There have been a great many studies concerning the effects of prolonged exposure to cold temperatures, but very few studies of seasonal acclimatization. In general, acclimation to cold temperatures elicits higher metabolic rates at all temperatures. Unfortunately this is not a consistent response and may vary not only

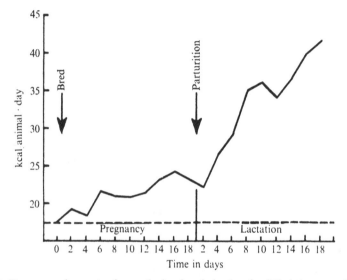

Fig. 8.4. Energy requirements of reproducing female bank voles (*Clethrionomys glareolus*) during gestation and lactation. The dotted line shows the normal energy requirements for a non-pregnant female of the same size (from Kaczmarski, 1966).

between species but within species (see Hart, 1971, p. 43 for review). However, for animals in the field we are dealing with acclimatization not acclimation. In several studies of acclimatization it has been pointed out that it frequently involves modification of heat conserving mechanisms, whereas acclimation involves increases in heat production. In fact, in certain species we find metabolism at any given test temperature may be lower in 'winter' animals than in 'summer' animals (Grodziński & Górecki, 1967; Hart, 1971). Results of acclimatization studies make any generalizations about increases or decreases in metabolism with season very difficult and probably species specific. In most cases the variation in metabolism is not greater than 10–15 per cent and varies with exposure temperature (Hart & Heroux, 1963; Pearson, 1962).

192

Habits. It has been pointed out that the habits of small mammals may influence the amount of energy they expend. McNab (1963) has suggested that croppers (mammals feeding on grasses or vegetative parts) have smaller home ranges than hunters (carnivores and seed eaters) and thus may spend less energy per day searching for their food. However, all indications are that metabolic rates relative to body size are similar in both groups. Thus any differences in energy expenditure would result from different rates of expenditure for activity and this will be discussed below.

Activity. Small mammals use considerably more energy in activity than rest and metabolism during activity may be 5–8 times as great as at rest (Wunder, 1970). Taylor *et al.* (1970) have shown that the net cost of running on the horizontal is a function of body size as shown in the following equation:

$$M_{run} = 8.46 \ W^{-0.40} \qquad (8.7)$$

where W is in g, MR in cc $O_2/g \cdot m$.

This factor, M_{run}, is the slope of the relation between metabolism and velocity of running. Thus, given body size, we can predict the change in metabolism of a mammal as it runs at different velocities. In addition to these factors of size and velocity of running, the kind of activity an animal is engaged in will influence metabolism. Whether an animal is running up or down hill modifies its metabolic expenditure somewhat. Intuitively one feels that running up inclines necessitates a higher energetic cost for mammals than running on the level. However, Taylor *et al.* (1972) have presented evidence that the cost of running uphill may be size dependent. This is further substantiated by the work of Wunder & Morrison (1974). Thus our capacity to estimate the physiological cost of activity is becoming fairly precise. However, in order to put these data into an ecological context, we need to know how much time an animal spends in activity and at what level of activity. These sorts of data are few.

We suggest much more information is needed on the time budgets of mammals in the field. Since most small mammals are nocturnal and secretive, such data will probably have to be generated via telemetric studies.

Duration of daily activity in small mammals (activity out of the nest) and the main patterns of their activity have been reviewed by Saint Girons (1966).

Small mammals

Behavioral thermoregulation. Behavioral thermoregulation is best developed in and important for small mammals rather than large forms. During exposure to cold (times of high energy expenditure), small mammals can modify their heat loss and thus conserve energy by huddling together and using a more favorable microclimate such as a nest. Daily activity may also be adjusted to take advantage of the least stressful time of day, especially regarding temperature (Pearson, 1960; Erkinaro, 1971). By storing food for winter small mammals gain two advantages. They will have food available during times of food shortage and will need to spend little time searching for food during adverse weather (Muul, 1968).

Underground nests of small mammals provide quite constant thermal environments (Daniel, 1964; Hayward, 1965*b*; Brown, 1968) on a daily basis; however, the temperature level may shift slightly from season to season. It has been shown experimentally that the presence of a nest can considerably reduce the energy expenditure of small mammals (Sealander, 1952; Pearson, 1960; Hudson, 1964; Trojan, 1970; Tertil, 1972).

Although there are no field studies, laboratory measurements indicate that small mammals may reduce metabolism during exposure to cold if they are allowed to huddle in groups. The magnitude of the decrease in metabolism due to huddling varies with the species studied (Ponugaeva, 1960) and may range from about 5–45 per cent. The decrease in metabolism is usually a function of the size of the group of animals, but, as group size increases above certain levels in various species, metabolism again increases due to antagonism and additional activity (Prychodko, 1958; Górecki, 1968; Fedyk, 1971).

Although the specific effects of nest insulation and huddling on metabolic rate are relatively small when taken as a percentage change, these factors are quite important when considering daily energy budgets since most small mammals may spend much of their time huddling in a nest. The effects of these phenomena are manifested in a daily energy budget by modifying the temperatures of exposure for the animals.

Energy of production

The theoretical aspects of the energy of production in small mammals have been discussed in Chapter 7 by Petrusewicz & Hansson. Production includes both energy which goes into growth and that which goes into reproduction.

194

Energy which goes into growth is difficult to measure and most investigators simply consider the increase in weight of an organism and the caloric value for that tissue. Such values can be expressed in several ways (kcal/g fresh weight, kcal/g dry weight, or kcal/g ash weight) and the value used depends on the question being asked. Energy values of growth and animal tissue have been summarized by Górecki (1965a, 1967), Sawicka-Kapusta (1970) and Fleharty *et al.* (1973). Body composition may change throughout the year and investigators should correct for this (see Fleharty *et al.*, 1973 for discussion).

The second pathway for production energy is that into reproduction. Few values exist for the energy content of litters and reproductive structures (see Kaczmarski, 1966; Fleharty *et al.*, 1973). However, for a correction, one need know only the energy content of litters and reproductive structures and the biomass of these.

Energy budget models

In dealing with energy flow in small mammals we are, at present, limited to looking at various components of flow and adding or integrating these together (see (1.1), (1.2)). In energy flow through small homeothermic animals, production (P) compared to respiration (R) consists of only about 2 per cent of assimilation (Grodziński & Górecki, 1967; Turner, 1970). Grodziński & French have recently reviewed all available data for 44 populations of small mammals and found that the mean production efficiency of small rodents is 2.3 per cent and of small shrews 0.7 per cent. Thus one of the major components of most energy flow models is an estimate of R.

Gessaman (1973) has reviewed the various models generated to date for estimating small mammal energy flow. These are summarized in Table 8.5.

Some of the earliest models of energy flow in small mammals simply used BMR × 2 as an estimate of energy requirements under field conditions (Golley, 1960). Following these early models some modifications were introduced. Pearson's (1960) model was one of the first to attempt to account for various ecological parameters and included the effects of huddling, nest insulation and a factor for activity. MacNab (1963) presented a model which described R for a deer mouse, although the model has general application to other mammals. Metabolism is described as a function of temperature, time at that temperature and whether the animal is active or not. For the purposes of the model a day

195

Table 8.5. *Models of mammalian energy budget and factors they include*

Authors of model	Body size	Temperature		Nest insulation	Huddling	Sex	Activity	Light/ dark	Season
		Outside	In nest						
Pearson (1960)	−	+	+	+	+	−	+	−	−
McNab (1963)	−	+	−	−	−	−	+	−	−
Grodzinski & Górecki (1967)	+	+	+	+	+	+	+	−	+
Trojan & Wojcieckowska (1969)	+	+	+	+	+	+	+	+	+
Chew & Chew (1970)	−	+	+	−	−	−	+	−	−
Newman (1971)	+	+	−	−	−	+	+	−	+
Randolph (1973)	−	+	+	−	−	−	+	−	+

is divided into a sinusoidal function of high temperature, low activity (daytime) and low temperature high activity (night time). The activity for activity energetics needs to be fed in as maximum to minimum following the sine-wave function. In addition, one needs, independently, to feed into the model the thermal conductance of each animal studied.

In addition to the models outlined in Table 8.5, Porter & Gates (1969) have approached animal energetics in a slightly different fashion introducing the new concept of climate space for animals.

It is our intent now to discuss in detail *only two* models of small mammal energetics and then to add some comments on the concept of climate space. One model was based on resting metabolic rate (RMR) and then another developed from average daily metabolic rate (ADMR).

RMR model

Early in the development of small mammal energetics, resting metabolism was used by some investigators for computing energy flow (Wiegert, 1961; Grodziński, 1961). Trojan (1970) modified the RMR approach to construct a more complex ecological model. The most complete model to date which is based upon RMR is that presented by Chew & Chew (1970); however, it still has two limitations. They have no simple way of estimating the cost of activity and they need to feed in metabolic data specific to the animal with which they are working.

Wunder (1975) has designed a generalized model to estimate R in small mammals given: (1) body size, (2) air temperature, (3) degree of running activity. The model takes the following general form:

$$R = \alpha M_B = M_{TR} + M_A, \tag{8.8}$$

where α is a coefficient to modify metabolism for the posture associated with activity (Schmidt-Nielsen, 1972a), M_B is basal metabolism, M_{TR} is metabolism associated with temperature regulation below thermoneutrality and M_A is metabolism due to activity. The mathematical form for the model is:

$$R = \alpha(3.8W^{-0.25}) + 1.05W^{-0.50}\,[(38 - 4W^{+0.25}) - T_A]$$
$$+ (8.46W^{-0.40})V, \tag{8.9}$$

where α is as above, W is weight in grams, T_A is ambient temperature and V is velocity of running in km/h.

197

Small mammals

This model coupled with estimates of ambient conditions in the field and estimates of time budgets can thus be used to estimate energy flux through an individual over time.

The ADMR model

Starting with measurements of average daily metabolic rate (ADMR) a model of daily energy budgets (DEB) was constructed for small mammals, mainly for small rodents. This model initially represented the cost of maintenance (i.e., respiration) of an animal of mean body weight (Grodziński, 1966; Grodziński & Górecki, 1967). Later, however, it was based on intraspecific functions of the allometric type describing the relationship between ADMR and body size in different species of voles, mice, and squirrels (Grodziński *et al.*, 1970; Hansson & Grodziński, 1970; Grodziński, 1971a; Drożdż *et al.*, 1971).

These budgets describe the respiration of adult animals, assuming that they do not change their body weight in the course of one day, i.e., that their production is equal to zero. Such DEBs represent the sum of the energy expenditure during the time the animal spends in its nest and in the period of activity out of the nest, together with additional costs of maintenance connected with female reproduction. Owing to the seasonal nature of reproduction and also to seasonal changes in activity and temperature outside the nest, these budgets were computed separately for different seasons, or at least for winter and summer days. Corrections added in these budgets to the empirical value of ADMR are limited to two alone, namely: (1) additional heat production for thermoregulation when the animal is active outside the nest, and (2) additional energy requirements of reproducing females (Grodziński, 1971a).

The daily energy budget of small mammals was recently developed by J. Weiner (Grodziński & Weiner, unpublished data) in the form of a general formula which utilizes interspecific equations describing relationships between ADMR and body size in small mammals, see (8.5) and (8.6). The general formula for a DEB based on the ADMR of small mammals has the form:

$$\text{DEB} = \text{ADMR} + \text{f} \left[\text{TC}(t_k - t_a) \right] + \text{CR}(\text{RP} \cdot \text{TR}), \qquad (8.10)$$

where f is the fraction of day spent outside nest, TC the thermo-conductance, t_k the lower critical temperature, assumed to be 20 °C, t_a the ambient temperature, CR the coefficient of respiration increase in a pregnant or lactating female (which on the average amounts to 0.61 – cf. p. 191 and Table 8.4), RP the fraction of reproducing females in a population and TR the duration of the breeding period as a fraction of one year. RP and TR may be assumed to amount to 0.2 and 0.5, respectively.

ADMR in (8.10) for rodents may be replaced by (8.5) while TC may be taken from (8.4) or from the following formula given by Hart (1971):

$$TC = 0.1094\ W^{-0.499} \approx 0.1097\ W^{-0.5}, \tag{8.11}$$

with TC in kcal/g·day·°C, W in g.

Upon substituting (8.5) and (8.11) together with constant coefficients into (8.10) and after transforming the equation and unifying various units we obtain an equation specific to rodents:

$$DEB = [2.437 + f(2.3278 - 0.1164\ t_a)]\ W^{-0.5} \tag{8.12}$$

with DEB in kcal/g.day, t_a in °C, W in g, and f dimensionless.

From (8.12) we may easily calculate the approximate respiration for each rodent species if we know its body weight W, the fraction of the day, f, that it spends in activity beyond the nest, and the temperature, t_a, of its environment. An analogous specific relation may also be written for insectivores starting with expression (8.6), although the data for assessing such corrections are inadequate in their case. Formula (8.12) involves the formal averaging of female reproduction costs over all the individuals in the population for an average day of the year. This is a simplification which admittedly facilitates the computation of a yearly energy budget for a whole population (see section on population models, p. 204).

The exponent in the equation (−0.50) gives values in kilocalories per unit of body weight in grams. If we simply change its sign we obtain the number of kilocalories per whole animal. Equation (8.10) written in its general form allows us to make adjustments to specific conditions in determining the DEB if only we know the necessary parameters, both physiological and populational.

Budgets representing only respiration or assimilation may easily be generalized to represent total consumption by adding the energy of feces and urine to the initial value (Drożdż *et al.*, 1971).

Daily and yearly energy budgets

Perhaps the simplest way to approach calculation of the daily energy budget (DEB) of an individual mammal is to envision the budget as the sum of its component parts. Thus DEB can be represented as:

$$DEB = R + P. \tag{8.13}$$

As mentioned above, for adults, except reproducing females, P is quite small relative to R. Thus the primary influence of P on DEB is through reproducing females and growing young. Grodziński & Górecki (1967) have suggested that P might be easiest to calculate on a population basis using a yearly time scale. However, R can be estimated for an animal using either the ADMR or RMR model. If the RMR model is used, R will be a function of the amount of time the animal is resting or active and the ambient conditions under which such behavior occurs. Since the model of (8.9) can give estimates of metabolism for very short periods of time, the R of a DEB can be envisaged as the sum of a variable metabolism for a day:

$$\text{Daily } R = \sum_{i=1}^{24} R. \qquad (8.14)$$

If P is calculated on a daily basis, then

$$\text{DEB} = \sum_{i=1}^{24} R + \sum_{i=1}^{24} P. \qquad (8.15)$$

For a seasonally cumulative budget one needs to know: (1) does the animal become torpid, and if so for how long, to what degree and under what ambient conditions; (2) how ambient conditions change in the field; (3) how the behavior of the animal may change from season to season (huddling, use of nest, moult, etc.); and (4) if the animal is a female, one needs to know how long she is pregnant and/or lactating. A yearly energy budget (YEB) can then take the following form

$$\text{YEB} = \sum_{i=1}^{365} \text{DEB} \qquad (8.16)$$

To calculate such a budget for day-by-day conditions would be quite tedious. Muul (1968) has constructed a seasonal energy budget for flying squirrels. Gebczynski *et al.* (1972) have done the same for a dormouse and Randolph (1973) presented a yearly model for short-tailed shrews (*Blarina*). Grodziński & Górecki (1967) have suggested that at least two seasons (summer and winter) should be considered if constructing a yearly budget.

It is interesting to note here that Mullen & Chew (1973) have compared estimates of metabolism for *Perognathus formosus* calculated from the model of Chew & Chew (1970) and from measurements of $D_2{}^{18}O$ in free-living animals. The results are quite similar. Using their estimates of temperature exposure, time of exposure and the activity correction

200

suggested by Chew & Chew for this 20 g mouse, we can compare their results with an estimate from the model presented in (8.9). We compared estimates for only 2 months, October and July. The estimate for October (using Chew & Chew's model) is 126.9 cc $O_2/g \cdot$ day and our estimate is 127 cc $O_2/g \cdot$ day. Their estimate for July is 66·3 cc $O_2/g \cdot$ day and our estimate is 61 cc $O_2/g \cdot$ day.

Climate space

One very important approach to modeling energy flow has received relatively little attention from students of small mammal ecology. This is the study of models of heat balance in terrestrial vertebrates approached with heat transfer equations (Birkebak *et al.*, 1966; Birkebak, 1966; Porter & Gates, 1969; Beckman *et al.*, 1971).

Porter & Gates (1969) have defined what they call the climate space of an animal. In essence, they have generated mathematical models, based on heat transfer theory and physiological data, for defining the environmental limits within which animals may function. They can then simulate changes in environmental parameters and predict animal limits and behavior. At present the models have been most refined for the desert iguana, *Dipsosaurus dorsalis*, a poikilotherm. However, personal communications with Porter indicate that models describing energy balance between mammals and their environment are possible.

One powerful advantage of such models is that they would allow one to stimulate environmental change and estimate changes in energy balance and thus in energy flow. Although not appropriate for model building at present this approach may be quite important in the next five to ten years.

Estimation of metabolic rate in the field

To date no studies estimating metabolic rate of small mammals in the field by means of the $D_2{}^{18}O$ technique have been responsible for generating models of energy flow. Mullen (1970, 1971*a,b*) has investigated field metabolism in two species of kangaroo rats, a pocket mouse, and the canyon mouse. Table 8.6 summarizes his results, showing only slightly higher metabolic rates in winter than in summer months.

Mullen & Chew (1973) compared estimates of energy flow in *Perognathus formosus* generated by indirect and direct techniques. The indirect technique was to use the model of Chew & Chew (1970) based on laboratory studies on metabolism and estimates of field time budgets. The other estimate was a direct integrated measure of metabolism by the $D_2{}^{18}O$ technique. Comparisons were calculated over several months and a fair agreement was found (10–20 per cent). Mullen & Chew suggest that one reason this agreement was so close is that they feel *Perognathus formosus* acts in the laboratory much as it does in the field, and a large component of the energy flow is due to thermoregulation. Thus they conclude that for species which are not especially active

Table 8.6. *Metabolism (expressed as cc $O_2/g \cdot day$) of free living small mammals determined by $D_2{}^{18}O$. All data recalculated from Mullen (1970, 1971a, b).*

Month	Perognathus formosus (20 g)	Dipodomys merriami (36 g)	Dipodomys microps (57 g)	Peromyscus crinitus (13 g)
January	–	61	–	129
February	168	107	105	220
March	162	146	96	–
April	151	100	148	–
May	–	–	–	–
June	–	–	133	–
July	77	–	70	–
August	83	44	–	–
September	100	73	–	–
October	160	122	12†	–
November	128	73	16†	129
December	–	98	86	–

† Suspected torpor.

indirect and direct estimates may be similar, but for species which may behave very differently in the field, estimates may be in error if calculated by the indirect model techniques.

In addition to the use of heavy water various radioisotopes have been studied as possible methods for estimating field metabolism. To date, however, none appear to be very promising (see Sawby, 1973, for discussion and review).

Other investigators have attempted to use biotelemetry of heart rate as a means to estimate metabolism in the field. At present none of the systems tried have been successful, as heart rate is only one of several parameters determining oxygen delivery to the tissues (see various chapters in Gessaman, 1973, for discussion).

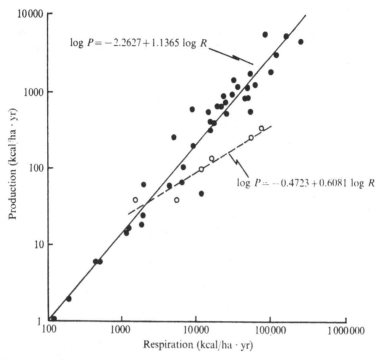

Fig. 8.5. Population production (*P*) described as a function of respiration (*R*) in rodents (solid line) and insectivores (broken line). Data points are for 44 populations of small mammals from Europe and North America. From Grodziński and French (unpublished data).

Table 8.7. *Effects of errors in estimates on energy budgets*

	Type of error	Size of error	Error as % *R*
(*A*)	In *R* components		
	(1) Estimating level of activity. (Largest error when level is low but estimated high.)	0.9–2.1 km/h	2–10
	(2) Amount of time running	—	5–15
	(3) Temperature when out of nest. (Depends on time out of nest: our limits 4–8 hrs.)	10 °C	6–12
	(4) Time out of nest. (Low error if little time spent running, high error if much time running.)	4–8 h	6–12
	(5) Temperature of nest. (Depends on amount of time in nest.)	5 °C	12–17
(*B*)	In *P* estimates		
	Depends directly on the degree of error in estimating number of animals breeding and breeding period, but will affect only *P*.		
(*C*)	In *N* estimates		
	These errors will affect DEB in direct relation to the magnitude of their error and will affect both *P* and *R*.		

203

Small mammals

Population models

To expand energy budget estimates to the population level we need simply to sum the DEB values for all individuals in the population (usually considered by cohorts of age/body size):

$$DEB_{pop} = \sum_{i=1}^{n} DEB \qquad (8.17)$$

where n = number in population.

In order to sum correctly we need to know: (1) the number of individuals in the population and the population age structure; (2) the levels of activity and conditions under which individuals are operating; and (3) the fraction of the population which is reproducing and length of the breeding season. Again the yearly budget for a population can be generated from a summation of the DEB_{pop}:

$$YEB_{pop} = \sum_{i=1}^{365} DEB_{pop}.$$

Summations of this sort have been used in several computations of energy flow through small populations in different ecosystems (Grodziński *et al.*, 1970; Chew & Chew, 1970; Grodziński, 1971*a*; Hansson, 1971*a*). As suggested above, production is more easily estimated separately and then added to respiration to give estimates of assimilation (see Chapter 7). In some 'desperate' cases P can be estimated as some small proportion of R using either of two formulae developed recently for small rodents and small insectivores (Fig. 8.5).

In order to assess the effects of errors in estimating various parameters involved in the calculation of energy budgets, we have listed the possible magnitude that these errors might have on energetic estimates (Table 8.7). These values were generated using the model in (8.9) and varying the parameters of time, temperature and level of activity. By comparing model estimates for R with values in the literature the model has an overall accuracy level of about 10–20 per cent (Wunder, 1975).

Most errors in estimating conditions which the animals are exposed to will only affect certain percentages of an R estimate. Errors in estimation of the fraction of the population breeding will affect the overall energy budget through mis-estimates of P and will depend directly on how far the breeding estimate is wrong. Errors in estimating the total population number will affect both P and R directly in relation to the magnitude of their error. Given the sorts of variance placed on population estimates (see Chapter 2) we suggest that this is the largest source of error in population energy budget estimates.

9. Elemental flow and standing crops for small mammal populations

J. B. GENTRY, L. A. BRIESE, D. W. KAUFMAN,
M. H. SMITH & J. G. WIENER

Following Lindeman's (1942) introduction of the concept of trophic transfer, there was rapid development in the field of bioenergetics of natural systems. The study of energetics usually involves the determination of biomass flow with caloric equivalents of biomass units given on a per unit weight basis. Since biomass can be expressed in units other than energy, i.e., the amount of an element, it is surprising that small mammals have been the subjects of many energy flow studies, but few on elemental cycling (Davis & Golley, 1963; Golley, 1967).

Since chemical limitation (e.g., oxygen depletion) is more obvious in aquatic environments and limnologists usually receive more training in chemistry than typical vertebrate ecologists, cycling processes have been more extensively studied in aquatic than in terrestrial ecosystems (Pomeroy, 1970). Rather, most mammal studies have been on nutrition and toxicology, giving us extensive information on these topics.

One goal of IBP was to study cycling of materials in natural systems throughout the world. Until recently, this goal was not comprehensively pursued by IBP scientists working with small mammals and little of their work has been published. Consequently, this review will present the methods for such studies, the importance of other types of ecological data to this field and recommendations for future work. In addition, we will use the limited data available to evaluate the importance of consumption, egestion, assimilation, emigration and immigration upon the elemental composition of small mammal populations that are fluctuating through time. These data will be modeled and discussed in terms of different elements.

Elemental analysis

The first step in cycling studies is to decide how a small mammal is to be captured and prepared for chemical analysis. After capture the animals should be weighed, dried and ground. Since zinc levels are found to be higher in animals captured in galvanized metal traps (Nabholz, 1973), and the grinding procedure can contaminate the samples for certain elements (Beyers *et al.*, 1971), these possibilities must be considered. Appropriate data such as age, sex, etc., should be recorded. The contents of the gastrointestinal tract should be removed before drying and processed separately from the carcass to allow for the calculation of the amounts of various elements with or without gastrointestinal contents. Animals can be dried in an oven at no more than 60 °C, but for many purposes freeze drying is preferable. Once ground into a dry powder, the animals are processed by such methods as tertiary acid digestion (Jackson, 1958), or other techniques suggested by Thiers (1957) to avoid contamination by trace elements. Under certain conditions, it may be advisable to wash the fur with dilute acid or water to eliminate dust as a contaminant. Two methods of detection have been used for general elemental surveys in mammals; the arc spectrograph (Briese, 1973; Nabholz, 1973; Sella, 1973) and the atomic absorption spectrophotometer (Beyers *et al.*, 1971). Other basic methods include activation analysis (Koch, 1960), and mass spectrometry (White, 1968). In addition, many chemical techniques are available for specific elements (Blaedel & Meloche, 1963), but are too numerous to elaborate upon here. Regardless of the method of detection, controls and standards should be processed with the samples.

Body composition

Estimation of standing crops of elements within the components of a natural system is a first step towards the elucidation of elemental relationships within the system. Since estimates for each biotic component are generally derived as products of the biomass and concentration of each element (Beyers *et al.*, 1971), knowledge of elemental concentrations is essential. Although considerable information on the chemical composition of animal bodies is available, most of it involves composition of various tissues and organs (e.g., Bowen, 1966; McDonald *et al.*, 1966) or gross chemical composition, such as nitrogen, ash and organic content (e.g., Dawson, 1970; Evans, 1973). These data are inadequate for estimates of elemental standing crops. Comparatively little work concerning elemental composition of whole body samples in mammals has been conducted.

Intraspecific differences

Many factors can vary the concentrations of elements within individuals and populations. These factors include body size, sex, reproductive condition and temporal, spatial and genetic variables.

Size-specific relationships in elemental concentration should be calculated for elemental cycling studies since field populations are characterized by constantly changing size structure. Size-specific changes are probably a function of age-specific changes. Most body and elemental components change during growth until chemical maturity is reached (Moulton, 1923; Sheng & Huggins, 1971). In addition, chemical maturity for Ca, N, K, Na, and Cl is attained at different ages (Bailey *et al.*, 1960).

Concentrations of 14 elements in old-field mice (*Peromyscus polionotus*) were analyzed during post-natal growth (Kaufman & Kaufman, unpublished results). Mice were analyzed at weekly intervals up to six weeks of age, when asymptotic weight is reached (Carmon *et al.*, 1963). There was a trend toward an increase in the live weight concentrations of N, P, Ca, K, Mg, Mn, Fe, Zn, Al, S, and Sr, while concentrations of Na, B, and Mo remained relatively constant at all ages (Table 9.1). Dry

Table 9.1. *Mean values of live weight and dry weight concentrations of Ca, Na and Fe in* Peromyscus polionotus *at different ages*

Age (days)	Live weight (g)	Live weight concentration			Dry weight concentration		
		Ca (pph)	Na (pph)	Fe (ppm)	Ca (pph)	Na (pph)	Fe (ppm)
0	1.59	0.100	0.100	59	0.569	0.584	338
7	3.39	0.235	0.093	57	1.050	0.418	257
14	5.32	0.429	0.093	63	1.497	0.326	221
21†	6.67	0.695	0.088	107	2.508	0.286	350
28	8.47	0.654	0.088	119	2.138	0.287	388
35	9.62	0.646	0.094	120	2.615	0.311	396
42	11.43	0.636	0.095	108	2.019	0.274	312

† Mice were weaned at this time.

weight concentrations of Na and K decrease with age but increase for all other elements, although Fe and Zn are higher at birth than at days 7 and 14 (Table 9.1). When variance is partitioned into age, sex, and genetic components, age has a significant effect on all elements except

207

Na for live weight concentration and S for dry weight concentration. Weaning has major effects on elemental concentration (Table 9.1).

Briese (1973) found significant changes in concentration of many elements with age in field trapped cotton rats (*Sigmodon hispidus*). Age had significant simple effects on Al, Cu, K, Mg, Mn, Na, Ca, P and Sr, while the interaction of age and season had effects on Fe, Na and Zn, age and sex on Zn, and age, season and sex on Ba, P, and Sr. Obviously age and body size are significant variables for studies of elemental concentrations in the laboratory and field.

Differences in elemental concentrations between sexes have been reported for *S. hispidus*. Sella (1973) found that female *S. hispidus* had significantly higher concentrations of Fe and Mn, while Briese (1973) found significant differences between the sexes for Ba and Sr and significant interactions between sex and/or season and size for Ba, Ca, P, Sr, and Zn. Some of the differences between males and females and between females at different sizes or seasons may be related to breeding condition (Briese, 1973). Evans (1973) found that breeding female *Microtus* had a higher fat content than the males, and Simkiss (1967) found Ca concentrations to be high in pregnant white rats but extremely low in lactating rats. Body fat is a significant covariate for several elements in *S. hispidus*, and factors that modify the energy budget of this species are probably also important to elemental concentrations (Briese, 1973).

A number of studies have documented temporal variation in concentrations of elements in the body as a whole. In an investigation of elemental composition of *S. hispidus*, Sella (1973) observed significant seasonal variation in the concentrations of all three elements studied (Mn, Fe and Zn). These differences were partially attributed to seasonal changes in the chemical status of the vegetation upon which the animals were grazing. Briese (1973), working with the same species, failed to find a seasonal difference for Mn but did for Al, ash, B, Fe, K, Mg, Mo, N, Na, Ba, Ca, P, and Zn. In addition, Nabholz (1973) found seasonal variation for Cu and Sr. Other studies have shown considerable seasonal differences in elemental concentration in short-tailed voles and common shrews (Hyvarinen, 1972; Evans, 1973).

Sella (1973) tested *S. hispidus* populations for differences in concentration of Fe, Mn and Zn. The two localities studied differed in the concentration of Fe and Mn in the soil and of Fe, Mn and Zn in the vegetation. Despite these differences only Zn differed between populations. In addition, Briese (1973) found a correlation between the concentration in the diet and in the body of *S. hispidus* for only 1 of 14 elements (B). Lack of correlation between concentrations in the body and the environment was also found in largemouth bass (Goodyear & Boyd, 1972). Apparently, variations in environmental concentrations of an element have no consistent effect on the animal's body concentration. Most vetebrates regulate elemental concentration within narrow homeostatic limits, and this may partially explain the emphasis that has been placed on studies of nutrition and toxicity in mammals.

Intraspecific differences in chemical composition may result from genetic differences between individuals. Two genetically divergent populations could exhibit substantially different elemental concentrations. A single gene effect on concentration of potassium in red blood cells of sheep has been demonstrated (Evans & King, 1955), and breed is a significant factor with regard to blood copper levels in sheep (Wiener *et al.*, 1969). Genotypic differences in gross body composition have been demonstrated among inbred lines of house mice (Dawson, 1970) and subspecies of deer mice (Hayward, 1965). However, most differences in elemental concentrations attributable to genetic differences between populations are difficult to distinguish from those caused by locality effects.

Interspecific differences

Beyers *et al.* (1971) hypothesized consistency in elemental concentrations across species except for certain elements that vary with body size (e.g., Fe). Considering the large number of intraspecific differences known to occur (Briese, 1973; Sella, 1973), this hypothesis is questionable. Nabholz (1973) analyzed nine species and found interspecific differences in concentration for 10 of 14 elements. As expected, elements such as Fe showed the largest differences among species. However, other elements not found by Beyers *et al.* (1971) to show species-specific differences did vary in Nabholz's study (e.g., K and Mg). Similar studies by our group on desert rodents (unpublished results) substantiate Nabholz's findings. Differences between species may be due to the comparisons being made between populations at different demographic stages rather than inherent specific variations. At any rate, differences among small mammal species are relatively small compared to those found among plants (Beyers *et al.*, 1971; Boyd & Walley, 1972; Nabholz, 1973), and the differences between the two forms are even more striking.

Prediction of standing crops

It is necessary to develop predictive equations relating the amount of each element to live body weight so that standing crops of elements can be estimated. Although numerous statistical differences exist, they are generally small and probably relatively minor compared to the errors involved in censusing techniques. We now have the information to calculate predictive equations for laboratory raised *P. polionotus* (Table 9.2) and for field trapped *S. hispidus* (Table 9.3) that can be extrapolated to natural systems.

Listed in Table 9.4 are data now available for 15 elements in 27 species. The relationship between the mean dry weight concentration of each element and mean dry body weight for each sample is given and it can be seen that nitrogen is the only element with a significant relationship between concentration and body weight. Regression equations relating amount of each element to body weight are also given in the table. The average concentrations given for all elements except N can be used to estimate standing crop values for use in field studies as follows:

$$E = C \cdot W, \tag{9.1}$$

where E is amount of an element, C is the mean concentration per unit weight and W is dry weight biomass in the same units. For example, let there be 10 small mammals on 0.1 ha weighing from 5 to 15 g each with a total weight of 90 g and a dry weight equal to 30 per cent of the live weight: to calculate the standing crop of Ca, then $W = 90 \times 0.30 = 27$ g dry weight, $C = 0.0315$ (Table 9.4) and $E = 0.0315 \times 27$ g $= 0.85$ g. In addition

$$S = E/A \tag{9.2}$$

where S is standing crop per unit area and A is area. For our example $S = 0.85$ g/0.1 ha $= 8.5$ g/ha. Extrapolations of E to animals quite different in weight from those species listed in Table 9.4 could be questionable.

Consumption and egestion

Accurate estimates of the amount consumed, egested and assimilated must be known in order to evaluate the role of small mammals in elemental cycling at the community level. Data on consumption rates and energy requirements come mainly from laboratory experiments with animals allowed only limited activity, which along with increased

Table 9.2. *Regression coefficients and intercepts for predicting amount in mg* (Y) *of each element from live weight in g* (X) *in* Peromyscus polionotus. *Formula was* $Y = a + bX$. $r^2 = $ *coefficient of determination. Values given* $\pm$ *2 standard errors*

Element	r^2‡	Slope ± 2 s.e.	Intercept ± 2 s.e.
Al	0.788	0.043 ± 0.003	−0.072 ± 0.027
B	0.837	0.002 ± 0.0001	−0.003 ± 0.001
Ca	0.716	7.718 ± 0.759	−12.602 ± 5.984
Fe	0.808	0.130 ± 0.010	−0.195 ± 0.078
K	0.915	2.953 ± 0.141	−0.419 ± 1.108†
Mg	0.748	0.269 ± 0.024	−0.533 ± 0.192
Mn	0.698	0.007 ± 0.001	−0.021 ± 0.006
Mo	0.919	0.001 ± 0.00004	−0.001 ± 0.0003
Na	0.794	0.915 ± 0.073	0.046 ± 0.574†
P	0.918	6.531 ± 0.301	−6.465 ± 2.410
Sr	0.575	0.003 ± 0.0004	−0.005 ± 0.003
Zn	0.823	0.050 ± 0.004	−0.072 ± 0.029

† Not significantly different from 0 at 0.05 level.
‡ All values significant at 0.01 level.

Table 9.3. *Regression coefficients and intercepts for predicting amount in mg* (Y) *of each element or component in a rat's body from its wet weight* (X). *General formula was in the form* $\log_e$ (Y = $\log_e$ a + b $\log_e$ X). *The transformation was necessary to improve* r^2, *the coefficient of determination* (*taken from Briese, 1973*)

Element	r^2‡	Slope ± 2 s.e.	$\log_e$ intercept ± 2 s.e.
Al	0.801	0.952 ± 0.055	−3.464 ± 0.245
B	0.790	1.219 ± 0.073	−7.760 ± 0.324
Ba	0.212	0.114 ± 0.025	−0.279 ± 0.114
Ca	0.809	1.918 ± 0.058	1.037 ± 0.260
Cu	0.774	0.920 ± 0.058	−5.322 ± 0.257
Fe	0.902	1.079 ± 0.041	−3.428 ± 0.183
K	0.898	0.920 ± 0.036	1.302 ± 0.160
Mg	0.819	0.870 ± 0.047	−0.943 ± 0.211
Mn	0.578	0.845 ± 0.082	−5.426 ± 0.373
Mo	0.652	0.044 ± 0.003	−0.134 ± 0.017
N	0.944	1.017 ± 0.028	−3.709 ± 0.128
Na	0.932	0.928 ± 0.031	0.122 ± 0.141†
P	0.930	0.949 ± 0.030	2.030 ± 0.134
Sr	0.650	0.886 ± 0.065	−4.652 ± 0.336
Zn	0.940	1.021 ± 0.030	−3.966 ± 0.133

† Not significantly different from 0 at 0.05 level.
‡ All values significant at 0.01 level.

Table 9.4. *Summary of dry weight concentration and amounts of 15 elements in small mammals.[a,b] Regression coefficients (b), intercepts (a) and coefficients of determination (r²) for average dry weight concentration against average dry body weight (g) and average amount of each element against average dry body weight (g). Mean values of dry weight concentrations (± 2 standard errors) are given. Form of regression formula was Y = a + bX*

	Dry weight concentration[c]				Amount[d]		
	r^2	Slope	Intercept	Mean ± 2 s.e.	r^2	Slope	Intercept
Al	0.02	−1.72	212.01	197.56 ± 32.69	0.69‡	169.46 ± 41.51‡	122.33 ± 411.28
B	0.02	0.03	4.89	5.15 ± 0.45	0.88‡	5.67 ± 0.80‡	5.50 ± 7.96
Ba	0.00	0.08	11.17	11.85 ± 5.13	0.31‡	15.36 ± 8.69‡	−30.64 ± 86.14
Ca	0.00	0.0008	3.41	3.15 ± 0.20	0.90‡	34.90 ± 4.38‡	−5.56 ± 43.38
Cu	0.00	−0.04	24.37	24.07 ± 3.66	0.61‡	24.26 ± 7.40‡	−13.23 ± 73.33
Fe	0.08	−4.47	339.03	303.23 ± 34.25	0.88‡	241.62 ± 34.41‡	340.81 ± 340.94
K	0.00	0.002	1.14	1.15 ± 0.07	0.96‡	12.02 ± 0.87‡	−3.21 ± 8.66
Mg	0.03	0.001	0.10	0.109 ± 0.010	0.87‡	1.21 ± 0.17‡	−0.76 ± 1.73
Mn	0.06	−0.51	19.30	15.00 ± 4.90	0.17†	6.69 ± 5.64†	43.36 ± 55.91
Mo	0.05	0.01	2.85	2.97 ± 0.15	0.92‡	3.21 ± 0.37‡	−2.28 ± 3.66
N	0.46‡	0.08‡	10.11‡	10.75 ± 0.27	0.96‡	123.11 ± 10.14‡	−147.74 ± 100.50†
Na	0.01	0.001	0.36	0.368 ± 0.019	0.97‡	3.92 ± 0.28‡	−1.59 ± 2.77
P	0.02	0.01	2.25	2.31 ± 0.12	0.91‡	24.83 ± 3.04‡	−21.77 ± 30.09
Sr	0.06	0.55	24.05	28.67 ± 5.49	0.60‡	38.79 ± 12.09‡	−75.84 ± 119.84
Zn	0.00	−0.33	109.31	106.67 ± 10.97	0.70‡	103.66 ± 25.91‡	12.78 ± 256.73

[a] *Blarina brevicauda, Peromyscus gossypinus, Ochrotomys nuttalli* (Beyers et al., 1971); *Sigmodon hispidus* (Briese, 1973); *B. brevicauda, Clethrionomys gapperi, Napeozapus insignis, O. nuttalli, Peromyscus leucopus, P. maniculatus, S. hispidus, Sorex cinereus, Zapus hudsonius* (Nabholz, 1973); *Ammospermophilus leucurus, Dipodomys agilis, D. merriami, D. microps, Eutamias panamintinus, Neotoma lepida, Onychomys torridus, Perognathus fallax, P. formosus, P. longimembris, Peromyscus boylii, P. californicus, P. crinitus, P. eremicus, P. maniculatus, P. polionotus, P. truei* (unpublished).

[b] 26 species (30 samples) analyzed for Ca, Fe, K, Mg, Na and Zn; all other elements analyzed in 25 species (27 samples) not including *B. brevicauda, O. nuttalli* and *P. gossypinus* from Beyers et al. (1971).

[c] Dry weight concentration in parts per hundred (pph) for Ca, K, Mg, N, Na and P and parts per million (ppm) for all others.

[d] Amount is in mg for Ca, K, Mg, N, Na and P and μg for all others.

† All values significant at 0.05 level.

‡ All values significant at 0.01 level.

stress and limited dietary choice makes extrapolation to natural conditions questionable (Golley, 1960; Wiegert, 1961; Odum *et al.*, 1962; Grodziński & Gorecki, 1967; Caldwell & Connell, 1968; Fleharty & Choate, 1973). Calculations for the flow of biomass have been made assuming that correction factors for activity, feeding habits and metabolism give satisfactory approximations for field situations (Golley, 1960; Odum *et al.*, 1962; Fleharty & Choate, 1973). For example, Golley (1960, 1967) assumed that the consumption rate of *Microtus* equaled twice the weight of the average stomach contents per activity period. This value was multiplied by the number of activity periods per day (assumed to be ten), to give the daily consumption rate. There is no experimental evidence to support the assumptions that an average stomach is half full, animals fill their stomach each activity period or that there are a given number of activity periods per day.

Other methods used to predict or measure metabolism and to indirectly calculate consumption rates in field situations are isotope excretion and metabolism of heavy water. Excretion rates of different radioactive isotopes correlate with oxygen consumption (Baker & Dunaway, 1969; Pulliam *et al.*, 1969; Wagner, 1970; Chew, 1971), but the corrections are dependent upon temperature and other factors which, probably, directly influence excretion rates (Mraz, 1959; Simkiss, 1967; Kostial & Momcilovic, 1972). Predictability in the field may be low because of changes in excretion rates (Wagner, 1970). Heavy water can also be used to determine field metabolism (Lifson *et al.*, 1955; Lee & Lifson, 1960; Lifson & Lee, 1961; Lifson & McClintock, 1966; Mullen, 1971*b*).

Both of these methods have drawbacks when applied to elemental cycling, since most methods used to determine consumption are concerned mainly with energy rather than biomass. Excretion rates and heavy water metabolism reveal little of the nutritional aspects of the diet of small mammals and the effect that dietary composition has on the amount consumed (Drożdż, 1969; Briese, 1973). Selective feeding by vertebrate herbivores has been well documented. Deer (Klein, 1962, 1970; Thomas *et al.*, 1964; Ullrey *et al.*, 1971), domestic sheep (Nicholson *et al.*, 1970), red grouse (Moss, 1972) and numerous other vertebrates exhibit feeding preferences for particular plant species. Digestibility (Ullrey *et al.*, 1971) and nutrient content (Klein, 1962; Gardarsson & Moss, 1970; Klein, 1970) appear to be important factors in selection of food. Effects of selective feeding and dietary shifts (Fleharty & Olson, 1969; Hansson, 1971*c*; Briese, 1973) on nutrition should be determined under natural conditions.

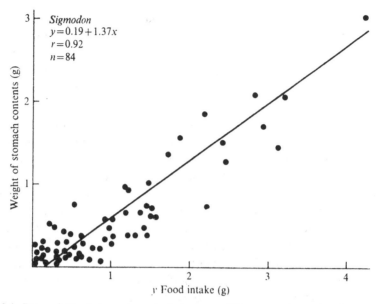

Fig. 9.1. Dry weight of stomach contents in grams (*X*) against dry weight in grams of amount consumed (*Y*) by *Sigmodon hispidus* fed insects, seeds or grass over a 4 h period (taken from Briese, 1973).

Briese (1973) elaborated on Golley's (1960) idea concerning stomach contents and found a linear relation between amount consumed and amount in the stomachs of *S. hispidus* after a 4 h period (Fig. 9.1). By trapping animals throughout a 24 h cycle the daily consumption was estimated (Table 9.5). Stomach contents were then analyzed for

Table 9.5. *Calculated daily consumption rates of cotton rats during each season (Briese, 1973). Consumption (I) in grams dry weight was calculated by summing the amount for each of the six 4 h periods during each day for each season. IA = consumption rate adjusted to a mean body size of 93.58 ± 5.33 g wet weight and LDW = lean dry weight of whole body*

Season	Sample size	$I + 2$ s.e.†	IA†	$I/LDW \pm 2$ s.e.†	IA/LDW†
Winter	96	11.30 ± 2.05^b	11.52^b	0.52 ± 0.07^b	0.52^b
Spring	77	9.06 ± 1.41^b	7.46^b	0.37 ± 0.05^{ab}	0.39^{ab}
Summer	73	9.18 ± 2.49^b	8.95^b	0.59 ± 0.11^b	0.59^b
Fall	51	3.96 ± 0.45^a	6.27^a	0.45 ± 0.06^a	0.42^a

† Any two means in the same column not sharing the same superscript are significantly different at 0.05 level.

elemental composition, and the amount of each element consumed per day determined for each season (Table 9.6). Estimating egestion is more difficult since nutrients are lost through multiple pathways. Analysis of the large intestine contents can be used to calculate a rough estimate of egestion (Briese, 1973; Table 9.6), but in most cases the estimates are minimal. Egestion can also be calculated by subtracting the amount of elements used for growth and reproduction from that consumed (Briese, 1973; Table 9.6). Other methods used to determine egestion rates have used different radioisotopes (Kitchings *et al.*, 1969).

Consumption and egestion rates are useful in determining which, if any, element might be limiting. In almost all cases, the amount of an element consumed will be greater than that utilized. A careful exam-ination of the difference between consumption and egestion of an element will reveal whether the animal is in a positive or negative state in relation to its environment. In the absence of precise data on egestion, the turnover ratio may be used to indicate which elements may be limiting, in that the lower the turnover ratio of the amount egested to that in the body, the more likely that element is to be limiting to population processes. Two elements may be limiting to *S. hispidus*; Ca in the fall and P in the spring and fall (Table 9.6). Aumann (1965) suggested that Na was limiting to *Microtus* populations, but Krebs *et al.* (1971), presented contradictory evidence. Sodium is not likely to be a limiting factor of *S. hispidus* populations (Table 9.6). On the basis of other types of data, Marey and coworkers (1966), also suggest Ca may be limiting in the fall, and Schultz (1964) hypothesized that Ca and P in the diet of lemmings were important in regulating their cycling. To ascertain whether a particular element is limiting using the turnover ratio, the lower limit of the amount in the diet that can still be assimil-ated must be determined. These data could then be used to set a lower limit on the turnover ratio for each element; values below this would suggest that the element could be limiting.

Population processes

Birth and death, two of the four factors which can change population number and size structure, introduce a dynamic aspect into elemental cycling studies. As previously discussed, an animal's body size can modify its elemental concentration. In addition, stage of growth and reproductive condition can change nutritional requirements and should be recorded and taken into account in models of elemental cycling

Table 9.6. *Amount of seven representative elements (measured in mg/day) involved in consumption, utilization for growth and reproduction, estimated amount egested, calculated amount egested and percentage of estimated amount egested divided by the amount contained in the average body of cotton rats during each season (Briese, 1973). Egestion equals consumption minus growth and reproduction. Calculated egestion was determined from the dry weight and mineral analyses of the large intestine contents*

Season	Element	Consumption	Growth	Reproduction	Egestion	Calculated egestion	Egestion/body × 100
Winter	B	0.3400	0.0004	0.0000	0.3396	0.1267	204.57
	Ca	51.98	1.65	0.00	50.33	25.81	5.56
	Cu	0.3300	0.0008	0.0000	0.3292	0.2289	83.13
	Fe	6.11	0.01	0.00	6.10	4.20	120.76
	P	57.63	1.09	0.00	56.54	51.13	9.12
	N†	565.00	6.24	0.00	558.76		21.47
	Na	24.95	0.15	0.00	24.80	17.87	32.56
Spring	B	0.2300	0.0003	0.0002	0.2295	0.0714	156.12
	Ca	46.80	1.08	0.74	44.98	31.06	4.56
	Cu	0.6100	0.0050	0.0005	0.6030	0.5747	155.81
	Fe	4.21	0.01	0.02	4.18	3.37	69.49
	P	50.40	0.71	0.74	48.94	51.49	6.58
	N†	450.00	5.81	5.83	438.36		12.96
	Na	18.46	0.17	0.21	18.07	18.10	17.41

Summer						
B	0.1800	0.0003	0.0001	0.1796	0.1820	246.02
Ca	38.22	2.08	0.83	35.31	12.36	5.22
Cu	0.2500	0.0068	0.0010	0.2422	0.1525	85.88
Fe	2.99	0.01	0.02	2.96	1.71	80.88
P	47.32	1.51	0.89	44.91	28.81	8.77
N†	455.00	8.41	4.95	441.64		19.64
Na	19.64	0.21	0.14	19.28	7.83	24.46
Fall						
B	0.0600	0.0030	0.0000	0.596	0.0181	124.17
Ca	9.36	2.43	0.32	6.61	3.22	1.83
Cu	0.2100	0.0025	0.0012	0.2063	0.0763	140.88
Fe	1.52	0.02	0.01	1.50	0.99	63.26
P	23.01	1.96	0.36	20.69	25.52	17.06
N†	253.50	9.22	1.58	242.69		17.74
Na	8.63	0.21	0.04	8.38	6.32	17.43
Mean for year						
B	0.2025	0.0003	0.0001	0.2025	0.1000	182.72
Ca	36.59	1.81	0.47	34.31	18.11	4.29
Cu	0.3500	0.0055	0.0007	0.3447	0.2550	116.48
Fe	3.71	0.01	0.01	3.68	2.57	83.59
P	44.59	1.32	0.50	42.77	39.24	10.38
N†	430.86	7.42	3.09	420.36		17.95
Na	17.92	0.18	0.10	17.63	12.51	23.72

† Nitrogen was not determined for the large intestine contents.

(Brody, 1945). Data need to be collected in the field on fluctuations in nutritional requirements of animals in different physiological states, and this information then needs to be incorporated into existing models.

The values for growth and reproduction given in Table 9.6 were calculated from elemental concentrations for pre- and postnatal developmental stages. The slope of the growth curve, empirically determined under field conditions, can be used to give incremental changes in biomass through time (Briese, 1973). The amount of each element used during the prenatal period is arbitrarily assigned to reproduction and that for the postnatal period to growth.

The remaining two factors, emigration and immigration, are usually assumed to be zero or equal, thus resulting in no net change in population number. Even if this were the case for mammal populations, another assumption is necessary, i.e., the biomass exchanged by these processes is equivalent in amount and elemental concentration. The latter is not a reasonable assumption for mammal populations (Briese & Smith, 1974). First, there may be a net movement in one direction for any one species or for species considered together. This condition would normally be expected because investigators choose areas with relatively dense populations to maximize the resulting data per unit effort, and net movement would usually be away from the area of highest density. Study areas are frequently chosen that have well defined habitat boundaries. Dispersing animals that cross these boundaries are more likely to die in the nonpreferred habitat, thus resulting in net one-way movements for each species (Briese & Smith, 1974). This net movement can be relatively large compared to the average standing crop, especially at certain times of the year. Movement at the edge of the study plot is an important variable and should not be ignored (M. H. Smith *et al.*, this volume).

Models

A model similar to those used in energy flow studies can be constructed for elemental cycling. Inputs and outputs for the study area are immigration and emigration. Consumption, egestion, mortality by predation, growth and reproduction are all important parameters for the population. We have chosen two elements (B and Ca) and one species to illustrate the technique and to emphasize the similarities and differences between elements (Fig. 9.2).

Even though the flows were expressed as a percentage of the standing crop, the consumption and egestion rates were quite different. Boron, which is a nonessential trace element (Bowen, 1966), has a high flux relative to the standing crop. In contrast, Ca which is an essential macro-element has relatively low values for consumption and egestion. Amounts for predation, growth and reproduction are similar for both of these elements. This is suggestive of some constant ratio between standing crops and secondary productivity similar to that described by Wiegert & Evans (1967) for the relationship between the energy of ingestion and secondary productivity in homeotherms. Further work is needed to test this hypothesis.

Immigration and emigration are essentially equal for both elements in *S. hispidus*, but there was a net input of both elements for other species (Fig. 9.2). These latter values could be important when considering elemental cycling for the entire study area. The models give a relative picture of the flow of elements through the *S. hispidus* population but do not integrate this species into the ecosystem or evaluate its potential importance.

Recommendations

Considering the state of development of elemental cycling in mammalian ecology, it is premature to make sweeping generalizations, but limited statements are appropriate. First, more data on whole body concentrations of elements in small mammals from other habitats and areas in the world are needed, and existing data need to be published in the open literature. Mammals of a greater size range should be examined to elucidate additional relationships between elemental concentrations and body size. Data on concentration should be combined with density and biomass estimates to calculate standing crops of elements. Geographical or habitat specific patterns in standing crop need to be determined. Some thought also needs to be given to the establishment of standards for elemental analysis. This might be accomplished by making some standard mammalian sample available to interested investigators. Chemical analysis should be conducted at both the elemental and molecular level of organization. Knowledge of the molecular state of each element will surely prove important in determining availability and flow in natural systems and importance as a potential limiting factor. More models of elemental flow through mammalian populations need to be constructed. The physico-chemical impact of mammals upon other components should also be examined in order to evaluate the overall importance of mammals in different ecosystems. Emphasis

219

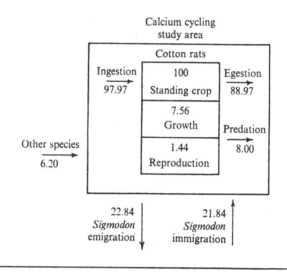

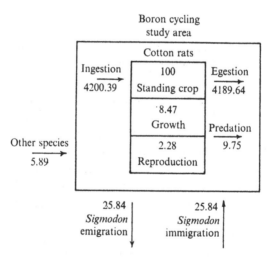

Fig. 9.2. The amount of Ca and B (as a percentage of the standing crop) cycled through an average population of *Sigmodon hispidus* (cotton rats) in one month. Standing crop was equal to 7.88 g Ca/ha and 0.001 g B/ha. The outer square represents the study area and the inner square represents a population of individuals. Data on *S. hispidus* densities and disappearance rates are from Provo (1962). Movement data are from Briese & Smith (1974). Concentrations of elements and the amount used for growth, reproduction, consumption and egestion are from Briese (1973).

should be placed on the regulation of flow within populations and the importance of mammals as potential regulators of elemental flow in ecosystems.

We thank Ms Sarah Collie, Glennis Kaufman, Jean Turner, Peggie Whitlock and Gloria Wiener and Dr Michael O'Farrell for help in preparing the manuscript. This effort was supported by contracts AT(38–1)-310 and AT(38–1)–819 between the US Atomic Energy Commission and the University of Georgia.

10. The role of small mammals in temperate forests, grasslands and cultivated fields

F. B. GOLLEY, L. RYSZKOWSKI & J. T. SOKUR

The rationale for the terrestrial ecosystem studies of the IBP is described in the volumes edited by Petrusewicz (1967). There have been two fundamental research approaches to small mammals within these studies.

One approach is that small mammals are considered as a compartment which receives inputs and has outputs (Fig. 10.1a). Research has

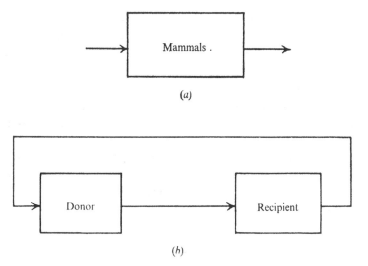

(a)

(b)

Fig. 10.1. Diagram of conceptual approach to mammal studies. (a) Mammals as a component receiving inputs and outputs; (b) mammals as a recipient compartment linked to a donor compartment.

focused on characterizing the compartment and on specific types of inputs and outputs; in most instances these involve food or energy without considering other components in the system. Many of the previous chapters are concerned with the characterization of the compartment in temperate regions and these data will not be summarized here.

Alternatively, small mammals are considered as one of a set of components linked through transfers. At the simplest level, the component receiving the action is the recipient, the other is the donor component (Fig. 10.1*b*). One transfer is direct, for example, the passage of food from the plant to the consumer; the other is feedback of the impact of the consumer on the consumed. Feedback is defined as the transfer of information which can alter the behavior of the donor compartment. Since transfer of information is not instantaneous, time is required for a feedback signal to travel from the recipient to donor so there is a time lag in the response to feedback. Steady-state or controlled system behavior is achieved by proper meshing of these signals as well as by the complexity of the circuitry.

Research on small mammals has used mainly the first approach or has investigated direct linear pathways between donor and recipient compartments. This focus has probably been due to the concentration on ecological energetics. Energy flows through systems in a somewhat linear fashion, and models of energy flow through food chains are conventionally linear. As a result, the research opportunities pointed out by Dinesman (1967) and Varley (1967) have been largely unexploited in IBP, and so we are relatively ignorant of the reciprocal action of feedback from mammals on plants and environment.

This lack of information makes it difficult to develop a thorough examination of the role of small mammals in temperate ecosystems. The object of this chapter is to develop a paradigm which accounts for the total behavior of the mammal in its ecosystem, to discuss the roles mammals play in the context of this paradigm, and finally to suggest research strategies for further study of the role of small mammals after IBP terminates.

The small mammal paradigm

Let us begin with an ecological system. This system is arbitrarily defined by specified boundaries (Fig. 10.2). The system may be a temperate forest, grassland or cultivated field. It will contain a variety of species populations which may number in the thousands, but which can be grouped functionally into three components; primary producers, consumers, and decomposers (Fig. 10.2). In addition, dead organic matter and soil are other ecosystem components. These components are linked by the transfer of energy, materials and information. This definition of components is conventional and can be found in Petrusewicz

(1967). Each component in the diagram can be expanded by identification of each functional subgroup. Small mammals are one such subgroup of the consumer component (Fig. 10.2); granivorous birds are another, as are social insects, sucking insects, etc. These subgroups can be expanded further by identification of the individual species populations, if that level of detail is desired.

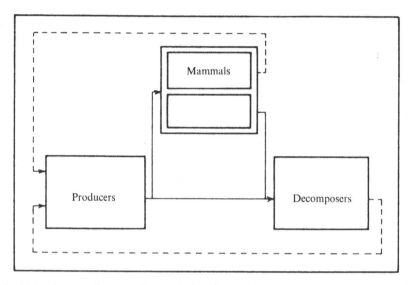

Fig. 10.2. Diagram of a generalized ecological system with three compartments. Mammals are one subcompartment of the consumers. Feedback loops are indicated by dashed lines.

This paradigm exhibits a major feature of all ecosystems. The main components, subcomponents, and species populations are linked together to form a system with transfers in both directions between all interacting components and stability is achieved through the interaction of the components, which are themselves under partial control. For example, in stable systems the consumer complex is internally controlled in such a way that one consumer population interacts with others in a web of internal relationships. Increase in numbers or activity of one population influences many other populations, so that the entire consumer complex functions as a whole. Further, the consumer complex acts to control the producer and decomposer components in the system (Golley, 1973).

Predator, parasite or disease populations are limited in their development by the diminution of energy as it passes through the food chain and by the accumulation of materials. Thus, there is a physical–chemical limit to the length of food chains. Elaboration within those limits is regulated biologically. In mature communities, there is great elaboration of food chain populations. Control is achieved by this variety – in populations there are more control units (predator, parasites and disease organisms) than units to be controlled, as one would expect from Ashby's (1956) Law of Requisite Variety.

We can begin to describe the linkages of small mammals to other components of the system and to other consumer subcomponents by tracing their feeding habits. There are numerous studies of this type – those of Whittaker (1966), Drożdż (1966) and Watts (1968) are recent examples. These studies show that small mammals have definite food preferences, mainly consuming green matter, fruits and seeds and/or insects. In turn, small mammals are consumed by a variety of predators and the wasted or unassimilated food, as well as their dead bodies, are transferred to the decomposer component of the system.

These direct transfers we have studied in much detail in IBP. The feedbacks are less well known. They include direct destruction of green leaves, shoots or stalks with an impact on photosynthesis; destruction of seeds with an impact on the number of progeny surviving the next growing season; the movement of seeds through the habitat in seed caches or by consumption and defecation; feeding on bark at the ground surface or on roots causing death of the plant; movement of nutrients in the system through defecation, through storage of food or by concentration of activity at a site; reconstruction of the habitat by burrowing, tunnelling and other earth-moving activities; serving as a reservoir for parasites and diseases of other consumers including man; serving as a buffer to predation of other consumers, and so on.

These impacts can be grouped into four main categories; (1) those concerned with destruction of a component by mammals; (2) those concerned with movement of materials or components by mammals; (3) those concerned with alteration of the environment; and (4) those concerned with other consumers, especially predators. Each of these categories involves direct effects on biological components, on the rates of transfers between components and on the environment. In addition, each of these impacts has secondary, tertiary and other indirect effects. However, before we examine these categories, we will consider the three types of temperate ecosystems, since each ecosystem controls or limits the nature of the impact of mammals within its system.

Cultivated fields are ecosystems artificially maintained at an early stage of succession, with simple structure, few possibilities of modifying the effect of climatic factors, high productivity, low energy cost in the production of a biomass unit and open cycles of mineral circulation. In these ecosystems, the cost of maintaining their stability is born by man. Man influences practically all of these ecosystem processes from the pedological processes of soil formation to energy flow, mineral cycling, and control of rodent, insect and predator populations.

Grassland ecosystems cover different successional stages up to climax communities of steppe or prairie. They have a more complicated structure than cultivated systems and are characterized by an especially deep horizon of soil organic matter. These ecosystems are controlled to a high degree by climatic factors, especially water, in the sequence of dry and wet seasons. Practically all above-ground primary production becomes dry and is shed when the growing season is over, and is partially decomposed before new growth begins. In late successional or climax grasslands, the energy flow of the production of a biomass unit (gross primary production to average standing crop) is moderate in comparison to the low costs in cultivated fields and high costs in forests. Mineral cycles within the ecosystem are well developed, with herbivores acting as an important influence on the rate of cycling. Use of grassland for pastures increases the impact of grazing animals which can lead to overexploitation, especially without management of the water regime. In pastures the impact of man on soil formation processes is rather small.

The forest ecosystems also cover different successional stages up to climax communities and have a very complicated structure. Forests change climatic factors producing distinct microclimates, and have high water storage, soil retention, etc. In climax forest ecosystems the energy flow of production of a biomass unit is high since the standing crop is very large. A major proportion of the live organic matter of these ecosystems is located in the above-ground vegetation. The retention of minerals is high. Influences of man on processes of soil formation are negligible.

The above characteristics of ecosystems frame the dimensions of the ecological role of small mammals. For example: (*a*) the influence of the burrowing activity of mammals on surface water runoff is quite different when the retention of water in the system is high, as in a forest, from that in grasslands where increased water storage through the effect of rodent burrows can be very important; (*b*) consumption of a large quantity of vegetation has a different importance when the bulk of primary production is shed each year and grazing promotes regrowth of the plant biomass, as in grasslands, as compared to forests; (c) the transport of minerals (e.g., gypsum, calcium carbonate, by rodents from deeper layers of soil profiles to the surface can be an important factor influencing soil chemistry in the grassland ecosystem characterized by a soil type like solonchak or solod, while in cultivated fields their burrowing activity has little significance on soil chemistry in comparison with man's input of mineral fertilizer; (d) the selection of seeds for food by rodents has a greater influence on the species composition of plant cover in forest ecosystems where seeds play important roles in propagation of vegetation than in grasslands where much of the regrowth is from roots.

The destruction of components by mammals

It is well known that small mammals are capable of destroying the vegetation. Elton's classic book, *Mice, Voles and Lemmings* describes numerous examples, and, as recently as 1967, Yugoslavia experienced widespread crop destruction by small mammals. Other recent examples from Europe include: (a) in Kazahstan, USSR, estimated cost of yearly damage to crops (1967–9) by *Citellus pygmaeus* amounted to 20–30 million rubles (Krylcov & Zaleskii, 1969); (b) in Bulgaria, estimated cost from crop damage (1967–8) by *Microtus arvalis* was 25 million leva (Straka, 1970); (c) in the Federal Republic of Germany, damage to forest plantations due to *Microtus agrestis* and *Clethrionomys glareolus* was observed in a recent year in an area of 10000 ha (Schindler, 1970); (d) in Italy, serious damage to tobacco cultures by *Pitymys savii* was recently observed (Murio & Lorito, 1969); (e) in Finland, costs from damage to orchards and ornamental plants by *M. agrestis* (1954–66) amounted to 2 million US dollars (Kanervo & Myllymäki, 1970); (f) the cost in Sweden (1962) of the damage to orchards due to *M. agrestis* amounted to 4–8 million US dollars (Stenmark, 1963). In addition, forest regeneration may be severely hampered by consumption of seeds by mice and shrews (Smith & Aldous, 1947; Ashby, 1959).

The majority of cases reported in the literature show that a significant impact from small mammals has occurred mainly in temperate agricultural fields or forestry plantations. Mature natural temperate forests and grasslands seem less likely to experience extreme rodent or shrew depredation. However, the data gathered from a variety of habitats through IBP show that a very low percentage of the primary production is consumed by small mammals. The average rate is about 3 per cent, with the only cases of very great consumption in an alfalfa field with a rodent density of 774/ha and on seeds produced in an old-field consumed by *Peromyscus* (Table 10.1). Thus, while the above reports suggest their

Table 10.1. *Energy consumption of small mammals as the percentage of available primary production consumed yearly in temperate ecosystems*

Ecosystem	Available primary production 10^6 kcal/ha · yr	Percentage consumption	Reference
Agricultural fields			
Rye field	40.7	0.5	Trojan (1969)
Alfalfa	39.8	0.8	Trojan (1969)
Alfalfa	38.8	1.4–21.4	Ryszkowski et al. (1973)
Grasslands			
Grass field	40.6	1.3	Trojan (1965)
Grass field	47.0	1.6	Golley (1960)
Old-field (seeds only) meadow	0.5	12.0	Odum et al. (1962)
Desert			
Desert shrub	2.4	5.5	Chew & Chew (1970)
Forest plantation	6.7	3.1	Gebczyńska (1970)
Forests			
Pine lichen	1.0	1.9	Ryszkowski (1969a)
Vaccinium–pine (40 years)	2.4	0.9–1.2	Ryszkowski (1969a)
Vaccinium–pine (140 years)	7.0	0.6	Ryszkowski (1969a)
Oak–pine	13.0	0.6–0.8	Ryszkowski (1969a)
Oak–hornbeam	2.1	4.6	Grodziński (1971b)
Mixed forest	16.2	0.6	Ryszkowski (1969a)
Beech forest	2.0	2.4–3.6	Hansson (1971a)
Spruce–grass	14.7–19.2	2.2–3.6	Hansson (1971a)
Circaceo–Alnetum	—	2.2	Aulak (1973)

potential destructive role, in most temperate systems small mammals destroy relatively little biomass. Their impact is more subtle.

The study of the consequences of destructive impacts on the system is difficult, and there are relatively few such investigations. Two research strategies are possible (Varley, 1967): mammals can either be excluded from an area and the vegetation in the exclosure compared with that available for attention from mammals or the state of the vegetation can be examined year after year together with the consumption by mammals. Variations in mammal consumption can be correlated with changes in vegetation and a causal relationship inferred.

Small mammals

The available studies of the role of mammals in the ecosystem indicate that they have an impact on the kinds of plants present in the system, their abundance, distribution, form and reproduction. For example, Batzli & Pitelka (1970) showed by enclosure experiments that *Microtus* kept the habitat more open and increased the number of plant species present. Where voles were excluded, grasses, the preferred food, became more dominant. Spitz (1968), also using enclosure techniques, found *Microtus* had little impact on alfalfa production yet strongly influenced the size of stems and the ratio of leaves to stems. Under vole influence, the quality of the alfalfa was reduced.

It is well known that in places of intensive rodent activity one can observe change in plant species composition. Kucheruk (1963) summarized the results of many authors on specific plant species composition within colonies of steppe rodents: characteristic plants for *Marmota* spp. colonies are: *Hyosciamus niger*, *Chenopodium foliosum* and *Sisymbrium* spp. For colonies of *Alticola strelzowi* indicator plants are *Crossilaria acicularis* and for colonies of *Lagurus* spp., *Carduus unicinatus* is characteristic. According to Skvorcova & Utehin (1969), 4 to 6 years after a colony of *Spalax microphthalamus* was abandoned the plant composition returned to its initial condition. In the case of a colony of *Citellus*, about 20 years were needed to restore the initial conditions (Formozov *et al.*, 1954).

Because of their high nutritional value seedlings are often preferred to older plants. Bark of trees, branches, stems and twigs are usually consumed by rodents only during winter, especially when the lower layers of snow are frozen. Injured young trees grow more slowly and Dinesman (1961) found that because of injuries made by *Microtus socialis* the height of trees was reduced (Table 10.2). A relatively small

Table 10.2.

Tree	Average height when uninjured (cm)	Average height when injured (cm)
Maple (*Acer negundo*)	123	106
Elm (*Ulmus pinnato-ramosa*)	123	105
(*Elaeagnus angustifolia*)	118	103

number of older injured trees die. Buchalczyk *et al.* (1970), estimated damages done by *M. oeconomus* to *Populus nigra*, *Salix purpurea* and *Alnus glutinosa* at 48 to 67 per cent of all surveyed trees, but only 6 to 7

230

per cent died because of the root injuries caused by the voles. In contrast, seedlings are destroyed by rodents in much greater proportion than older trees. *A. flavicollis* destroyed 78 per cent of the total number of seedlings in a deciduous forest. Sviridenko (1940) classified trees in the following categories according to damage done by rodents to seedlings in deciduous forests: (*a*) 80 to 100 per cent of seedlings destroyed; oaks, elms, maples, lindens; (*b*) 50 to 60 per cent destroyed; ashes, rowan; (*c*) 0 to 20 per cent; hazels and bird cherries. Such information can be found quite often in the literature, but there is a complete lack of information on the ratio of plant biomass consumed to biomass destroyed in the case of trees.

The destructive activity of rodents at peak density has also been compared with that of fungi and insects, according to Pivovarova (after Dinesman, 1961); This is shown in Table 10.3. The percentage of seedlings destroyed by rodents was quite high in comparison to the activity of fungi and insects except in the case of maple.

Table 10.3.

Species of seedlings	Number of seedlings/ha		Percentage injured by rodents	Percentage injured by fungi and/or insects
	Beginning of spring	End of autumn		
Hornbeam (*Carpinus*)	20 000	4 600	32	18
Maple (*Acer*)	7 200	1 660	25	44
Oak (*Quercus*)	8 800	2 800	41	27

The effect of grazing on trees and shrubs is often serious because the plant has a long period of production and damage or destruction not only affects future production but also requires a long time for repair or replacement of the damage. This impact can be especially serious in orchards where the individual tree is valuable (Tahon, 1969).

In herbaceous systems and grasslands, the consumption or destruction of foliage by small mammals interupts the predominant pathway of vegetation to the decomposers through the litter. Mammals can thus act as an analog to a fire in reducing organic material to a more easily decomposed form which short circuits the long process of decay. Because of this, we suspect that in certain grassland systems rodent outbreaks may be beneficial rather than destructive. Without rodents, the dead organic matter might accumulate and tie up essential nutrients to such an extent that primary production would be altered.

Small mammals

These examples illustrate the options open to the vegetation in response to destruction of the foliage. First, primary production may increase if consumption is light and interferes with maturation of the plant. Alternatively, there may be a switch between plant parts or there may be a switch between plant species away from those which are preferred food toward less desirable parts or species. Over time palatability may change with selection for organic compounds that are toxic to the herbivore or with development of structural protective appendages such as spines. Whittaker & Feeny (1971) in their review of chemical interactions between organisms discuss the limitations of herbivores in detoxification of organic compounds. Apparently palatability is a compromise between the selection of foliage for nutrients and avoidance of toxins. In addition, the body size and the form of the digestive tract influence the relation of the animal to toxic materials. Large mammal herbivores may be able to tolerate larger doses of toxin than small mammals, and ruminants may be able to detoxify plant materials better than nonruminants. These interactions between plants and animals have been the subject of detailed study by animal scientists (Longhurst *et al.*, 1968), but there is little information for the small mammals we are considering here. It will be important to find out if the devastating consumption of foliage in cultivated fields by small mammals is due to the selection against plant toxins in the crop and thus, an increase in its palatability to these herbivores.

Throughout the coniferous forest regions, small mammals may significantly reduce forest regeneration through direct destruction of the seed crop (Smith & Aldous, 1947; Spencer, 1955). Ashby (1959) also states that small mammals prevent regeneration of deciduous forests in Great Britain through consumption of seeds and seedlings, although Tanton (1965*a*) feels that the importance of small mammals in seed destruction is overestimated. Dinesman (1961) summarized results of many authors on this topic as follows: In forests with *Apodemus flavicollis* all seedlings of maple are eaten while elm and hazel survive. In Kaukaz forests, because of feeding activity of *A. flavicollis* seeds of beech are completely eaten. Zhukov (1949) found that the number of oak seedlings depends on the rodent density (Table 10.4).

Mice and shrews can consume large quantities of seed. Radvanyi (1966) found that small mammals consumed 49 per cent of *Picea glauca* seed distributed in an Alberta forest when mouse densities were about 15/ha. Abbott (1961) estimated that white-footed mice (*Peromyscus maniculatus*) consumed 260 seeds per animal daily, and the redback vole

Table 10.4.

Crop of acorns	Number of seedlings/ha	
	Without rodents	At peak density of rodents
Very poor	280	138
Poor	1 880	920
Moderate	9 870	4 830
Good	108 600	52 900

(*Clethrionomys gapperi*) consumed 232 seeds per day when provided with white pine seed in feeders. In a similar type of study in a Tennessee oak–hickory forest, white-footed mice were estimated to consume 17 per cent of the pine seeds in feeders, in contrast to 64 per cent by short-tailed shrews (*Blarina brevicauda*) (Mathies *et al.*, 1972). Gashwiler (1970) using an exclosure technique showed that 22 to 44 per cent of conifer seeds in Oregon were destroyed by mice and shrews, depending on the species of conifer. He concluded that mice and birds could adversely influence forest regeneration after the forest was clear-cut for timber. Bramble & Goddard (1942) using enclosures found that survival and germination of seeds was much higher when protected from mammals and birds in the scrub-oak barrens of Pennsylvania.

Rodents also collect and store large quantities of seed and other foods. Sviridenko (1957) reviewed records for several mammals as shown in Table 10.5.

Table 10.5.

Species	Food stored	Amount stored
Sciurus vulgaris	Acorns	650 seeds
Eutamias sirbiricus	Various seeds	up to 8 kg
Citellus undulatus	Various seeds and grasses	up to 6 kg
Glis glis	Nuts	up to 12–15 kg
Spalax leucodon	Potatoes	up to 12 kg
M. arvalis	Herbs, grasses	up to 1.4 kg
M. socialis	Seeds	up to 1.2 kg
M. nivalis	Grass	up to 2.0–2.3 kg
Arvicola terrestris	Roots	up to 1 kg
Apodemus flavicollis	Acorns	up to 4 kg
Cricetus raddei	Seeds, herbs, fruits	up to 7 kg

233

In order to develop a paradigm of mammal–seed interactions, we require information on seed selection by small mammals, seed palatability, and distribution of seeds and mammals in time and space. The authors do not know of any information on the full spectrum of seeds available over a year in any temperate community. Such information should include the size and numbers of seeds available monthly at the surfaces where mammals feed. We must also know the range of seed sizes selected by small mammals, assuming they select seeds as discussed by Rosenzweig & Sterner (1970) and are not entirely opportunists. If small mammals are limited to seeds of a specific size, then the plant may escape seed predation by mammals by decreasing or increasing seed size. Naturally, the plant also has limited options since the energy stored in the seed must be sufficient for germination after the winter period and the energy required to produce the seed cannot be excessive. Here is an opportunity for mutualistic selection where the proper interaction between plant and predators results in adequate reproduction of the plant and adequate food for the mammal. We see no reason that Janzen's (1969) concept of the influence of seed predators on the survival and species of tropical trees cannot be applied to temperate forests and small mammals. However, the difference between tropical and temperate environments will undoubtedly have an effect on the plant and animal strategies used in these interactions. The intrusion of a long winter period with plant dormancy but continued animal activity may act against the production of few large seeds per tree and toward many small seeds dispersed over relatively large areas, possibly with protective chemical compounds.

These two modes of destruction of plant components by mammals (i.e., foliage feeding and seed predation), can be diagrammed using the symbols in Fig. 10.2. The two different kinds of consumption in one case have a direct effect on the production capacity of the community and in the other on the ability of the producers to replace themselves. The immediate impact of foliage feeding is of major importance in annual communities where storage in roots and stems is unimportant. In other communities foliage consumption can be compensated for from storage tissues. In other words, in forests, periodic outbreaks of foliage feeders reduces production (Varley & Gradwell, 1962) but does not necessarily destroy the forest.

In contrast, seed predation has a time lag in its impact since the effect is on the number of mature plants which, in turn, affect the environment of the mammal. In the annual community, excessive seed

234

consumption could reduce plant density in the next season. In the forest, there may be tens of years available for regeneration. The strategies for assuring community stability over time and space where the environment is severe for part of the year and this severity is quite variable from year to year must be quite complicated. Obviously, this line of research has been neglected in IBP and should be considerably increased, if for no other reason than its practical significance.

Movement of materials

Small mammals are mobile, and through their movements they transport seeds, organic materials, and chemicals in the ecosystem. While we expect that many seeds are transported by animals, the authors are unaware of any quantitative data showing the role of small mammals in this process. The significance of such transport is also unknown. Caches of seeds collected by mammals may result in clumps of seedlings and mature plants, as observed by West (1968) in Oregon. He estimates that as many as 50 per cent of bitter-bush and 15 per cent of ponderosa pine seedlings may result from rodent caches. Presumably, the pattern of tree spacing, location of seed in sites favoring germination, and improvement of germination might be of advantage to the plants.

The movement of chemical materials is even more subtle and difficult to examine. Ants and termites accumulate elements near their colonies (Gentry & Stritz, 1972) and it is suspected that small mammals also do this. Dinesman (1967) mentions the luxurious flora around mammal burrows which could result from enrichment of the soil, or concentration of seeds at one location, or both. Also, within a colony of *Marmota* spp. the plants were observed by Zimina *et al.* (1970) to be taller. The same observations were reported by Formozov & Voronov (1939) within colonies of *Lagurus lagurus*; here plants such as *Stipa*, *Festuca* and *Artemesia* were larger. Finally, Kucheruk (1963) observed that *Xanthium strimarium* L. within and outside of a colony of *Meriones tamariscinus* was strongly influenced by the presence of the rodents; as shown in Table 10.6.

Kucheruk (1963), after Richards, points out that feces of rodents influence growth of *Azotobacter* in soils. In mountain pastures in the region of Tian Shan and Pamir about 5 to 15 per cent of the area is characterized by changed plant cover because of the activity of *Marmota* spp. (Zimina *et al.*, 1970). Movement of chemicals may also be accelerated by cutting leaves, stems and fruits which then act as input to

Table 10.6.

	Within colony			Outside colony		
	Mean	Minimum	Maximum	Mean	Minimum	Maximum
Number of stems	5.5	1.0	14	2.2	1.0	7.0
Height (cm)	193.5	160.0	220	116.5	95.0	145.0
Maximum diameter of stem (cm)	3.3	1.8	5.0	2.0	1.2	2.7

litter. Stark (1973) describes the effect of *Tamiasciurus* on the Jeffrey pine ecosystem as an increased litterfall of 18 $g/m^2 \cdot yr$ which adds 137 g Ca, 284 g Na, and 186 g K per quarter hectare to the litter component. This effect may be important in a number of forested systems. Dinesman (1967) also describes how salt is leached from mounds in saline soil with a consequent improvement in the soil's ability to support plant growth. These chemical effects could be of special significance in systems with low fertility and should be more thoroughly investigated.

Alteration of the habitat and other impacts

It is also well known that small mammals alter the substrate through digging, construction of burrows, tunnels, runways and other constructions. These actions not only enrich the substrate as mentioned above, but also may enhance water movement. In both cases, the role of the mammal is to improve growth conditions for the vegetation. Voronov (1953) has made some quantitative estimates of burrowing activity and structure of nests for a variety of mammals (Table 10.7). The amount of

Table 10.7.

Species	Length of corridor (cm)	Volume of nest (cm³)	Amount of removed soil (m³)
Mole	6000	2096	0.077
Badger	4600	398000	2.896
Fox	2700	440000	2.103
A. sylvaticus	100	351.2	0.0012
A. flavicollis	100	2280	0.0043
C. glareolus	95	283	0.0011
A. terrestris	1500	4886	0.053
M. arvalis	40	3299	0.0037

soil removed in nest construction obviously can be quite great. Kuznetzov (1970) estimated that 11 to 12 moles (*Talpa caucasica*) dislodge 0.15 m^3 of soil/ha·day in deciduous forests. In mountain pastures of Tian Shan and Pamir, Zimina *et al.* (1970), estimated that marmots (*Marmota* spp.) dislodge 18–120 m^3 of earth/ha. Under laboratory conditions, *Apodemus sylvaticus* may dislodge during two hours 1–3 kg of earth. The weight of the dislodged earth is 50 to 150 times heavier than the body weight of this mouse (Dufour, 1971). *Pitymys subterraneus* may move to the surface of the soil about 12000 kg of earth/ha and the length of their corridors and nests under the surface is about 28 m (Novikov & Petrov, 1953). In a year of peak numbers, *Microtus socialis* may dislodge on the surface 2.1 m^3 of earth/ha, and the dislodged soil covers 0.004 per cent of the whole territory (Hodashova, 1960). Voles at a density of 300 to 400 individuals/ha dislodge up to 10 m^3 of earth/ha during one year from a depth of 10–40 cm. Rodents such as *Ellobius* spp., *Myospalax* spp., and *Spalax* spp., living their whole life in the soil, dislodge 3–15 m^3 of earth/ha·yr, while rodents such as *Marmota* dislodge 18–120 m^3/ha·yr (Zimina, 1970).

Dislodged earth is one of the factors influencing microrelief of the soil which in turn influences surface water runoff. Small mammals which dislodge the earth with the help of the head (*Talpa, Spalax*) make small circular embankments with the hole in the middle. Voles, using the hind legs to dislodge the earth, make semilunar embankments. These different shaped mounds influence flow of surface water. Due to water sinking into the holes, the moisture of the soil close to the burrow is higher than at the same depth in burrowed soil. Kaczmarczyk (unpublished data of the Department of Agroecology at Turew, Poland) found the following percentage positive differences in moisture due to burrows of *M. arvalis* in alfalfa fields, compared to the moisture of unaffected soil at the same depth:

1 day after rain = 1·5 per cent
4 days after rain = 1·1 per cent
20 days after rain = 0·8 per cent.

Rodents may also influence the soil volume and density. Within a colony of *Citellus pygmaeus* the volume weight of dry steppe (solonetz) soil at the colony was 1.29 and in unaffected soil 1.44. Because of the decrease in volume weight of the soil, it is friable and the water current is increased from 0.27 mm/min in unaffected soil to 5.07 mm/min within the colony (Abaturov & Zubkova, 1969). However, the most friable soil is in new colonies, while in very old colonies the soil is even less friable than is unaffected areas.

Movement of soil by rodents may also have an impact on the distribution of chemicals in the system. In dry steppe *Citellus pygmaeus* (sousliks) dislodge 1 500 kg of soil/ha·yr from depths of 40–200 cm. In this mass of dislodged soil, Abaturov *et al.*, (1969) found: SO_4, 14 kg/ha; Na, 5 kg/ha; Cl, 3 kg/ha; Ca, 77 kg/ha; Al, 122 kg/ha; Fe, 49 kg/ha; S, 18 kg/ha. Because of very deep ground water tables, there is no capillary water uptake and burrowing activity of sousliks is the main source of renewal of minerals in the upper layers of the soil (Abaturov *et al.*, 1969). In spruce forests, moles dislodge up to 19 000 kg of earth/ha during one year from depths of 10–40 cm. The following chemicals were found: Si, 47 kg; Fe, 47 kg; Al, 139 kg; Ca and Mg, 36 kg (Abaturov, 1972).

Because of burrowing activity, small mammals mix horizons of the soil profile. An increase of organic matter below horizon A of the soil profile due to burrowing activity of rodents was observed by Kucheruk (1963). In *Spalax* colonies the humus horizon may be extended by 20 cm in depth. At the same time, the content of Ca and Mg in humus horizons is increased, which, in turn, influences pH conditions. Also, changes have been observed in the ratio of humin and fulvo acids, indicating an increase of decomposition processes in the soil due to rodent activity (Hodashova, 1970).

Interaction with other consumers and predators

Small mammals feed on and are fed upon by a variety of other consumers. Insects fall into the appropriate size range which can be utilized by small mammals and form an alternative food to seeds. In certain instances, insect populations may be under heavy pressure by this type of predation. For example, Buckner (1964, 1966*b*) describes shrew predation on larch sawfly larvae. Shrews locate larvae by smell and can consume a major proportion of the larval population. Small mammals can also act as a more available food to carnivores. For example, Ryszkowski *et al.* (1973), showed that hares and *M. arvalis* played reciprocal roles in food of foxes. When voles comprised a major portion of the foxes' diet, hares were less important.

In these instances, the role of small mammals is interpreted in the context of the consumer complex rather than as feedback to other components or the environment. For system stability, it is necessary that the consumer complex be internally stable in order to effectively control other system components. We have restricted our discussion of the role of small mammals mainly to the impact on producers or the substrate but their role within the consumer complex may also be of considerable significance.

Conclusions

The impact of small mammals on vegetation and consumer populations is well known to biologists in a qualitative way, through observations of destruction or severe damage to the other components of the community. Further, the rates of flow of energy and matter to mammals in the consumed foliage or seeds is relatively well known through the research of IBP. We know very little of the impact of these flows on the donor compartments, nor can we accurately quantify the feedback flux directly. Thus, on a purely descriptive level our knowledge of the impact of small mammals on the community is very limited. This lack of knowledge not only concerns the actual feeding or action of the mammal, but also the quantitative response of the vegetation to the feeding.

However, in spite of the meager published information we will propose several hypotheses concerning the role of small mammals in temperate ecosystems. First, in terms of material and energy transfer, small mammals are components of ecosystems which, when compared with invertebrates, have a high expenditure of energy per unit of biomass produced. For this reason, small mammals are efficient in the processes of mineralization of organic matter and much less efficient in the storage processes of the ecosystem. Second, side effects of feeding such as destruction or wasting of food may be equal to or more important than direct consumption. Plant debris left on the surface or transferred into burrows may influence such diverse factors as soil water storage and soil temperature. The results of burrowing activity are even more striking since the chemicals transported by these movements can be an important source of nutrients, especially under dry conditions.

Considering the characteristics of the ecosystems under survey, we expect that the greatest impact of rodents will be in grassland ecosystems. In this system, decomposition of the annual primary production is one of the key ecosystem processes. Accumulation of organic matter in the litter can depress production by tying-up nutrients in the inactive biomass. The bioenergetic character of small mammals with a high ratio of respiration to production makes their impact analogous to fire and outbreaks of rodents under these conditions function as part of the control mechanism of mineral cycling in the ecosystem. In this context, the meaning of control or stability at the population level may be quite different from that at the ecosystem level of organization.

The potential role of small mammals in temperate ecosystems is diagrammed in Fig. 10.3. In very intensive agriculture, there is no

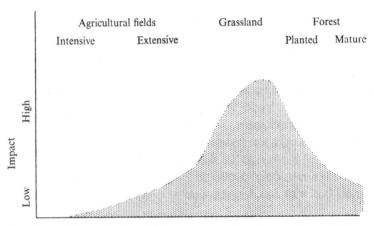

Fig. 10.3. Expectation of the impact of small mammals on temperate ecosystems.

suitable habitat for small mammals and their impact is essentially zero. For example, in Belgium the last widespread outbreaks of *Microtus arvalis* were observed in 1937, 1945/6 and 1948. Since then, no widespread outbreaks have been observed (Bruel, 1969) due to the energetic control operations associated with intensive agriculture, and very high densities have been observed only in small areas of poorly managed, extensive pastures (I. Tahon, personal communication).

At the other extreme, in mature forest the impact of rodents on the system is negligible. The large storage of nutrients in the tree biomass, the uptake through the well-developed root system, and the control of the microclimate all result in rodents and other small mammals having a less significant direct impact. Rather, in this system small mammals have their greatest impact in the consumption and destruction of seeds and seedlings.

Further research is needed for an adequate description of the role of small mammals in temperate ecosystems. We need to know the frequency distribution, abundance and the time pattern of foods, seeds, insects and foliage available to the mammals and the chemical and physical

240

structure of these foods. To complete the analysis, information is also needed on the feeding behavior of the mammal, the size and shape of food particles that can be consumed and the time and pattern of searching and feeding. With these data, we will be able to construct hypotheses which can be tested by field experiments utilizing enclosures or other techniques. These studies still will not answer the questions about the mutualistic response of mammal and vegetation or mammal and predator nor give insight into how such interactions evolve and develop. For this analysis, we need detailed study of the consequences of feeding and activity over time in the context of the plasticity of response of the plants and animals. This latter analysis can only be done with both the action and feedback considered at the same time, since the impact is to both compartments. It is evidence of how much we have learned through IBP that this research strategy which now seems so obvious, was not adopted by IBP six years ago.

The authors wish to thank the participants at the synthesis meeting for their many constructive comments concerning this paper. Ken Ashby and Robert Chew were especially helpful in pointing out several logical problems and providing additional examples of the impact of small mammals.

11. The role of small mammals in arctic ecosystems

G. O. BATZLI

Few regions of earth have been so romanticized as the Arctic. Nearly everyone knows of the tribulations of early arctic explorers and of the cultural wisdom of the Eskimos which enabled them to survive rigorous environmental conditions, and nearly everyone has heard tales of the mass movements of lemmings. Although there is often less fact than fancy in the stories told, one fact is clear; arctic ecosystems are cold dominated.

It may seem trivial in a scientific paper to call attention to such an elementary fact, but it is the length and severity of the cold season that determines a great number of characteristics of arctic ecosystems (Sater, 1969). Indeed, the definition of the Arctic is usually based upon temperature patterns; for instance, the region of the Northern Hemisphere where the mean temperature during the warmest month of the year (usually July) is less than 10 °C (Britton, 1957; Dunbar, 1968). The biological significance of this isotherm rests upon its rough correspondence with the northern boundary of coniferous forests. For the purposes of this review then, arctic ecosystems include cold dominated, circumpolar terrestrial regions with vegetation of low stature; regions usually called tundra.

Because of other factors, particularly elevation and proximity to the sea, arctic tundra does not always occur at the same latitude. Nevertheless, it usually occurs north of the Arctic Circle, and has several physical attributes besides low temperatures which also constrain biological activity. Day length varies from 0 hours in the winter to 24 hours in the summer. While not intense, solar radiation during summer is continuous, and it melts the winter snow mantle and the upper layer of soil. The resulting surge of net primary production is comparable to that for herbaceous communities in temperate regions (0.5–3.0 g/m²·day), but the annual rate of production is low owing to the short growing season (Bliss, 1962). Below the shallow active layer of soil lies a permanently frozen zone (permafrost) which may be up to 500 m deep and can dramatically influence morphological features of the surface (Black, 1954; Péwé, 1957). Finally, although cold weather and hard-packed snow prevent many vertebrates from inhabiting the tundra during winter, it is the insulative property of snow that allows small mammals to remain active (Formozov, 1946; Pruitt, 1970).

The main objectives of this chapter are to evaluate and to document the importance of small mammals in arctic ecosystems by examining their roles.

Evaluation of roles

One need only consider the position of small mammals in arctic food webs to suspect that they will be influential members of the ecosystem. Several authors have diagrammed tundra food webs (Pitelka, 1957*a*; Tikhomirov, 1959), and small mammals are in the center of the web (Fig. 11.1). Small mammals consume vegetation and soil arthropods and

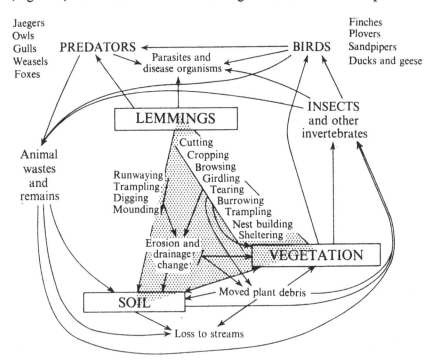

Fig. 11.1. Diagram of food web and lemming activities in coastal tundra of northern Alaska (from Pitelka, 1957*a*).

are in turn consumed by avian and mammalian predators. Other activities of small mammals, such as burrowing, establishing trails and depositing waste materials, also affect the vegetation and soils.

244

Clearly then, small mammals influence three major components of arctic ecosystems; the soil and vegetation on which they depend for food and shelter and the predators for whom they are a resource. Any evaluation of the significance of small mammals must be concerned with the degree to which the characteristics of these important components are interdependent. Insofar as these relationships can be stated quantitatively, our understanding of ecosystem dynamics will be enhanced.

Because information for any one site is hopelessly incomplete, this review will attempt to integrate many studies done at different sites. Particular emphasis will be placed upon habitat selection, population dynamics, nutrition and energetics. If sufficient information is available on these features of small mammal populations, the impact of small mammals on their environment can be estimated. In order to facilitate the process, groups of closely related species will be considered together.

Distribution of arctic small mammals

Four groups of small mammals occur regularly in arctic tundra; microtine rodents, sciurid rodents, hares and shrews. Unfortunately, their distribution does not always correspond to the distribution of tundra (Ognev, 1928–50; Hall & Kelson, 1959). Those forms which are tundra-specific evolved in central Eurasia during the early Pleistocene and migrated to North America (Hoffmann & Taber, 1968).

Macpherson (1965) analyzed the geographic distribution of Canadian arctic mammals and concluded that the only small mammals which were tundra-specific and widespread before the last glaciation (Wisconsin) were the arctic hare (*Lepus arcticus*) and the varying (collared) lemming (*Dicrostonyx groenlandicus* [= *torquatus* ?] and *D. hudsonius*). These forms are still largely confined to the arctic tundra, as is the hoofed lemming (*D. torquatus*) in Eurasia (Ognev, 1948, vol. 6). The mountain hare (*L. timidus* [= *othus*]), however, is closely associated with woody vegetation and usually only reaches the southern parts of the tundra (Ognev, 1940, vol. 4; Pavlenen, 1971).

Before the Wisconsin glaciation, brown lemmings (*Lemmus trimucronatus* [= *sibiricus* ?]) and arctic ground squirrels (*Spermophilus* [= *Citellus*] *undulatus* [= *parryii*]) were tundra-specific but confined to Eurasia and far western parts of North America. Now both occur in arctic and alpine tundra on the mainland west of Hudson Bay, and *Lemmus* occurs in the Canadian archipelago east to Baffin Island (Macpherson, 1965). Modern Eurasian relatives *Lemmus lemmus*,

245

Small mammals

Lemmus obensis (= *sibiricus*), and *Spermophilus undulatus* show similar habitat affinities.

The masked shrew (*Sorex cinereus*) and the red vole (*Clethrionomys rutilus*) were primarily woodland forms before the Wisconsin and still are, although they also inhabit portions of the tundra. *Clethrionomys rutilus* is holarctic and joined by a congener *C. rufocanus* in western Eurasia, but the form of woodland shrew which reaches Eurasian tundra is *S. araneus*. Finally, Macpherson (1965) notes that two other species, the tundra vole (*Microtus oeconomus*) and the arctic shrew (*S. arcticus* [= *tundrensis*]) first inhabited North American tundra at the end of the Wisconsin. In Eurasia *M. oeconomus* [= *ratticeps*] has a much wider distribution reaching into temperate regions. *Sorex arcticus* also occurs in Eurasian tundra but the taxonomy of this and closely related shrews is uncertain (Hoffmann, 1970; Meylan & Hausser, 1973).

Three other forms of *Microtus* not mentioned by Macpherson penetrate the arctic tundra; *M. miurus* in Alaska and *M. gregalis* (possibly conspecific with *M. miurus*) and *M. middendorffi* in Siberia (Bee & Hall, 1956; Shvarts, 1971). Even though other small mammals sometimes go into the arctic tundra they are usually uncommon and associated with exceptional circumstances, such as marmots (*Marmota caligata*) occuring in rocky areas (Shvarts, 1963; Childs, 1969).

Only species which regularly occur in tundra will be considered for this review. Obviously, more is known about some than others and the review will be uneven. Microtine rodents will be emphasized because they reach the highest densities and receive more study. Coverage of the literature from the USSR will be incomplete owing to difficulty in obtaining translations.

Habitats and population dynamics

The intensity of relationships between small mammals and their environment is directly related to their density and activity patterns. Quantitative estimates of impact will be no better than our knowledge of population dynamics and dispersion. Unfortunately, most of the work on tundra populations has not been done using mark–recapture techniques, and few investigators estimated absolute densities. Never-the less, a large literature about tundra rodents, particularly microtines, exists and can be usefully summarized.

Arctic microtine populations display periodic increases and declines (cycles) on a 3- to 4-year basis as do their relatives at lower latitudes

246

(Krebs & Myers, 1974). The causes of these cycles remain obscure and controversial, and no attempt to analyze the causes will be made here. Rather, the main characteristics of the populations will be described so that inferences may be drawn about their roles. Three features – population density, reproductive activity and population structure – will be stressed because these will determine the intensity of microtine activity.

Lemmus

Members of the genus *Lemmus* reach the highest densities of all tundra rodents (Table 11.1). Some taxonomic work (Sidorowicz, 1960*b*, 1964) relegates all forms of *Lemmus* to a single species, but the ecological circumstances of the forms differ.

The Norwegian lemming (*L. lemmus*), because of its occasional massive emigrations from alpine habitats into agricultural regions, is mentioned in historical chronicles dating from the Middle Ages (Marsden, 1964). Ecological studies of this form date back to Collett (1878, 1895). As Elton (1942) pointed out, Collett first suggested that emigration resulted from over-population of normal habitats, an idea supported by modern work. Elton also reviewed much of the older literature on cyclic fluctuations of microtine and other wildlife populations.

Movement patterns of Norwegian lemmings have been described in detail at locations in Norway (Clough, 1968) Sweden (Curry-Lindahl, 1962; deKock & Robinson, 1966; Bergström, 1967), Finland (Kalela *et al.*, 1961; Koponen *et al.*, 1961; Myllymäki *et al.*, 1962; Aho & Kalela, 1966; Kalela *et al.*, 1971), and the USSR (Koshkina, 1962). Two kinds of movements occur, both of which involve changes in habitat. Seasonal changes over short distances occur every year, but the longer emigrations are associated with high density populations only. Details of the longer movements will not be considered here because Norwegian lemmings move far from the arctic-alpine tundra.

Kalela *et al.* (1961, 1971) describe seasonal changes in habitat, and the following account is taken largely from their work. Vegetation on the mountains of northern Europe changes rapidly with elevation. Alpine fells, heaths and snow beds at the highest elevations give way to bands of willow thickets, birch forest and coniferous forest at progressively lower elevations. Except during times of high population density, lemmings are restricted to habitats at the upper edges of the birch forest and above.

Small mammals

Studies of foraging patterns revealed that winter activity is concentrated under good snow cover in alpine habitats, particularly snow beds (*Deschampsia–Anthoxanthum* type, *Salix–Cassiope* type and *Ranunculus* type), grass–sedge heaths (*Juncus–Festuca* type and *Carex* type) and dwarf shrub heaths (*Phyllodoce–Vaccinium* type). These habitats flood when the snow melts, and the lemmings move into areas with better drainage, especially willow thickets and peatlands. Summer activities are concentrated in moist areas with fresh vegetation, including a well developed layer of bryophytes, throughout the season. The lemmings favor areas where thickets (*Salix–Betula*), bog (*Sphagnum*), fen (mixed bryophytes) and brookside meadow (*Ranunculus acris* type) intermingle. Regions with damp moss carpets in the upper birch forest may also be heavily utilized, and may even be the optimal habitat when populations are low (F. A. Pitelka, personal communication).

The ecological situation of *Lemmus obensis* and *L. trimucronatus* differs from *L. lemmus* in several ways. Although the southern portions of their ranges reach into alpine tundra, these forms are primarily inhabitants of coastal plains where grasses (*Arctophila*, *Dupontia*) and sedges (*Carex*, *Eriophorum*) dominate the wet lowlands and polygonized ground, and decumbent shrubs and forbs (*Salix*, *Dryas*) are important on ridges and hummocks (Bee & Hall, 1956; Britton, 1957; Pitelka, 1957a; Tikhomirov, 1959; Schamurin *et al.*, 1972).

Seasonal movement patterns of *L. obensis* and *L. trimucronatus* resemble those of *L. lemmus* and appear to be responses to patterns of flooding and snow depth. In winter the animals congregate in lowlands, particularly polygon troughs and meadows where snow depth is greater. At melt-off these areas become flooded and lemmings are more concentrated in uplands. As the tundra dries relative densities increase in the lowlands, apparently because these habitats contain the highest densities of preferred foods (Batzli, 1973). No long emigrations have been witnessed for these forms, although unusual activity may occur during some peak years (Thompson, 1955a).

There are few good estimates of the absolute densities of *L. lemmus* in its normal habitats. Data from northern sites suggest maximal densities of 100–250/ha (Table 11.1) while minimal densities are very low, perhaps 0.1/ha. At southern sites, alpine habitats are extremely patchy and are shared by several species of microtines which may outnumber *Lemmus* in some habitats. Overall peak densities of *Lemmus*, calculated by weighting the densities in various habitats by the amount of habitat available, may be only 25/ha.

Table 11.1. *Maximal densities estimated for microtine rodents in tundra habitats*

Species	Density (no./ha)	Location	Source
Lemmus lemmus	25†	S. Norway	Hagen, Skar & Östbye (unpublished results)
	150†	Central Norway	Clough (1968)
	100–140†	N. Finland	Kalela *et al.* (1971)
	250	Kola Penninsula, USSR	Koshkina (1970)
L. obensis	700–2000	Yamal Penninsula, USSR	Karaseva *et al.* (1971)
	100–350	Taimyr Penninsula, USSR	Vinokurov *et al.* (1972)
L. trimucronatus	90–175	N. Alaska	Thompson (1955*b*)
	75–175	N. Alaska	Schultz (1969)
	100–200	N. Alaska	Maher (1970)
	75	N. Canada	Krebs (1964)
Dicrostonyx groenlandicus	2–3	Devon Island, Canada	Speller (1972)
	20–30	N. Alaska	Feist (unpublished results)
	1–30	N. Alaska	Pitelka & Batzli (unpublished results)
	25	N. Canada	Krebs (1964)
	35–40	S. Hudson Bay	Shelford (1943)
	25	S. Hudson Bay	Brooks (1970)
D. torquatus	250	Taimyr Peninsula, USSR	Vinokurov *et al.* (1972)
Clethrionomys rutilus	75–110*†	Kola Penninsula, USSR	Koshkina (1965)
C. rufocanus	30*†	N. Finland	Kalela (1957)
	5†	S. Norway	Hagen, Skar & Östbye (unpublished results)
Microtus oeconomus	15†	S. Norway	Hagen, Skar & Östbye (unpublished results)
	160*	N. USSR	Schwarz *et al.* (1969)

† Density calculated for this paper from data of authors cited.
* Estimates for subarctic forest–tundra or taiga.

More density estimates have been made for *L. trimucronatus* in northern Alaska than for any other region, although some of these were based upon concurrent observations (Table 11.1). The range of maximal densities is 75–200/ha, and minimal densities again may be as low as 0·1/ha. In this region, peak densities are usually reached just before snow melt after which populations decline because of widespread destruction of habitat and the intense predation which follows (Pitelka, 1973). Preliminary results relating data from Pitelka's snap trapping to live trapping data indicate average peak densities of 100–150/ha over all habitat types.

A recent report (Karaseva *et al.*, 1971), of peak densities for *Lemmus obensis* is considerably higher than that reported for another location, 700–2000/ha as against 100–350/ha (Table 11.1). The higher density was based upon extrapolation from a series of 100 m² sites to one hectare which produced values an order of magnitude greater than other estimates for the genus. Because lemmings routinely move 30–50 m between captures in live traps, it seems likely that Kareseva *et al.* were sampling much larger areas than 100 m² and greatly overestimated the densities.

All *Lemmus* spp. appear to have a winter and a summer reproductive season (Table 11.2). Procuring specimens for analyses is extremely

Table 11.2. *Summary of breeding seasons for microtine rodents in arctic and subarctic habitats during peak years. Some data estimated to to nearest 5 per cent*

	Percentage pregnancies in mature females†					
Species	Dec–Feb	Mar–Apr	June 1	July 15	Sept 1	Source
Lemmus lemmus	?	+	10	77	12	Koshkina & Khalansky (1962)
	?	+	70	57	0	Koponen (1970)
L. trimucronatus	?	+	0	80	0	Krebs (1964)
	33	29	0	70	0	Mullen (1968)
L. obensis	+	+	?	57	?	Shvarts *et al.* (1969) Karaseva *et al.* (1971)
Dicrostonyx	+	+	0	90	0	Krebs (1964)
groenlandicus	?	+	10	85	0	Speller (1972)
Clethrionomys *rutilus*	—	—	75	40	0	Koshkina (1965)
C. rufocanus	—	—	90	69	0	Kalela (1957)
M. oeconomus	—	—	60	58	0	Tast (1966)

† Because only ¾ of the pregnancies are usually detected (the ova may not implant until the 5th day) a value of 75% or above indicates almost all females were pregnant.
+ Indicates animals pregnant but no data on the percentage pregnant.

difficult during winter, so much of the evidence is circumstantial, based upon age distributions in the spring or changes of population density over winter. One study in which samples were obtained throughout the winter indicated fairly continuous breeding by about 40 per cent of the females from December through April (Mullen, 1968). All breeding ceases at times of snow melt (spring) or freezing (autumn). Apparently these are difficult times for lemmings, owing to cold without a protective snow mantle in the autumn and flooding in the spring (Fuller, 1967). Maximal breeding occurs during early or mid-summer when nearly all the mature females are pregnant (Table 11.2).

Obviously, the changes in breeding patterns affect the age structure of the populations. On the basis of size, growth rate, moult and behavior, lemmings can be divided into four age classes (Table 11.3). The nestlings

Table 11.3. *Criteria for age classes of lemmings* (Lemmus *and* Dicrostonyx). *Values given are means and based upon laboratory or field observations* (*Thompson, 1955b; Bee & Hall, 1956; Quay & Quay, 1956; Hansen, 1957; Koponen, 1970; Batzli et al., 1974*)

Age class	Age (days)	Weight (g)	Growth rate (g/day)	Moult	Behavior
Nestling	0–14	3–13	0.8	Juvenile pelage develops	Leave nest and weaned at end of period
Juvenile (J)	15–29	14–28	1.0	Moult into post-juvenile pelage	Independent, but not usually reproductive
Subadult (SA)	30–59	♀29–39 ♂29–44	0.4 0.5	Moult into pre-adult pelage	Usually reproductive, but not at maximal rate
Adult (A)	60+	♀40+ ♂45+	Slower and more erratic	Moult into adult pelage	Maximal reproductive rates

do not leave the nest and have not been sampled systematically, so they are not considered in the discussion of age structure.

The best way to observe the influence of pauses and resurgence of breeding on the age structure of populations is to compare age structure at the beginning and the end of a breeding season. The only suitable data which exist for arctic populations pertain to the summer breeding season (Table 11.4). Although scanty, the data are consistent, and a clear pattern emerges. At the beginning of the breeding season, after a pause at snow melt, most of the *Lemmus* populations are adult, a structure typical of declining populations. By the end of the breeding season most of the populations are juveniles or subadults, a structure typical of a growing population. These seasonal trends probably occur every year and are superimposed on the long-term population dynamics of the 3- to 4-year cycle.

Small mammals

Table 11.4. *Population structure of microtines in tundra habitats given as percentage in three age classes at snow melt (May–June) and freezing (September). Some data estimated to the nearest 5 per cent because they were taken from graphs. Numbers in parentheses indicate percentages of males and females in population*

Species	Spring J	Spring SA	Spring A	Autumn J	Autumn SA	Autumn A	Source
Lemmus lemmus ♂♂	0	6	94	27	57	16	Koshkina & Khalansky
♀♀	1	10	89	30	44	26	(1961)
♂♂	0	32	68	45	41	14	Koponen (1970)
♀♀	2	39	59	33	41	26	
Lemmus trimucronatus ♂♂ (55)	5	35	60	(55)			Pitelka (1957b)
					89	11	
♀♀ (45)	10	45	45	(45)			
♂♂	15	11	74	24	56	20	Krebs (1964)
♂♂ (45)	2	15	83	(47) 75	15	10	Batzli (unpublished results)
♀♀ (55)	2	16	82	(53) 62	18	20	
Dicrostonyx groenlandicus ♂♂	21	5	74	20	52	28	Krebs (1964)
♂♂ & ♀♀	22	29	49	70	18	12	Speller (1972)

Dicrostonyx

From nearly every point of view, the collared or hoofed lemmings of the genus *Dicrostonyx* show the greatest degree of adaptation to arctic winters of all the small mammals (Irving, 1972). Not only do these species occur earliest in the tundra fauna of North America (pre-Wisconsin) and Eurasia (Cromerian), but also their range extends into the high arctic where other small mammals do not occur (Hoffmann & Taber, 1968; Hall & Kelson, 1959). Morphological adaptations include shortened tails, lack of pinnae and, in winter, enlarged claws and white pelage. In addition, collared lemmings show a remarkable resistance to hypothermic collapse (Ferguson & Folk, 1970).

The ranges of *Lemmus* and *Dicrostonyx* are largely overlapping, except in western Europe and the high arctic (Ognev, 1948, vol. 6; Hall & Kelson, 1959). As might be expected, the species are ecologically separated. *Dicrostonyx* prefers upland heaths, higher and drier tundra, where forbs and woody vegetation are more common (Watson,

1956; Pitelka, 1957a; Krebs, 1964). Even in areas where *Lemmus* is not present and *Dicrostonyx* moves into grass–sedge lowlands, densities are highest on ridges and high center polygons (Brooks, 1970; Speller, 1972). These habitat preferences seem to be related to dietary requirements (dicot leaves) and the need for dry burrow sites.

Maximal densities for *Dicrostonyx groenlandicus* usually fall well below those for *Lemmus* (Table 11.1). In the high arctic, densities may only reach 2–3/ha in favorable habitat, while at lower latitudes peak densities may be an order of magnitude higher. The higher densities may only be reached occasionally, at some locations perhaps once in 20–30 years (Pitelka, 1973). One report indicated that *Dicrostonyx torquatus* sometimes reaches densities comparable to *Lemmus*, but the method of estimation was not described (Vinokurov *et al.*, 1972).

Breeding seasons for *Dicrostonyx* are similar to *Lemmus* in that a summer and a winter season are separated by pauses in late spring and early autumn (Table 11.2). Data from age distributions (more juveniles) suggest that spring breeding in *Dicrostonyx* lasts later into May than in *Lemmus* (Table 11.4). During the height of summer breeding, nearly all females are pregnant. Shifts in population structure from predominantly adult individuals in the spring to predominantly young in the autumn parallel those described for *Lemmus*.

Other microtines

When *Clethrionomys* and *Microtus* do occur on the tundra, it is usually at more inland or southern regions, and their distributions overlap those of lemmings. In North America *Clethrionomys*, like *Dicrostonyx*, is most abundant in well-drained habitats, particularly those containing shrubs. *Microtus oeconomus* prefers low, wet areas dominated by grasses and sedges, perhaps even more so than *Lemmus*. *M. miurus* seems to require both dry soil and nearby water and is commonly found on stream banks and lake shores, particularly where willows are spaced so that an understory occurs (Bee & Hall, 1956; Pitelka, 1957a). In some regions of Alaska, such as the tundra of the interior north slope (Umiat) or the western coast (Cape Thompson), all five species of tundra microtines occur (Pitelka, 1957a; Pruitt, 1966).

In alpine regions in northern Europe several species of microtines overlap the normal habitat of *Lemmus lemmus*, and the species composition varies. In 1970 populations of *Lemmus*, *M. oeconomus*, *C. rufocanus*, and *C. glareolus* peaked in an alpine region of southern

Small mammals

Norway (Hagen, Skar & Östbye, unpublished results). Alpine meadows and thickets of northern Finland harbor *Lemmus*, *M. oeconomus*, *M. agrestis* and *C. rufocanus* populations (Tast, 1968). In forest–tundra of the Kola Penninsula, USSR, *C. rufocanus*, *C. rutilus* and *L. lemmus* all occur (Koshkina, 1966), while *L. obensis*, *D. torquatus*, *M. middendorffi*, *M. gregalis*, *M. oeconomus* and *C. rutilus* all regularly inhabit southern tundra and forest–tundra of the Yamal Penninsula (Pjastolova, 1972).

Most of the population studies of microtines other than lemmings have not been done in tundra but in the transitional forest–tundra or in the taiga where densities of some species appear to be higher (Table 11.1). There are few density estimates for tundra. The distributions of these species in Eurasia are complex, and they respond to one anothers' densities in a manner suggesting competitive exclusion (Linn, 1954; Schwarz, 1963; Tast, 1968). Perhaps for these reasons, I found no clear exposition of differences in habitat preference where these microtines intermingle, although most Eurasian forms show preferences similar to their close North American relatives.

A major difference between lemmings and other arctic microtines seems to be the ability of lemmings to breed in the winter. Apparently only *M. gregalis* and *M. middendorffi* begin breeding before May (Shvarts *et al.*, 1969, Table 11.1). Autumnal populations, made up largely of juveniles and subadults from the late summer breeding, over-winter but grow very little until spring when both growth and reproduction commence. The overwintering population, which composes the whole population in May, is rapidly replaced by younger animals during the summer and may form only 30 per cent or less of the population by mid-July. A similar pattern has been observed for *Microtus oeconomus* in Alaska (F. A. Pitelka, personal communication).

Other small mammals

Only one of the other arctic small mammals seems to have had much work done on its population dynamics in the tundra; the arctic ground squirrel *Spermophilus undulatus*. Because they require elaborate burrows and maintenance of visual contact, permanent ground squirrel colonies are restricted to well drained sites with low vegetation, e.g., glacial moraines and knolls in alpine regions and stream banks and old beach ridges in arctic tundra (Mayer, 1953; Bee & Hall, 1956; Tikhomirov, 1959; Murray & Murray, 1969).

254

Densities within permanent colonies of squirrels appear to be stable owing to territorial defence of suitable burrows by males and females. Carl (1971) reported densities of 41 and 49 adults for two successive Junes in a breeding colony on a 30 ha site. By October, when hibernation began, the adult populations had declined to 24 and 20, but many of the young produced in July remained (91 out of 141 for the first year and 43 out of 180 for the second). Many of these young overwinter on the breeding colony site, but most were expelled after emerging the following May. Thus, in June the population was 100 per cent adult, but from July through October only 20–30 per cent adult.

Arctic hares (*Lepus arcticus* and *L. timidus*) have seldom been studied in North America because of their general scarcity (Bee & Hall, 1956; Aleksiuk, 1964; Bergerud, 1967), although large herds consisting of hundreds of individuals, have been reported on Ellesmere Island (W. A. Fuller, personal communication). Detailed population studies have not been done on *L. timidus* in Eurasian tundra either, but large herds of 100–500 have also been reported occasionally (Tikhomirov, 1959). Because hares are more common in taiga and alpine situations, most of the ecological observations have been made there, where densities may reach 0.25/ha. (Nyholm, 1968; Pavlenen, 1971).

The literature on arctic shrews is similar to that on hares; most is concerned with taxonomy and distribution. *Sorex arcticus* and *Sorex cinereus* occur in most habitats of more inland and more southern tundra of North America but at lower densities than microtines (Bee & Hall, 1956; Pruitt, 1966). *S. cinereus* occurs in wet meadows or *Eriophorum* tussocks while *S. arcticus* is more characteristic of drier habitats such as hillsides and ridges. Much of the work on *S. arcticus* and *S. araneus* in Eurasia has been directed toward interesting winter adaptations which include a reduction in body size (Mezhzherin, 1964; Hyvärinen, 1969). Nearly all population studies of these shrews have been done in lower latitudes in taiga or dcciduuus forest where densities may reach 10–20/ha (Yudin, 1962; Buckner, 1966a, 1969).

Nutrition and energetics

A detailed knowledge of food habits is a prerequisite for adequate nutritional studies of natural populations. Most studies of food habits in the Arctic have been done indirectly by observing signs of grazing in the field or by testing palatability of food items in the laboratory. Estimates of the percentage of diet contributed by various food items require analyses of stomach contents, and few such estimates exist for arctic small mammals. Whatever the technique, all observations agree that arctic rodents are herbivorous.

Food habits of microtines seem to be related to habitat preference. Those which prefer moist habitats (*Lemmus* spp., *Microtus oeconomus*) eat primarily monocots and mosses while those in drier habitats, where dicots are more abundant, prefer dicots (*Dicrostonyx* spp., *Clethrionomys* spp., *M. miurus*) (Ognev, 1948, vols. 6–7; Bee & Hall, 1956; Tikhomirov, 1959; Kalela *et al.*, 1961; Tast, 1966; Stoddart, 1967; Brooks, 1970).

Analyses of stomach contents indicate that substantial shifts in diet occur with location and with season (Table 11.5). Shifts within a species may be in part a reflection of availability of forage. Thus, *Lemmus lemmus* in forest–tundra, where grasses and sedges are less available, takes more mosses (Koshkina, 1961), and *Lemmus trimucronatus* takes more mosses in winter when monocots are less available (Batzli, 1973).

Dietary composition does not simply mimic vegetational composition, however. When two species live side by side, for instance *Lemmus trimucronatus* and *Dicrostonyx groenlandicus* in upland tundra, their food habits may be extremely different (Table 11.5).

Ground squirrels seem to be more catholic in their tastes than most other small mammals. They readily eat a wide variety of monocots and dicots and may take insects (Mayer, 1953; Tikhomirov, 1959). Arctic hares may take herbs in the summer, but most of their diet consists of the leaves and stems of woody plants, particularly willows (Nyholm, 1968; Pulliainen, 1972). No detailed studies of the food habits of arctic shrews have been done, but, no doubt, they are insectivores like other members of the genus *Sorex* (Buckner, 1964).

Once the dietary composition has been determined, the question of rates of consumption can be addressed. Consumption rates can be observed directly in the laboratory, but more often they have been estimated from a knowledge of energy requirements and the percentage of metabolizable energy in the diets. Metabolizable energy is often equated with assimilated energy, but for technical reasons that usage is misleading (Batzli, 1974). Whatever its name, metabolizable energy can be measured by deducting the energy lost in feces and urine from the energy consumed.

Energy requirements and rates of consumption have been measured for several arctic small mammals, though the animals were usually not from arctic populations. Some results are summarized in Table 11.6.

Table 11.5. *Percentage composition of diet for microtine rodents in tundra as determined by stomach content analyses*

Species	Mosses	Sedges	Food type Grasses	Dicots	Miscellaneous	Source
Lemmus lemmus	*(Dicranum, Pleurozonium, Polytrichum)*	*(Carex, Eriophorum)*	*(Deschampsia, Nardus, Festuca)*	(berries)	(graminoids)	Koshkina (1961)
Tundra with bog May–June	15	55	30	0	0	
Tundra with meadow September	30	0	55	10	5	
Forest–tundra September	60	0	25	10	5	
Lemmus trimucronatus	*(Polytrichum, Dicranum, Calliergon)*	*(Carex, Eriophorum)*	*(Dupontia, Poa)*	*(Saxifraga, Stellaria, Draba)*	(liverworts, lichens)	Batzli (1973)
Upland tundra June	24	30	38	6	2	
Marshy tundra June	14	20	56	7	3	
Low polygons July	24	26	35	13	2	
Low polygons January	38	37	18	5	2	
Dicrostonyx groenlandicus	*(Polytrichum, Dicranum)*	*(Carex, Eriophorum)*	*(Dupontia)*	*(Salix, Potentilla, Dryas)*	(lichens)	Batzli (unpublished results)
Upland tundra June	4	4	8	83	1	

Following Grodziński (1971*b*), average daily metabolic rates have been related to body size (*W*) using a power function with an exponent of 0.5. All of these functions are based upon metabolism under conditions of thermal neutrality. A multiple regression which corrects for temperature conditions is also presented for *Lemmus*.

Comparison of the power functions indicates that after corrections for body size are made, microtines which live primarily on tundra (*Lemmus* and *Dicrostonyx*) have higher metabolic rates than those which live primarily in the subarctic (*Microtus, Clethrionomys*). Shrews, of course, have the highest metabolic rate of all. These high rates may be required to maintain body temperature under arctic thermal regimes. In arctic hares, however, larger body size and low thermal conductance allow maintenance of body temperature with low metabolic rates (Wang *et al.*, 1973). Presumably this would be adaptive in the barren high arctic where food availability can be extremely low.

All but one of the values for metabolizable energy in Table 11.6 were obtained for food which did not represent natural diets in the arctic. The exception is the value for *Lemmus* which is based upon two independent studies (Melchior, 1972; Batzli, 1973) and therefore seems reliable. The low digestibility of food by *Lemmus* may be a peculiarity of arctic monocots. or it may be a peculiarity of the lemming, but microtines at lower latitudes digest monocots much more efficiently. Drożdż *et al.* (1971) report 55 per cent digestibility of grass by *Arvicola terrestris*, and I have found the value to be 50 per cent for *Microtus californicus*. One result of low digestive efficiency is the greatly increased rates of consumption shown in the table for *Lemmus* as compared to other microtines.

As can be seen from the data presented above, *Lemmus* is the most abundant small mammal in much of the circumpolar tundra, and its populations have received the most study. For these reasons, I have calculated an energy budget for *Lemmus* during a peak year. Although *L. trimucronatus* is the form for which I have the best data (Tables 11.1–11.6), most of the data refer to the snowless period of June to September, and much remains to be learned about processes under the snow. Nevertheless, Table 11.7 represents my view of a typical population during a cyclic peak from the end of a growing season (September) through a winter of population build-up, a peak in spring and a decline during the following growing season. Several assumptions were required but all were relatively conservative, so, if there is error, the energy budget will be underestimated.

Table 11.6. *Rates of metabolism and consumption rates for arctic small mammals. ADMR is the function for average daily metabolic rate at temperature shown. DEB is daily energy budget*

Species	ADMR (kcal/day)	Temp. (°C)	Wt (g)	DEB (kcal/day)	Percentage metabolizable energy in diet	Consumption (kcal/day)	Source
Lemmus trimucronatus	$1.28W^{0.75} - 0.45T + 6.40$ or †$3.16W^{0.5}$	15	40	‡ 20.0	33	60	Collier et al. (1974); Melchior (1972); Batzli (1973)
Dicrostonyx groenlandicus	†$2.86W^{0.5}$	17–20	43	‡18.9	66?	29	Morrison & Teitz (1953)
Clethrionomys rutilus	†$2.00W^{0.5}$	22	25	‡10.0	69?	14	Morrison & Teitz (1953)
summer	$2.53W^{0.5}$	15	22	13.5	86	14	Grodziński (1971)
Microtus oeconomus	$2.75?W^{0.5}$	15	28	16.0	80	20	Grodziński (1971)
summer	‡$2.43W^{0.5}$	20	22	14.1	67	21	Gebczynska (1970)
winter	†$1.92W^{0.5}$	20	21	‡11.4	70?	16	Gebczynska (1970)
Spermophilus undulatus	†$2.12W^{0.5}$	22	450	‡45.0	67?	68	Morrison & Teitz (1953)
Lepus arcticus	†$2.23W^{0.5}$	20	3000	‡122.0	—	—	Wang et al. (1973)
Sorex cinereus	$3.74W^{0.5}$	15	3.8	8.4	83	10	Grodziński (1971)

† Calculated for this paper from data of authors cited.
‡ Based upon nonreproductive animals.
? Percentage metabolizable energy computed from data of authors cited.

Table 11.7. *Population density, structure, energetics and consumption (kcal) for* Lemmus trimucronatus *on one hectare during a peak year. See text for discussion of assumptions. PEB is population energy budget*

Month	Mean temp. at ground (°C)	Juveniles		Subadults		Adults		Totals		Population consumption
		No.	PEB	No.	PEB	No.	PEB	No.	PEB	
September	0	27	10368	13	6903	5	4023	45	21294	63882
October	−5	4	1729	20	11687	6	5279	30	18695	56085
November	−10	0	0	8	4788	12	10638	20	15426	46278
December	−15	5	2511	7	5103	13	14384	25	21998	65994
January	−20	11	6969	10	7434	14	15990	35	30393	91179
February	−20	20	9688	15	10318	15	15268	50	35274	105882
March	−20	32	17162	21	15553	17	19576	70	52291	156843
April	−15	50	24300	30	20514	20	21078	100	65892	197676
May	−10	53	24725	60	37107	37	33967	150	95799	287397
June	0	5	1920	25	14973	70	68106	100	84999	254997
July	+5	7	2539	5	3398	38	42014	50	47951	143853
August	+5	5	1817	5	3398	15	15630	25	20845	62535
Totals			103728		141176		265953		510857	1532571

The densities and age structure of the population are conjecture for October through April, but it is conjecture based upon knowledge of the situations in September and May and some knowledge of the timing and intensity of winter breeding (Mullen, 1968). Breeding is assumed to occur from December through April, during which time 40 per cent of adult females are reproducing, and from June through August, reaching a peak of 100 per cent of adult females reproducing in July. Breeding intensity for subadult females is assumed to be half that of adult females (Batzli, unpublished results).

Mean temperature values at ground level were based upon extrapolations from normal weather at Barrow, Alaska (Climatological Data, US Department of Commerce) and information on micrometeorology contained in Kelley & Weaver (1969) and US IBP Tundra Biome Reports (Brown & West, 1970; Brown & Bowen, 1971, 1972). Average daily metabolic rates are calculated for the appropriate temperature and body size for each age class. Average weights for the age classes (20 g for juveniles, 35 g for subadults, 60 g for adult females and 75 g for adult males) were based upon data for Barrow populations (Batzli, unpublished results). Metabolic requirements for reproducing females were increased by 80 per cent following work on *Microtus* in temperate latitudes (Trojan & Wojciechowska, 1967; Migula, 1969). Consumption rates for each month are simply the total population energy budget multiplied by three. The regression for determining ADMR at different body sizes and ambient temperatures (Table 11.6) was calculated from data on oxygen consumption in a large chamber with food and bedding but no nests. Following Collier *et al.* (1974), lemmings were assumed to be resting at thermal neutrality (in the nest) 50 per cent of the time (RMR $= 0.7073 \ W^{0.76}$ kcal/day).

Results of the calculations indicate moderately high energy flow through a high lemming population (511 Mcal/ha·yr) when compared to high populations of microtines at lower latitudes; 170 Mcal/ha·yr for *Microtus pennsylvanicus* (Golley, 1960); 375 Mcal/ha·yr for *M. oeconomus* (Gebczynska, 1970), 214–814 Mcal/ha·yr for *M. arvalis* in agricultural regions (Trojan, 1969; Ryszkowski *et al.*, 1973). However, the consumption rate for *Lemmus* is considerably greater (1533 Mcal/ha·yr) than for *Microtus* (250–1051 Mcal/ha·yr). Energy flow through *Lemmus* is greater than might be expected from its density owing to larger body size and higher metabolic rate, but it is primarily the low digestibility of forage that produces the very high consumption rate.

Small mammals

No consideration of nutrition can be complete without mention of diet quality, but so little is known about the nutritional requirements of natural populations of herbivorous rodents that little can be said. It may be that nutritional quality influences food selection by *Lemmus*, but the proposition remains unproved. An important role has been proposed for nutrients in the population dynamics of *Lemmus* (Pitelka & Schultz, 1964; Schultz, 1969; Bunnell, 1973). However, since the influence of *Lemmus* on vegetation is clearer than the reverse, a review of the subject will be delayed for the next section.

Environmental impact

Vegetation

The immediate effect of small mammals on their forage is clear; the standing crop is reduced. But the long-term effects are not so clear. Moderate grazing may actually stimulate production of new shoots by perennials (and all arctic plants are perennials), while heavy grazing may eliminate a species, thus altering composition of the plant community. Which of these alternatives occurs will probably depend on the species involved and on the density of the grazer.

Many observations indicate that high populations of *Lemmus* seriously disrupt vegetative cover. For instance, Koshkina (1962) reported that during the build up of a high population under snow *L. lemmus* destroyed 60–90 per cent of moss cover and clipped virtually all monocot shoots in their habitat. Others have reported consumption of 30–70 per cent of moss cover in alpine winter habitats of Finnish Lapland (Kalela *et al.*, 1971; Kalela & Kaponen, 1971). In southern Norway the distribution of *Lemmus* is extremely patchy, but deKock *et al.* (1968), concluded that overgrazing contributes to the collapse of lemming populations. Curry-Lindahl (1962) in Sweden did not think that destruction of food plants was such a serious matter, however.

Lemmus obensis and *L. trimucronatus* depend less on mosses and more on the green stem bases of monocots in the winter. During a winter of high population virtually all of the monocot stems are clipped (Pitelka, 1957*b*; Tikhomirov, 1959). Although few measurements of available forage under the snow have been made, some sampling results for wet meadow tundra indicate that 15–30 g/m^2 dry weight of green stem bases are present in September, then decline to 10 g/m^2 by January and remain at that level the rest of winter if grazing is light (Dennis & Tiezsen, 1972; Bunnell, 1973).

If we allow an average value of 4·6 kcal/g (Batzli, unpublished results) the monocot stem bases represent 460×10^3 kcal/ha of available food in January. During the rest of winter until snow melt the diet of *L. trimucronatus* consists of 55–65 per cent monocots. (Table 11.5; Batzli, unpublished results) Energetic requirements during this time (January to June) are 280×10^3 kcal/ha and consumption is 839×10^3 kcal/ha when a high population occurs (Table 11.7). Since about 60 per cent of consumption consists of monocot stem bases, about 503×10^3 kcal/ha or all (109 per cent) of these must be consumed. Of course, these figures will vary depending upon habitat and population densities, but it is no wonder that biologists report widespread destruction of habitat and unusual movements associated with peak populations near Barrow (Thompson, 1955*c*; Pitelka, 1957*b*; Pitelka & Schultz, 1964).

During the course of the summer following peak densities, lemmings depress the standing crop of monocots about 50 per cent below that in exclosures. In areas where the sod has been uprooted in order to gain access to rhizomes, it may be depressed 90 per cent (Pitelka & Schultz, 1964), and such uprooting can occur on one third of the habitat. Mosses are similarly devastated (Peiper, 1963). Recovery of vegetation proceeds in the following summer in areas where only clipping occurs, but uprooted areas recover more slowly.

The effects of microtines on vegetation are not limited to temporary reductions of standing crop and stature. Tikhomirov (1959) reported that lemming grazing suppresses flowering in cotton grass (*Eriophorum angustifolium*) and tundra grass (*Dupontia fischeri*). Using controlled grazing by voles, *M. oeconomus* and *M. middendorffi*, Smirnov & Tokmakova (1971, 1972) found that moderate grazing increased production of *Eriophorum* and *Carex* by stimulating new shoot growth. They estimated that optimal densities would be 30–50 voles/ha above which productivity would be decreased. Grazing also seemed to suppress flowering of cotton grass in the following summer. Flowering of *Eriophorum* in northern Alaska also varies widely, and F. A. Pitelka (personal communication) reports that flowering is heaviest during the summer before a lemming peak. On the other hand, Tast & Kalela (1971) reported that *Eriophorum* inside and outside of exclosures showed no differences in flowering. They presented no data for the summer following a peak population, however.

Other aspects of microtine activity besides grazing have effects on vegetation. Building of burrows and runways can keep significant portions of the tundra virtually free from vegetation, particularly during peak populations when trails can cover 20 per cent of the surface and holes may be as common as $1/m^2$ in heavily utilized habitats (Tikhomirov, 1959). Bare areas created by digging out and deposition of soil are invaded by plant species which favor disturbed sites, thus creating a vegetational mosaic. Finally, the deposition of urine and feces may have a fertilizing effect which produces more robust plants. Unfortunately, most of these effects are not well documented except in Tikhomirov's monograph (1959).

Long-term effects of microtines on arctic vegetation have not been well documented either. Workers on the Yamal and Taimyr Penninsulas of the USSR. suggest that microtine activity enhances the dominance of monocots over mosses and aids in the creation of grass–sedge meadows (Tikhomirov, 1959; Pjastalova, 1972). Exclosures which have been in place for 18 years near Barrow, Alaska, show different responses in different habitats. In upland habitats, where grazing is prevented, lichens and mosses become well developed at the expense of monocots. In lowland habitats, mosses may develop at the expense of monocots which show a larger amount of standing dead material than surrounding grazed areas. In some areas, particularly in wet *Eriophorum* sites, monocot growth seems more vigorous within the exclosures. None of these trends have been quantitatively documented, however, nor have the effects of the small exclosures themselves been determined. Nevertheless it appears that in the long run microtine activity disrupts the ground layer of mosses and lichens more than it depresses monocot. growth.

Of the non-microtine small mammals, only the ground squirrel has significant effects on vegetation. Squirrels generally occur in areas which are dominated by *Dryas* and shrubs. Under the influence of foraging and digging, however, herbs invade, and grasses (*Calamagrostis, Poa*) become particularly common (Tikhomirov, 1959; Carl, 1971).

Microtopography and soil

Anyone who has observed polygons riddled with holes and runways must be impressed with the potential of lemmings for modifying microtopography. In addition, lemming holes and runways seem to be associated with all stages of development of hummocks on well-drained soils in northern Alaska. Finally, lemming runways are often associated with frost cracks in the soil, particularly in polygon troughs. The degree

264

to which lemmings are merely selecting suitable habitat and the degree to which they are responsible for alterations of topography is not clear, but their burrows and pathways must at least alter drainage patterns, thereby affecting the thermal regime of the soil (Pitelka, 1957*a*; Tikhomirov, 1959; Batzli, unpublished results).

Another way in which lemmings affect the thermal regime of the soil is by their winter clipping. Nearly all of the remaining clippings are inedible dead material and are left on the surface of the tundra. This reduction in stature of the dead vegetation apparently reduces its insulating effect, for the depth of thaw becomes deeper than in unclipped areas (Pitelka & Schultz, 1964). In low areas, the spring melt waters move these clippings about and windrows are left at pond edges as the water recedes. These dense mats decrease depth of thaw, decrease shoot growth and may even contribute to the formation of small peaty mounds (Tikhomirov, 1959).

Heavy clipping may also be viewed as adding large amounts of organic matter to the soils where, because it is moist, decomposition will occur more rapidly. This together with urine and feces creates a rapid accumulation of nutrients at the soil surface. As nutrient concentrations were high in vegetation during a summer of peak populations, it seems that some of these nutrients must leach into the soil solution, and be absorbed rapidly by the plants. (Peiper, 1963; Schultz, 1969; Bunnell, 1973). Concentrations were low in vegetation the next summer; apparently nutrients which are not leached return to the soil more slowly. Nutrient concentrations slowly built up during the following summers until another lemming peak occurred.

Concentrations of nutrients in the above-ground vegetation were correlated with nutrient levels in the top 10 cm of soil (Bunnell, 1973). It is here that most (75–90 per cent) of the below ground biomass of the plants is located (Dennis & Johnson, 1970), and one might expect rapid uptake of nutrients from this level. Samples taken from exclosures showed no cylic fluctuations in nutrient levels (Peiper, 1963; Schultz, 1969). Although the reasons for this pattern of change are not clear, cycling of nutrients in the soil and vegetation does seem to be linked to lemming cycles, and lemmings may control the rates of nutrient exchange between soils and vegetation in the tundra.

Finally, it should be pointed out that the digging of rodents may also alter the structural characteristics of soil. For instance, the digging of lemmings produces aggregates which increase particle size and improve drainage (Tikhomirov, 1959).

Small mammals

Although more patchily distributed than lemmings, ground squirrels have an equally large potential for altering local soil characteristics where they occur. They build elaborate burrow systems with up to 50 openings, each of which may be 25 cm in diameter and surrounded by dirt piles. These openings and tunnels continually shift, and abandoned portions of the system may collapse. Further, the squirrels excavate simpler burrow systems, called duck holes, and shallow pits on the boundaries of territories. The effects on microtopography and drainage are clear although they have not been quantified (Tikhomirov, 1959; Carl, 1971).

Predators

Only a few major predators of small mammals breed regularly in tundra ecosystems, the snowy owl (*Nyctea scandiaca*), the short-eared owl (*Asio flammeus*), several species of jaegers (*Stercorarius* spp.), the arctic fox (*Alopex lagopus*) and two species of weasels (*Mustela rixosa* [= *nivalis*] and *M. erminea*). All of these are circumpolar in their distribution. Although other predators occur on the tundra (gulls, various raptors, red foxes, wolves and bears), these are much less dependent on tundra small mammals than the species mentioned above.

Avian predators are migratory and particularly dependent upon small mammal populations during the breeding season. Once they have nested, their food supply must be gathered relatively close to the nest. Both the snowy owl and pomarine jaegers are large birds and specialize on microtines for food during their breeding season (Pitelka *et al.*, 1955). Since they consume large numbers of microtines (4–7 adult lemmings/ day for snowy owls and 3–4/day for jaegers) it is not surprising that the density of microtines determines the breeding success of these birds (Pitelka *et al.*, 1955; Maher, 1970; Gessaman, 1972). When densities of microtines are low, these birds do not breed at all. Rather, they become local vagrants living off whatever they can capture, and small birds become more important in their diet. The breeding success of shore birds, for instance, may be reduced by high predation rates at times when lemming populations have declined (U. Safriel, unpublished results).

Foxes and weasels can often turn to other prey (ground squirrels and birds) during a summer of low microtine densities, but during winter microtines are the only available prey. Thus, not only breeding success, but also winter survival of mammalian predators may be dependent on microtine populations, and both breeding densities and winter densities of these predators have been correlated with lemming densities (Macpherson, 1969; Maher, 1967; MacLean *et al.*, 1974).

The information on predators cited above is based upon observations in North America, but similar fluctuations in response to microtine densities have been reported in Eurasia (Curry-Lindahl, 1962; Shvarts, 1963; Mysterud, 1970; Tast & Kalela, 1971). Therefore, density of predators in the tundra in general appears to be closely linked to microtine density, and predation rates of other prey, particularly small birds, may also be affected. In return, high predation rates may contribute to lemming declines and delay the recovery of low microtine populations, thus influencing the course of microtine cycles (Maher, 1967; Pitelka, 1973; MacLean *et al.*, 1974).

Summary and conclusions

It is clear from the discussions above that small mammals, particularly microtine rodents, play a pivotal role in tundra ecosystems. Whereas peak microtine populations in temperate grasslands may only consume 1–35 per cent of available food (Golley, 1960; Batzli & Pitelka, 1970; Gebczynska, 1970) during a particular season, lemmings may consume nearly all of their accessible winter forage.

The effects on the vegetation, microtopography and soil are striking. Productivity of vegetation and nutrient concentrations in both vegetation and soil fluctuate in rhythm with microtine cycles, and these fluctuations appear to be caused by microtine grazing. This may be the only case of herbivore control of nutrient flux through a terrestrial ecosystem. In addition, changes in topography (formation of hummocks) and vegetational composition (reduction of mosses and lichens) seem to be associated with small mammal activity.

Reproductive activity and density of predator populations also cycle in response to microtine density. Indirect effects may be felt by other tundra inhabitants, particularly breeding birds, when predators shift their diet following a microtine decline.

A word of caution must be added. The distribution and population dynamics of small mammals in the arctic is relatively poorly known. While the conclusions I have drawn almost certainly apply to low coastal tundra which supports high densities of lemmings, their application in other types of tundra may be debated.

Although this paper is largely based upon published reports in the open literature, significant aid was given by participants in IBP programs who generously shared their thoughts and data. I particularly wish to thank F. A. Pitelka, S. F. MacLean, B. Collier, N. C. Stenseth, B. M. Fitzgerald, D. D. Feist, E. Östbye, S. W. Speller, S. S. Shvarts and E. Haukioja. Contributions of unpublished data gathered for IBP are recognized in the text by citing the author's name. Drs F. A. Pitelka, W. A. Fuller and R. M. Chew made helpful comments on the manuscript for which I am grateful.

My own research was supported by the National Science Foundation (Grant No. GV–33852) as a part of the US IBP Tundra Biome Programme. Support of field work was also provided by the Naval Arctic Research Laboratory, for which I thank J. F. Schindler, Director, and L. S. Underwood, Assistant Director for Science.

12. The role of small mammals in tropical ecosystems

T. H. FLEMING

Compared with temperate and arctic areas, tropical regions of the world are relatively rich in numbers and species of small mammals. The continental United States and surrounding islands, an area of approximately 7.4×10^6 km^2, for example, contains about 319 species of mammals, 267 (84 per cent) of which can be considered 'small mammals' (i.e., members of the taxonomic orders Marsupialia, Insectivora, Chiroptera, Rodentia, and Lagomorpha) (Burt & Grossenheider, 1964). By contrast, the Central American country of Panama, which includes an area of about 7.7×10^4 km^2 – only one-hundredth the area of the United States – contains at least 205 mammal species, of which 166 (81 per cent) are small mammals (Handley, 1966). In a restricted portion of Zaire, the Lake Kivu region, Dieterlen (1967a) reported a total of 120 species of mammals of which 45 to 50 were rodents. At least 161 species of mammals, including 107 species of small mammals (66 per cent), inhabit the lowland tropical forest of Malaya, an area of somewhat less than 1.3×10^5 km^2 (Harrison, 1962). Finally, at the community level, tropical forests in Panama contain twice as many species of mammals as temperate forests in the eastern United States (Fleming, 1973a).

Despite, or perhaps because of, the richness of tropical small mammal faunas, relatively little is known about the functional roles played by these mammals in their ecosystems. In addition to serving as important links in the flow of energy and in mineral cycling, tropical mammals play important roles as seed predators and pollination and seed dispersal agents. Furthermore, their activities as economic pests and/or reservoirs for infectious diseases can make them of direct concern to man and his welfare. For these reasons it seems clear that we need a much more extensive knowledge of patterns and processes (Wiens, 1973) of tropical small mammals than is presently available. The purpose of this paper, therefore, is to review the existing data on the ecology of tropical mammals and to point out the kinds of information that are currently unavailable. In this review, emphasis will be placed on the community structure, functional roles, and population dynamics of small mammals in lowland tropical ecosystems. Particular stress will be placed on forest ecosystems because they have been most extensively studied in many tropical areas.

Community structure of tropical mammals

Taxonomic composition

The taxonomic, and hence in part the trophic, composition of tropical small mammal communities varies from one zoogeographic region to another. In the Neotropical (Central and South America) and Australasian (Australia, Tasmania, New Guinea and surrounding islands) regions, rodents and marsupials are the numerically dominant non-volant forms whereas in the Ethiopian (Africa and Madagascar) and Oriental (tropical Asia and associated continental islands) regions, rodents and insectivores predominate. Because of a paucity of detailed information about the food habits of most tropical mammals, it is difficult to tell whether these taxonomic differences have resulted in trophic differences between the regions. In the New World, marsupials (Didelphidae) tend to be strongly insectivorous (Enders, 1935; Fleming, 1972) so that they probably partially fill the 'terrestrial insectivore' food niche in lowland Neotropical communities. When more data on food habits are available, it will probably turn out that there are few major differences in the trophic composition of the non-chiropteran portions of tropical mammal communities as demonstrated for Panamanian and Malayan communities by Fleming (1973a).

Bats are important members of tropical mammal communities but their trophic impact on ecosystems has seldom been considered by mammalian ecologists. A recent study by Wilson (1973) indicates that the trophic, as well as the taxonomic, compositions of tropical bat faunas, and presumably bat communities, differ from each other in several respects. In all zoogeographic regions, bats that consume insects in flight ('aerial insectivores') are trophically most important. However, frugivorous bats are trophically important only in the Neotropical and Australasian regions, and nectarivorous bats are important primarily in the Neotropical region. Although probably influenced to some extent by the different phylogenetic histories of the bat faunas, these trophic differences may also reflect important regional differences in the abundance and availability of fruit and flower resources suitable for consumption by bats.

Species richness

The number of small mammals known to inhabit various tropical habitats is shown in Table 12.1. In communities in which both bats and other species have been sampled, total species counts range from 46 to 60 or more. When just the diversity of rodents is examined, rather consistent results are found in most areas: tropical habitats, with the probable exception of those at higher altitudes, usually contain 10 to 16 species of rodents (Table 12.1). While the data upon which this conclusion is based come from a variety of different habitats, the number of samples is still small so that these results may be biased. It would not be surprising, for example, to find that fewer rodent species coexist in grasslands than in forests. This is probably the case in the grasslands of Central America, where less than 10 species are likely to occur together. Sheppe (1972) mentions the potential coexistence of only five species of rodents on the seasonally-inundated floodplains of Zambia. Also, when a larger number of lowland wet forests have been studied, the upper limit to the number of coexisting species may increase. There are currently few data available on the diversity of rodents in the 'classic' rain forests of South America, Africa, and Asia. Until data from these areas are available, strong conclusions about the richness of tropical rodent communities cannot be made.

Although, again, few areas have been adequately sampled, it appears that the number of species of sympatric marsupials or insectivores in tropical habitats is always less than the number of rodent species. One to six species are usually found in a given habitat (Table 12.1).

Estimates (probably underestimates) of the species richness of tropical bat communities are available primarily for the Neotropical region. These data (Table 12.1) indicate that 27 to 40 or more species can be found in one habitat, which means that fully two-thirds of the species of small mammals in some tropical habitats are bats. Liat (1966) reported 33 species of bats inhabiting the lowland forests around Kuala Lumpur, Malaysia. These data clearly indicate that, because of their richness in number of species and individuals, bats deserve considerably more attention from the standpoint of biomass and energy flow than they have received to date.

Relative abundance

In addition to species richness, another important aspect of community structure is the relative abundance of each species. Are tropical

Table 12.1. *Composition of tropical small mammal communities at various locations and habitats*

Country	Location and latitude	Habitat† and elevation	No. species of small mammals					Source
			Marsupials	Insectivores	Bats	Rodents	Total	
Costa Rica	Guanacaste Prov. (10° N)	Dry tropical forest (45 m)	6	0	30+	10	46+	Fleming (1973a), Organization for Tropical Studies (unpublished data)
Costa Rica	Heredia Prov. (10° N)	Wet tropical forest (100 m)	6	0	40+	14	60+	Fleming (1973a), Organization for Tropical Studies (unpublished data)
Panama	Canal Zone (9° N)	Dry tropical forest (50 m)	6	0	27+	16	49+	Fleming (1970, 1972), Fleming et al. (1972)
Panama	Canal Zone (9° N)	Moist tropical forest (5 m)	6	0	31+	16	53+	Fleming (1970, 1972), Fleming et al. (1972)
Brazil	Teresopolis (22° S)	Moist subtropical forest (850 m)	3+	0	N.D.	14+	17+	Davis (1945)
Zaire	Lake Kivu (2° S)	Lower montane wet forest (2200 m)	0	4	N.D.	14	18+	Rahm (1967)
Zaire	Lake Kivu (2° S)	Subtropical wet forest (1500 m)	0	4	N.D.	14	18+	Rahm (1967)
Zaire	Lake Kivu (2° S)	Banana plantation (1800 m)	0	1	N.D.	16	17+	Rahm (1967)
Zaire	Lake Kivu (2° S)	*Pennisetum* grassland (1460 m)	0	1	N.D.	12	13+	Dieterlen (1967a)
Uganda	Queen Elizabeth Park (0°)	*Imperata* grassland (976 m)	0	4	N.D.	9	13+	Delany (1964b)
Uganda	Mayanja forest (0°)	Subtropical moist forest (1200 m)	0	N.D.	N.D.	14	14+	Delany (1971)
Ivory Coast	Lamto Research Station (7° N)	Savanna (<100 m)	0	1	N.D.	14	15+	Bellier (1967)
Malaya	Kuala Lumpur (3° N)	Moist tropical forest (c. 500 m)	0	2	N.D.	14+	16+	Harrison (1969)
Malaya	Ulu Gombak (3° N)	Moist tropical forest (578 m)	0	2	N.D.	14+	16+	Rudd (1965)

† Based on the Holdridge (1947) classification. N.D. Indicates no data available.

communities characterized by a lack of dominance (i.e., no species that are extremely common) as suggested by MacArthur (1969), or is numerical representation inequitably partitioned among species?

Before presenting pertinent data, I wish to stress that the determination of relative abundances of small mammals is an extremely difficult task and one that is fraught with a multitude of potential sampling errors. As should be clear from other contributions in this volume, many variables can influence trapping results; these include the kinds of traps and baits used, size limitations of the traps, placement of the traps (on the ground or in trees), species-specific behavior towards traps, and many others. The operation of one or more of these biases during a trapping study will undoubtedly obscure to some extent the ecological situation that is being investigated. For example, when relative abundances are based on trapping results, it is likely that the abundance of highly trappable species will be overestimated while the abundance of trap-shy species will be underestimated.

Dieterlen's (1967*a*) study of rodent populations in the Lake Kivu region of Zaire can be used to illustrate the extent to which trapping results can differ from actual conditions. Dieterlen estimated the abundance of rodents by two methods: conventional snap trapping and hand capture in small areas (40 m × 20 m) surrounded by corrugated fencing. If we assume the hand collections more nearly approximate actual abundances than snap trapping, then the latter method overestimated the relative abundance of 6 out of 12 species in two grassland habitats. Trapping greatly overestimated the abundance of *Dendromus insignis* and underestimated the abundance of *Lophuromys aquilus*. These results, in conjunction with other kinds of biases mentioned above, indicate that care must always be exercised when basing estimates of relative abundance on trapping data.

Although the data were obtained by a variety of sampling methods, rather similar estimates of the relative abundances of small mammals have been obtained in a number of different tropical habitats (Fig. 12.1). Data from nine localities indicate that communities of small mammals usually contain one or two relatively common species and many uncommon species. In the communities shown in Fig. 12.1, the three most common species account for an average of 69 per cent (range, 55 per cent to 85 per cent) of the individuals captured.

Other, less thorough, studies lend support to the idea that there is a high degree of numerical 'dominance' (i.e., an inequable distribution of relative abundances) in most tropical rodent communities. For example,

273

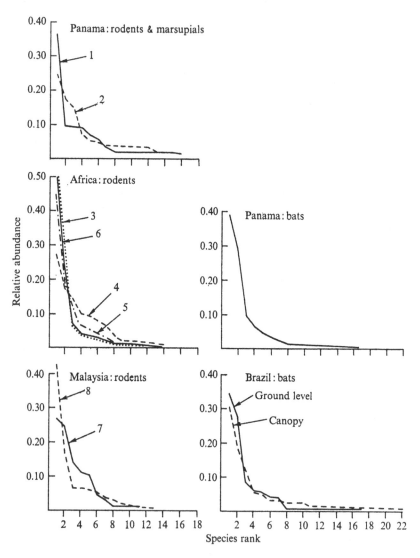

Fig. 12.1. Relative abundances of small mammals in various tropical habitats as follows: Panama, dry tropical forest (1) and moist tropical forest (2) (Fleming, 1970, 1971, 1972); Africa, lower montane wet forest (3), subtropical wet forest (4), banana plantation (5), and subtropical moist forest (6) (Delany, 1971; Rahm, 1967); Malaysia, moist tropical forest (7, 8) (Harrison, 1969; Rudd, 1965). The Panamanian bat data are from the dry tropical forest (Fleming, 1973a), and the Brazilian bat data are from the wet tropical forest (Handley, 1967).

Fleming (1973*c*) found that two species, *Heteromys desmarestianus* and *Liomys salvini*, accounted for 94 and 96 per cent of the rodents trapped in Costa Rican wet and dry tropical forests, respectively. Likewise, one or two species accounted for more than 65 per cent of the individuals caught in 32 of 41 trapping periods in Malawi (Hanney, 1965). One species, *Praomys jacksoni*, accounted for more than 66 per cent of the individuals trapped in montane forest habitats whereas *Mastomys natalensis* was the dominant species in cultivated areas. Finally, Delany (1964*a*) found one or two species 'common' and many species apparently 'uncommon' at many localities in Uganda.

Few tropical bat communities have been sampled quantitatively, and the usual sampling technique, the Japanese 'mist net', is probably quite selective in the species it captures (see Fleming *et al.*, 1972). Nevertheless, available data seem to indicate that, as in the case of rodents, there is a high degree of numerical dominance in tropical bat communities (Fig. 12.1). Although these Neotropical communities contain over 30 species of bats, only one to three species can be considered 'common.' Interestingly, Handley (1967) found that although the distribution of relative abundances at ground level and in the canopy was similar, different species were dominant at the two levels. In the canopy *Artibeus lituratus* and *A. jamaicensis* were the two most common species, but at ground level these species ranked third and fourth in abundance, respectively, being replaced by *Carollia perspicillata* and *C. subrufa* as the dominant species.

Biomass

There are few estimates of the biomass per unit area of tropical small mammals in the literature. Available data include biomass estimates for two localities in Panama (my own calculations based on densities presented in Fleming, 1970, 1971, 1972), three localities in Africa (Bellier, 1967; Dieterlen, 1967*a*; Poulet, 1972*b*), and one in tropical Asia (Harrison, 1969).

In two Panamanian forests, average wet weight biomass, calculated by multiplying average weights of trapped species by their estimated average densities, ranged from 5562 to 7597 g/ha (Table 12.2). Rodents (including squirrels and the agouti, *Dasyprocta punctata*) comprised 72 per cent and 83 per cent of these totals in the dry and moist forests, respectively. Biomass undoubtedly fluctuates seasonally owing to seasonal density changes, but density estimates for most species are not

Table 12.2. *Wet weight biomass estimates for populations or communities of tropical small mammals*

Location and habitat	No. of species	Average density (no./ha)	Average biomass (g (wet wt)/ha)	Source
Panama: dry tropical forest	11 (rodents)	18.9	4025	Fleming (1970, 1971, 1972)
	5 (marsupials)	3.5	1538	
	16 (total)	22.4	5562	
Panama: moist tropical forest	9 (rodents)	11.3	6304	Fleming (1970, 1971, 1972)
	4 (marsupials)	2.1	1293	
	13 (total)	13.4	7597	
Ivory Coast: savanna with palms	11+ (rodents)	—	460 (burned) to 916 (unburned)	Bellier (1967)
	>1 ? (insectivores)	—	11 (burned) to 86 (unburned)	
Lake Kivu, Zaire: *Pennisetum* grassland	5 (rodents) >1 ? (insectivores) }	370	14000–15000	Dieterlen (1967a)
Senegal: thornbush savanna	1 rodent	3–4 ?	393–599 after 'normal' rains; 42–64 after 'sub-normal' rains	Poulet (1972b)
Malaya: primary forest	11 (rodents)	4.9	840	Harrison (1969)
secondary forest	4 (rodents) 1 (tree shrew) }	7.1	810	Harrison (1969)
scrub	5 (rodents)	3.2	282	Harrison (1969)
grassland	4 (rodents)	5.5	324	Harrison (1969)

precise enough to warrant the calculation of seasonal changes in community biomass. At the population level, however, biomass of *Liomys adspersus*, *Oryzomys capito*, and *Proechimys semispinosus* varied from 208 to 629 g/ha, 0 to 159 g/ha and 125 to 1099 g/ha, respectively, in the dry tropical forest (Fleming, 1971). Gliwicz (1973) estimated that the average biomass of a population of *Proechimys semispinosus* on a small island in Gatun Lake, Panama Canal Zone, was 2900 g/ha.

Relatively precise biomass estimates from Africa are available only for grassland or savanna habitats (Table 12.2). Using a hand capture technique, Dieterlen (1967*a,c*) estimated that combined densities of rodents and insectivores in *Pennisetum* grassland were as high as one animal per 27 m², which extrapolated to a biomass of 14500 to 15000 g/ha. Other African biomass estimates are not this high. For example, Bellier (1967) reported that the biomass of rodents varied from 40 to 980 g/ha on burned savanna and from 120 to 1645 g/ha on unburned savanna; biomass of insectivores varied from 0 to 60 g/ha and from 15 to 280 g/ha, respectively, in the two habitats. Biomass also varies seasonally in the gerbil, *Taterillus pygargus*, which inhabits the dry thornbush savanna in Senegal (Poulet, 1972*b*). After 'normal rains', biomass of this species averages 393 to 599 g/ha whereas it averages 42 to 64 g/ha after 'subnormal' rains (Table 12.2). Delany (1964*b*) sampled a variety of habitats in Uganda and noted that biomass of rodents, expressed as g/trap unit, was highest (10.1 g/t.u.) in *Imperata* grassland and averaged 2.6 g/t.u. in several other habitats (e.g., burned and unburned grassland, scrub, semi-deciduous forest). Finally, Hanney (1965) recorded highest biomass, expressed as g/100 trap nights, on wet plateaus in Malawi with averages ranging from 1200 g/100 t.n. in cultivated areas on the Nyika River to 350–500 g/100 t.n. in *Brachystegia* woodland; in dry areas his maximum yield was 250 g/100 t.n.

These few biomass estimates, which are based on a variety of different data-gathering techniques, make it extremely difficult to generalize about the size of standing crops of tropical small mammals. Many more studies conducted in a variety of habitats and using standardized techniques are needed before general trends become apparent. At this point it seems safe to conclude that the sizes of standing crops do vary between habitats and, seasonally, within habitats. Although the subject has received little attention to date, another generalization that may emerge is, as in the case of relative abundances, the distribution of biomass is highly inequable in tropical small mammal communities. For example, in the Panamanian moist and dry tropical forests two

277

Small mammals

species (the agouti, *Dasyprocta punctata*, spiny rat, *Proechimys semi-spinosus*, or a squirrel, *Sciurus granatensis*) account for 66 per cent and 43 per cent of the total biomass, respectively. In both forests the agouti alone accounts for 29 to 52 per cent of the biomass owing to its relatively large size (c. 4000 g) although it sometimes occurs at densities of less than one individual/ha. The paca (*Agouti paca*), which is the poorly-known nocturnal counterpart of the agouti in Neotropical forests, undoubtedly also represents a large portion of the rodent biomass because of its large size (c. 10000 g); however, its population densities and biomass are yet to be estimated. Whether or not the Panamanian situation regarding the distribution of biomass among species is typical of other tropical communities needs to be determined.

Spatial distribution of feeding habits

Wherever spatial heterogeneity permits, species of small mammals are distributed three-dimensionally in their habitats. In dealing with the ways in which small mammals are distributed in tropical habitats, it is useful to partition species into different size classes, food habits, and spatial adaptations. In this way a roughly three-dimensional 'niche matrix' can be constructed to visualize the ways in which species distribute themselves spatially and trophically.

Table 12.3 is an attempt to indicate the spatial distribution patterns of the small mammals (mostly rodents) in three tropical habitats, one each in the Neotropical, Ethiopian, and Oriental regions. In constructing these tables, I have defined spatial adaptations (i.e., terrestrial, scansorial, or arboreal) on the basis of where a species obtains most of its food; scansorial species feed both on the ground and off the ground in shrubs and trees. As indicated earlier, details of the food habits of most tropical mammals are poorly known so that the species in the three habitats have been assigned to general feeding categories whose limits are defined in Fleming (1973*a*). Dietary information used in characterizing the species in Table 12.3 comes from Delany (1964*b*, 1971), Dieterlen (1967*a*), Fleming (1970, 1972), Harrison (1954, 1962), and Harrison & Liat (1950).

In the Panamanian community, a majority of the species are either terrestrial or scansorial. The ground-dwellers tend to be mostly vegetarian, with the larger species (i.e., *Proechimys*, *Agouti*, and *Dasyprocta*) concentrating on fruit and large seeds. Most of the scansorial species, including the marsupials, are omnivores, whereas two of the three arboreal species are vegetarian.

278

Table 12.3. *Spatial distribution of small mammals in three tropical habitats by body size and food habits, which include: F, fruit; G, seeds; H, vegetation; I, insects, and O, a mixture of food items. Data for Panama are from Fleming (1970, 1972); for Uganda, from Delany (1971 and personal communication); for Malaya, from Harrison (1954, 1962, 1969).*

Spatial adaptation	<100	Weight (g) 100–1000	>1000
(A) Panama: dry tropical forest			
Arboreal		†*Caluromys*: O	*Coendou*: H
		Diplomys: F	
Scansorial	†*Marmosa*: O(I)	†*Metachirus*: O	†*Didelphis*: O
	Oryzomys concolor: O?	†*Philander*: O	
	Nyctomys: F?	*Sciurus* (2 spp.): O	
		Tylomys: F?	
Terrestrial	*Liomys*: G	*Sigmodon*: H	*Agouti*: H, F
	Oryzomys capito: O	*Proechimys*: H, F	*Dasyprocta*: H, F
	Nectomys: O?		
	Zygodontomys: O		
(B) Uganda: moist subtropical forest			
Arboreal	*Grammomys*: H	*Heliosciurus*: ?	
	Graphiurus: H		
	Hylomyscus: H		
	Thamnomys: H?		
Scansorial	*Praomys*: O	*Oenomys*: H?	
Terrestrial	*Hybomys*: H	*Aethomys*: H	*Thryonomys*: ?
	Lophuromys (2 spp.): I	‡*Malacomys*: O	*Atherurus*: H
	Mus (2 spp.): O, I	*Cricetomys*: ?	
(C) Malaya: moist tropical forest			
Arboreal	*Rattus cremoriventer*: F		
	Flying Squirrels (several		
	spp.): H, F		
Scansorial		*Callosciurus* (2 spp.): G, F	
		Sundasciurus: G, F	
Terrestrial	*Rattus exulans*: O	*Rhinosciurus*: I	*Atherurus*: H
	R. whiteheadi: O?	*Rattus mülleri*: O	
		R. bowersi: O	
		R. rajah: O	
		R. surifer: ?	
		R. sabanus: O	

† Marsupials.   ‡ Aquatic.

Small mammals

The rodent fauna of the Mayanja forest in Uganda contains 5 arboreal species and 10 ground-dwelling or aquatic forms. As in Panama, the arboreal species are vegetarians whereas a diversity of feeding habits is found among the terrestrial species. There are apparently fewer scansorial species in this community than in the Panamanian one owing to the absence of marsupials.

The Malaysian community is rich in squirrels and rats of the genus *Rattus*. The distribution of species resembles that in Africa: there is a greater number of arboreal or terrestrial forms than scansorial forms. Compared with the Panamanian forest, the Malaysian forest contains fewer scansorial omnivores but a greater proportion of terrestrial omnivores. Furthermore, there is a greater diversity of medium-sized rats in the Malaysian community than in either the Panamanian or Ugandan communities.

Bats contribute importantly to the structure of tropical mammal communities but have seldom been studied in conjunction with other small mammals. As is well known, the diversity of food habits in tropical bats is much greater than in temperate bats. For example, of the 27 species netted in a Panamanian dry tropical forest, 9 were insectivorous, 11 frugivorous, 3 omnivorous, 2 carnivorous, and 1 each specialized on fish and blood (Fleming *et al.*, 1972). Compared to the bat fauna of lowland Malaya, the Panamanian fauna contains a greater proportion of frugivores and nectarivores and a lower proportion of insectivores (Fleming, 1973*a*). When they become better known, bat communities in tropical Africa will probably more closely resemble those in tropical Asia than tropical America because of the closer taxonomic and trophic resemblance of Ethiopian and Oriental bat faunas (Wilson, 1973).

Functional roles of tropical mammals

Energy flow

Almost nothing is known about rates and patterns of energy flow through populations or communities of small mammals in the tropics. To my knowledge only two studies address themselves to this ecological question. Poulet (1972*b*) measured net primary productivity and seed consumption by the gerbil *Taterillus pygargus* in the dry thornbush savanna of Senegal, West Africa. Of the 40 to 60 kg (dry weight) of seeds produced/ha·year, *Taterillus* consumed 2.1 to 6.6 kg/ha or about

280

9 per cent of the annual seed crop. This level of seed consumption was considered a preliminary estimate (Poulet, 1972*b*).

For a population of the echimyid rodent *Proechimys semispinosus* living on a 16 hectare island in Gatum Lake, Panama Canal Zone, Gliwicz (1973) estimated that annual production was 97×10^3 kcal/year, cost of maintenance 1973×10^3 kcal/year, and assimilation 2028×10^3 kcal/year. Her calculated values yield an efficiency of production (= production/assimilation) of 4.73 per cent, somewhat higher than the efficiency of the temperate-zone beech forest rodents (2.65 per cent) studied by Grodziński *et al.* (1969). No estimates were made of the proportion of annual plant production these frugivorous rodents consume.

Since no other estimates of energy flow in small tropical mammals are apparently available, it is impossible at this time to make meaningful temperate–tropical energetic comparisons. Estimates of energy relationships in temperate mammals, summarized in Chew & Chew (1970) and Grodziński (1971*a*), indicate that small mammals consume from less than 1 per cent up to 14 per cent, typically less than 6 per cent, of the plant food available to them. Since this 'available food' is itself only a small portion of the net primary productivity of any habitat, it is clear that small mammals account for only a small proportion of the energy flow in ecosystems. Because there is no reason to believe that tropical mammals are more 'efficient' at consuming available plant material than more northern (or southern) forms, tropical mammals can also be expected to account for only a small percentage of the energy flow through their ecosystems.

In partial support of this expectation, Karr (1973) has calculated that the energy needed to maintain observed densities of tropical bird communities is similar to that needed to support temperate bird communities. Thus, other things being equal, temperate and tropical bird communities should consume roughly equivalent amounts of plant or insect material, at least during the breeding season when bird densities in the temperate zone are highest. Whether or not 'energy compensation', as Karr (1971) calls this phenomenon, also occurs in temperate and tropical mammals must await further investigation.

Other functional roles

In addition to contributing to energy flow and nutrient cycling, small mammals play extremely important roles as seed predators and/or

281

dispersal agents and as pollination agents in tropical forest ecosystems. These three roles appear to be partitioned among several taxonomically unrelated groups of mammals. Seed predation and/or dispersal is effected by marsupials, bats, primates, rodents, and certain carnivores (e.g., in the New World, the coati (*Nasua narica*), kinkajou (*Potos flavus*), olingo (*Bassaricyon gabbii*), and the tayra (*Eira barbara*)). On the other hand, bats appear to be the principal pollinating agents among tropical mammals.

According to Janzen (1971*a*) '. . . seed predation [is] the cost of reliable dispersal and [is] directly analogous to a juicy fruit or complex exploding pod.' He continues by saying that it is difficult to separate the act of seed predation from dispersal for a given seed type (e.g., a palm nut) because, when handled by mammals, some seeds are usually killed and eaten whereas others may be cached and perhaps neglected, to germinate later. The general importance of predators and dispersers is that they help to determine the 'seed shadow' of a particular plant, which in turn can determine whether or not a particular plant will leave any offspring and, if so, where the offspring will be located. These actions can ultimately help determine the density of a particular species and also the plant species composition of the habitat (Janzen, 1970).

Janzen's detailed studies (Janzen, 1971*b*, 1972; Wilson & Janzen, 1972) of the interactions between seed predators, dispersal agents and plants in Costa Rican dry tropical forests have provided an overview of the importance of mammalian dispersal agents to tropical plant communities. For example, the leguminous vine *Dioclea megacarpa* suffers some 'predispersal mortality' from one of its main dispersal agents, the variegated squirrel *Sciurus variegatoides*, which eats 0 to 43 per cent (but usually less than 10 per cent) of the seeds while they are in the 'milk' stage. However, this mortality is more than compensated for when squirrels and other rodents such as the agouti and paca carry mature, fallen seeds away from the parent vine. Seeds that fail to be dispersed by these mammals suffer intense predation by the bruchid beetle *Caryedes brasiliensis* or, if they escape notice by the beetles, seedlings underneath parent vines suffer high mortality from the larvae of a noctuid moth. Thus only those seeds that are carried away from conspecific plants stand a reasonable chance of germinating and producing a new individual.

A similar situation obtains in other species of tropical trees such as the palm *Scheelea rostrata* and the Panama tree, *Sterculia apetala*. Both of these species suffer intense seed predation either by two species of bruchid beetles (in the case of *Scheelea*) or the cotton stainer bug, *Dysdercus fasciatus* (in the case of *Sterculia*), unless mammalian dispersal agents such as squirrels, agoutis, pacas, and probably other rodents (e.g., *Proechimys semispinosus* and *Liomys adspersus* or *L. salvini*) carry the palm nuts away from the parent plant or squirrels and monkeys remove *Sterculia* pods from the vicinity of the parent before the pods dehisce. In these and other species of plants whose seeds are destroyed by the larvae of host-specific bruchid beetles, the probability of a seed germinating increases the farther it is removed from a conspecific plant (Janzen, 1970).

Using data collected on Barro Colorado Island, Panama, Smythe (1970) has discussed the evolutionary implications of mammalian dispersal agents on the timing of the fruiting seasons of tropical trees. Those species producing small seeds which pass through the guts of animals such as bats, primates, and other scansorial or arboreal frugivores tend to have prolonged fruiting seasons which maximize the exposure of seeds to dispersal agents. On the other hand, those species producing larger fruits whose seeds cannot be swallowed whole are more seasonal in their fruiting and depend on the scatterholding behavior of larger mammals such as the agouti and paca for their dispersal via burial.

Because of their relatively high mobility, bats are important both as seed dispersal and pollinating agents in tropical forest ecosystems. Although Baker (1972), Faegri & van der Pilj (1966), and van der Pilj (1968) have reviewed the general aspects of bat–plant interactions, there is little detailed knowledge about the services that bats provide for plants. For example, how far do bats carry seeds in their guts before voiding them? How far is pollen carried before it is deposited on another flower and what is the probability that the site of deposition will be a conspecific individual?

Radio-tracking and banding studies on Trinidad (Williams & Williams, 1970) and in Costa Rica (Laval, 1970; Fleming *et al.*, 1972; Heithaus *et al.*, unpublished results) indicate that foraging distances are probably correlated with body size in plant-visiting bats of the family Phyllostomatidae. The large omnivorous species *Phyllostomus hastatus* feeds at distances of up to ten kilometers from its roost, and recapture distances of nearly two kilometers have been recorded for a variety of smaller species. These data suggest that bats are capable of moving pollen and/or seeds several kilometers from their point of origin.

In the dry tropical-riparian forest ecosystem of western Costa Rica, at

283

least 41 species of plants (5 to 10 per cent of the shrub and tree flora) are visited regularly for nectar and pollen or fruit by at least 12 species of phyllostomatid bats (Heithaus *et al.*, unpublished results). Twenty-one species of pollen and 21 species of seeds and fruit pulp have been collected from these bats; one species, *Manilkara achras*, is both bat-pollinated and bat-dispersed. Mixed-species pollen loads occur on many bats, particularly on individuals of *Phyllostomus discolor*, which suggests that bats may visit several different flower species in one night. This behavior could serve to reduce the 'efficiency' of pollen exchange for a given plant species. Also, owing to differences in species-specific foraging behavior, seed dispersal distances and the habitats in which seeds are deposited will probably vary depending on the identity of the seed vector. For example, if seeds of the abundant riparian shrub *Piper tuberculatum* are ingested by *Artibeus jamaicensis*, they will likely be carried a longer distance, perhaps to a new habitat, than if they are ingested by *Sturnira lilium*, a species that apparently forages relatively short distances and primarily along a single river system. Seeds of *Piper* are, therefore, perhaps more likely to end up in the 'correct' habitat if they are dispersed by *Sturnira* rather than by *Artibeus*.

While information is beginning to accumulate, at least in the New World, on the activities of bats as pollinating and/or seed dispersal agents, the energetic impact of plant-visiting and insectivorous bats on the primary and secondary productivity of tropical ecosystems has yet to be studied. Because many tropical bats are relatively poor laboratory subjects (see Rasweiler, 1973), it may be difficult to obtain the physiological information needed for energetic calculations. Therefore, progress in this area of tropical ecology will probably be particularly slow.

Population dynamics of tropical mammals

Compared to the large amount of work that has been done in the temperate zone, there is relatively little detailed information about the population dynamics of tropical small mammals. A perusal of the data in Table 12.4, which summarizes most of the available information on the demographic parameters of tropical rodents, will show that there is virtually no demographic data for African species (for a review of the literature see Delany, 1972), a very limited amount for neotropical species, and scattered information for Australasian species. This section of the paper will discuss three aspects of these and other

284

published data: (1) population densities and fluctuations, (2) reproduction and productivity, and (3) mortality rates.

Population densities and fluctuations

As is probably the case in temperate zones, population densities of tropical forest-dwelling rodents tend to be lower than those of grassland species. Results of a number of studies of forest-dwelling species indicate that typical densities range from 3 to 25 rodents/ha (Table 12.4). Hanney (1964) estimated the density of *Lophuromys flavopunctatus*, a forest-dwelling montane species, to be about 5 individuals/ha on the Zomba Plateau in Nyasaland. In contrast, Dieterlen (1967c) reported total densities of several species of rodents and shrews in *Pennisetum* grassland in Zaire to be as high as 278 individuals/ha. Other habitats, including fields of mixed grasses and shrubs or swampy second growth, near Lake Kivu contained small mammal densities as high as 435 individuals/ha. Although Dieterlen's density estimates are probably too high because of the unconventional method he used to gather his data, his results, along with other studies cited by him, suggest that small mammal densities can be very high in and around cultivated areas in Africa. *Rattus exulans* attains a density of 90 individuals/ha in grassland on Ponape Island in the central Pacific (Table 12.4).

There are few data on densities of rodents living in other tropical habitats. Poulet (1972b) reported that densities of *Taterillus pygargus* ranged from less than 1 to about 9 individuals/ha in thornbush savanna in Senegal. On the subtropical Hawaiian atoll of Kure, *Rattus exulans* sometimes reaches a density of 111/ha (Table 12.4).

Population densities of forest-dwelling marsupials in Panama and Queensland, Australia, are similar to those of sympatric rodents. Minimum densities of *Didelphis marsupialis*, *Marmosa robinsoni*, and *Philander opossum* in a Panamanian dry tropical forest were 0.1 to 1.3, 0.3 to 2.3, and about 0.5/ha, respectively (Fleming, 1972). In a subtropical Queensland rain forest, the density of *Antechinus stuartii* averaged 12.4/ha with a range of 3.0–24.9/ha (Wood, 1970) (cf. Table 12.4).

In addition to average density values, it is of interest to know the range of variation of tropical mammal densities. Are tropical populations, on the average, more 'stable' (i.e., less variable in density) than temperate populations? According to data presented in Table 12.4, populations of many forest species are relatively stable. Excluding those

Table 12.4. *Demographic parameters of tropical rodents*

Family and species	Locality and habitat	Median density and range (no./ha)	Length breeding season (months)	No. litters per year	Average litter size	Annual productivity	Age at sexual maturity (months)	Annual probability of survival	Source
HETEROMYIDAE									
Liomys adspersus	Panama (9° N): dry tropical forest (50 m)	8.2 (5.4–11.0)	6	1.4	4.0	5.6	c. 3	0.28	Fleming (1971)
L. salvini	Costa Rica (10° N): Dry tropical forest (45 m)	6.2 (4.0–8.4)	6	1.8	3.8	6.8	3	>0.18	Fleming (1974)
Heteromys desmarestianus	Costa Rica (10° N): wet tropical forest (100 m)	13.8 (9.9–17.8)	>10	2.9	3.1	9.0	8	c. 0.26	Fleming (1974)
CRICETIDAE									
Oryzomys capito	Panama (9° N): dry and moist tropical forests (5 & 50 m)	2.3 (0.3–4.3)	12	6.1	3.9	23.8	1.5	<0.05	Fleming (1971)
ECHIMYIDAE									
Proechimys semispinosus	Panama (9° N): dry and moist tropical forests (5 & 50 m)	3.0 (0.4–5.6)	12	4.7	2.5	11.6	c. 6	0.36	Fleming (1971)
	Panama (9° N): island, moist tropical forest (50 m)	8.5 (7.3–9.7)	12	3.7	3.2	12.0	6		Gliwicz (1973)
MURIDAE									
Rattus rattus	Venezuela (10° N): dry tropical forest (<100 m)	3.8 (1.5–6.1)	12					c. 0.05	Gomez (1960)
R. rattus argiventer	Ponape Island, Pacific Ocean (7° N): grassland, forest and palm plantation (<100 m)		12	3.2	3.8	12.2	3–4		Storer (1962)
R. rattus	Oahu, Hawaiian Islands (27° N): *Prosopis* forest (15 m)	25 (20–64)	10	2.3	5.1	11.8		<0.05?	Tamarin & Malecha (1971, 1972)
R. rattus jalorensis	Malaya (3° N): moist tropical forest and palm plantation (c. 500 m)		12	3.5	4.2	14.7		0.35	Harrison (1951, 1955, 1956)

Species	Location / habitat								Reference
R. rattus argiventer	Malaya (3° N): wasteland (c. 500 m)		12	1.1	5.5	6.2		0.26	Harrison (1951, 1955, 1956)
R. exulans	Ponape Island, Pacific Ocean (7° N): grassland, forest, and palm plantation (<100 m)	c. 90? in grassland	12	3.9	2.5	9.8	3–4	0.36	Storer (1962)
R. exulans	Oahu, Hawaiian Islands (27° N): *Prosopis* forest (15 m)	15 (0–30)	≥10	4.3	4.0	17.2		<0.05	Tamarin & Malecha (1971)
R. exulans	Kure Atoll, Hawaiian Islands (28° N): beach magnolia community (3 m)	111 (49–185)	9	1.2	3.8	4.4	2	0.45	Wirtz (1972)
R. exulans	Malaya (3° N): moist tropical forest (c. 500 m)		12	4.4	4.2	25.6		0.08	Harrison (1951, 1955, 1956)
Rattus mülleri	Malaya (3° N): moist tropical forest (c. 500 m)		12	2.3	3.8	9.2		0.23	Harrison (1951, 1955, 1956)
Rattus whiteheadi	Malaya (3° N): moist tropical forest (c. 500 m)		12	4.7	3.0	14.1		0.10	Harrison (1951, 1955, 1956)
Rattus rajah	Malaya (3° N): moist tropical forest (c. 500 m)		12	1.7	3.3	5.6		0.28	Harrison (1951, 1955, 1956)
Rattus diardii	Malaya (3° N): town (c. 500 m)		12	4.6	5.2	23.9		0.10	Harrison (1951, 1955, 1956)
Rattus bowersi	Malaya (3° N): moist tropical forest (c. 500 m)		12	1.7	4.0	6.8		0.26	Harrison (1951, 1955, 1956)
Rattus sabanus	Malaya (3° N): moist tropical forest (c. 500 m)		12	2.4	3.1	7.4		0.14	Harrison (1951, 1955, 1956)
Rattus canus	Malaya (3° N): moist tropical forest (c. 500 m)		12	2.8	3.0	8.4		0.19	Harrison (1951, 1955, 1956)
Rattus fuscipes	Queensland (28° S): subtropical moist forest (610 m)	16 (12–25)	≥4		4.4		6	?	Wood (1971)
Melomys cervinipes	Queensland (28° S): subtropical moist forest (610 m)	11 (7–16)	≥9		1.8		>6	?	Wood (1971)
Chiropodomys gliroides	Malaya (3° N): moist tropical forest (c. 500 m)		12	2.3	2.2	5.5		0.35	Harrison (1951, 1955, 1956)

species (i.e., *Oryzomys capito*, *Proechimys semispinosus*, and *Rattus exulans* on Oahu) whose densities sometimes fall below 1/ha, the mean ratio of highest/lowest observed density was 2.7 ($N = 8$) with most values ranging between 2 and 4.

Results of two other studies not included in Table 12.4 support the idea that populations of some tropical rodents are relatively stable. Harrison (1956) reported that populations of various species of *Rattus* in forests around Kuala Lumpur, were stable in size and age structure in a 2½ year period between 1948 and 1950. Trapping results from five areas in Uganda that were visited four times a year for three years (Rahm, 1967) suggest that population levels of a number of species, including *Lophuromys aquilus*, *Oenomys hypoxanthus*, *Praomys jacksoni*, *Thamnomys dolichurus* and others, were relatively stable over the period of study.

In those rodent populations that do vary in size, fluctuations tend to occur on a seasonal basis and probably often reflect seasonal changes in reproductive activity and recruitment. Population peaks usually occur in the wet season in the Panamanian, Costa Rican, and Hawaiian species that have been studied (Fleming, 1971, 1973*b*; Tamarin & Malecha, 1971) whereas peak populations of *Rattus fuscipes* and *Melomys cervinipes* in Queensland occur in the 'dry' season (Wood, 1971). Dry season population peaks also occur in rodents and insectivores inhabiting the savanna in Ivory Coast (Bellier, 1967) and in *Taterillus pygargus* in Senegal (Poulet, 1972*b*).

Reports of rodent plagues in the tropics are summarized in Fleming (1971), Dieterlen (1967*c*), and Wood (1971). Most plagues appear to occur where man has replaced native vegetation with cultivation. Under such conditions, unusually large numbers of rodents can build up and cause extensive damage to crops. In Malaya, the wood rat, *Rattus tiomanicus*, which presumably occurs at relatively low densities in undisturbed forests, sometimes reaches densities of 250 to 500/ha and it can become an economic pest in mature oil palm plantations (Wood, 1969). It is not known whether populations of any tropical rodents vary cyclically in size as populations of temperate and arctic microtines are reported to do.

Reproduction and productivity

There is considerable variation in the breeding patterns of tropical mammals, and the timing of reproductive activities can vary considerably, even among species living in the same habitat. In lowland Panama, for example, about one-half of the mammals studied to date (23 of 45 species) are seasonal breeders (Fleming, 1973*b*). The highest percentage of species with pregnant or lactating females occurs in the dry season (January through April), and most of the seasonal breeders time their reproductive activities so that young are weaned at the end of the dry season or early in the wet season. Among small Panamanian mammals, most cricetid and echimyid rodents probably breed year-round, but seasonal breeding occurs in marsupials, bats, squirrels, and heteromyid and dasyproctid rodents. Elsewhere in the Neotropics, breeding is reported to be seasonal in the cricetid rodents but year-round in the echimyid rodents *Proechimys dimidiatus* and *P. iheringi* inhabiting two forests in Brazil (Davis, 1945).

As expected for a continent as topographically and climatically diverse as Africa, there is considerable intra- and interspecific variation in the breeding patterns of small African mammals. Species such as *Taterillus pygarus* in Senegal and *Mastomys natalensis* in the drier parts of Malawi and Tanzania that live in strongly seasonal habitats reproduce seasonally (Chapman *et al.*, 1959; Hanney, 1965; Poulet, 1972*b*). Other species that are seasonal breeders in Malawi or Nyasaland include *Lophuromys aquilus, Beamys major, Thamnomys dolichurus, T. cometes, Rhabdomys pumilio, Mus triton,* and *Tatera leucogaster* (Hanney & Morris, 1962; Hanney, 1964, 1965). Hanney (1965) stated that in Malawi maximum reproduction in most species occurs in the wet season or the early part of the dry season; males of at least three species (*Saccostomus campestris, Tatera leucogaster* and *Lophuromys aquilus*), become infertile for part of the dry season.

Year-round breeding is reported in several species: *Mastomys natalensis* (although there are probably seasonal peaks) in Sierra Leone (Brambell & Davis, 1941), *Lophuromys flavopunctatus* (in which seasonal peaks also occur) and *Praomys morio* in Uganda (Delany, 1971), *Praomys jacksoni* (which has seasonal peaks) and *Acomys cahirinus* in Malawi (Hanney, 1965), and in the squirrels *Funisciurus anerythrus* and *Tamiscus emini* (both of which have seasonal peaks) in the equatorial forest in Zaire (Rahm, 1970). In his study of the reproductive patterns of rodents in the Lake Kivu region of Zaire, Dieterlen (1967*b*) noted that when data from all species were combined, there was a positive relationship between rainfall and the percentage of females pregnant or lactating. Although most species bred year-round, reproductive activity was highest in November through June, the period of greatest rainfall. In the seven most common species there were two pronounced breeding peaks; one in January through April and another in June. A similar correlation between rainfall and the occurrence of pregnancies in forest-dwelling murid rodents has been found in Gabon and Zaire (Dubost, 1968; Rahm, 1970).

The timing of reproduction is also variable in small mammals in tropical Asia, subtropical Australia, and on various Pacific islands. In a limited survey of the reproductive condition of females of 26 species in North Borneo, Wade (1958) noted pregnant females in June, July and August but not in April and May. From these and other observations, she concluded that breeding is seasonal in most species of mammals in the evergreen rain forest and that the onset of breeding corresponds with the annual period of minimum precipitation; a similar situation to that in Panama. Rudd (1965) observed little reproductive activity in several species of small mammals (mostly rodents) over a seven month period in Selangor State, Malaya. In his extensive studies of *Rattus* and other rodents around Kuala Lumpur, Harrison (1951, 1952, 1955) reported that although no distinct breeding seasons occur in *Rattus diardii*, *R. exulans*, *R. sabanus*, and *R. whiteheadi*, there appeared to be an inverse relationship between the prevalence of pregnancy and rainfall. In these species the proportion of pregnant females per three month sample period was relatively low, varying from 11 to 30 per cent, throughout the year. Males of some of the forest species (i.e., *Rattus mülleri*, *R. sabanus*, *R. rajah* and *R. whiteheadi*) underwent a reduction in testis weight, and presumably fertility, from October to March, whereas no such trend was evident in certain grassland species (e.g., *Rattus argiventer* and *R. exulans*).

Two common rodents of the subtropical rain forest in Queensland, *Rattus fuscipes* and *Melomys cervinipes*, are both seasonal breeders, but the length of the breeding season is variable from year to year (Wood, 1971). In the former species, breeding began in November and ended in February, April, or June or July in three different years. Some females of the latter species were reproductively active in every month but the main season occurred in September through May. *Antechinus stuartii*, a common dasyurid marsupial in the same habitat, also breeds seasonally with females becoming pregnant in September (Wood, 1970).

The breeding patterns of *Rattus rattus* and *R. exulans* on islands in the Pacific Ocean vary geographically. On Ponape Island (7° N) both species breed throughout the year, but *R. rattus* has a breeding peak in the dry season (January through March) whereas *R. exulans* has a peak

in July through September (Jackson, 1962). On the Hawaiian island of
Oahu (27° N), reproduction in both species is reduced in the winter i.e.,
October through April (Tamarin & Malecha, 1972), and on Kure atoll
of the same archipelago, *R. exulans* breeds mainly in May through
September (Wirtz, 1972).

At our present state of knowledge it is difficult to generalize about the
timing and intensity of reproduction in tropical mammals. The reproductive activities of many rodents, including the seasonal breeders in
Panama and Borneo and some of the year-round breeding *Rattus* in
Malaya, appear to be inversely correlated with rainfall, whereas many
African rodents breed primarily in the wet periods of the year. Regardless of geographical location, it appears that wet–dry seasonality does
affect the intensity of reproduction in most tropical mammals, even
those that breed throughout the year. Although, in most cases, the
proximate causes behind the reproductive seasonality of tropical
mammals have not yet been elucidated, it seems reasonable to postulate
that their reproductive activities are timed so that young are produced
at the most favorable times of the year, whether this means optimum
food supplies, dry nest sites, freedom from predators, or possibly other
factors.

Annual productivity, defined as the number of offspring produced per
female per year, is relatively low in tropical rodents. In 39 species for
which there are reasonably complete data productivity averages 9.8 and
ranges from 1.1 (in *Hybomys univittatus* in Uganda) to 26 (in *Rattus
exulans* in Malaya) (Tables 12.4, 12.5). Additional data used in the
above calculations include annual productivity estimates of 4.8 in
Lophuromys flavopunctatus (Hanney, 1964) and 13.3 in *Taterillus
pygargus* (Poulet, 1972*b*). Delany's (1971) estimates of productivity are
based on the number of previous litters, as determined by counts of
placental scars, and probably underestimate yearly productivity. When
annual productivity is recalculated without Delany's data, the mean
value is 11.3 ($N = 30$; range, 4.4–25.6). If annual litter production in
Mastomys natalensis were known, then the upper limit of productivity
would certainly be higher, because litter size in this species is large.
(According to Chapman *et al.* (1959), the average value is 11.2 with a
range of 3–16.)

There may be a phylogenetic basis to some of the observed variation
in productivity of tropical rodents. Fleming (1971) noted, for example,
that cricetid and murid rodents are usually more productive than
heteromyid rodents. However, since murid rodents provide the bulk of

the data, no meaningful phylogenetic comparisons can yet be made among tropical rodents.

Geographic variation in productivity is evident in *Rattus rattus* and *R. exulans* (Tamarin & Malecha, 1972; Table 12.4). Yearly productivity in *R. rattus* on Oahu was similar to that on Ponape but somewhat lower

Table 12.5. *Reproductive data for some African rodents. Data for Uganda are from Delany (1971), data for Malawi from Hanney (1965)*

Species	Av. litter size	Av. no. of previous litters (range)	Av. no. of previous young (range)
(*A*) Uganda			
Lophuromys flavopunctatus	2.2	0.51 (0–5)	1.12 (0–11)
Praomys morio	3.3	0.42 (0–6)	1.39 (0–19.8)
Hylomyscus stella	3.2	0.53 (0–6)	1.68 (0–19.2)
Aethomys kaiseri	3.5	(0–1)	(0–3.5)
Grammomys dolichurus	2.7	(0–5)	(0–13.5)
Hybomys univittatus	2.3	(0–1)	(0–2.3)
Malacomys longipes	3.5	(0–2)	(0–7.0)
Oenomys hypoxanthus	2.9	(0–1)	(0–2.9)
Thamnomys rutilans	2.0	(0–2)	(0–4.0)

Species	Av. litter size (range)	Max. no. of litters per season	Max. annual productivity
(*B*) Malawi			
Thamnomys cometes	2.9 (2–5)	2	5.8
Aethomys chrysophilus	3.2 (2–5)	1–4	6.4
Praomys jacksoni	4 (–)	≤3	12
Rhabdomys pumilio	4.5 (2–7)	≤4	18
Mus triton	6 (5–7)	≤3	18
Acomys cahirinus	3.2 (2–5)	≤3	9.6

than in Malaya. Productivity in *R. exulans* on Oahu was higher than that on Ponape or Kure atoll but lower than in Malaya. Causal factors behind this variation, which implies geographic variation in mortality rates, have not yet been identified.

Differences in body size could possibly account for some of the observed variation in productivity. In general, we should expect an inverse relationship between average adult female weight and annual productivity if Smith's (1954) general observations of a negative relationship between body size and intrinsic rate of natural increase holds true for rodents. When data for the species in Table 12.4 are plotted (Fig. 12.2), there is a slight but statistically insignificant negative relationship ($r = -0.22$, $P > 0.05$) between female weight and annual productivity. Removing the three most productive species (one cricetid, *Oryzomys capito*, and two Malayan murids, *Rattus exulans* and *R. diardii*) from the data would flatten the curve even more.

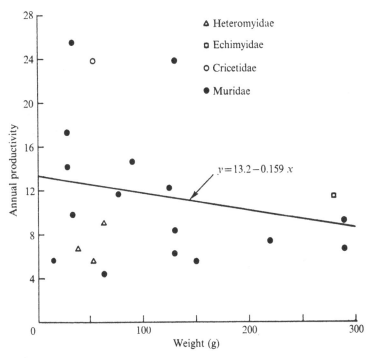

Fig. 12.2. Relationship between adult female weight and annual productivity (as defined in text) in 20 species of tropical rodents. Sources of data are listed in Table 12.4.

Habitat differences probably also influence the observed variation in productivity. Temperate grassland species generally tend to be more productive than forest species, but a paucity of data for tropical grassland forms makes it impracticable to compare tropical grassland species with forest species at this time. It is perhaps noteworthy, however, that two of the three highly productive species in Fig. 12.2 are nonforest species: *Rattus diardii* lives in houses as does *R. exulans* which also lives in grassland and scrub. The prolific African rodent *Mastomys natalensis* is also closely associated with man (Rosevear, 1969). The data on island populations of *Rattus rattus* and *R. exulans*,

293

along with data on *R. rattus jarak* from the island of Jarak in the Straits of Malacca (Harrison, 1951), suggest that low productivity and hence low turnover rates might be characteristic of rodents living on tropical or subtropical islands.

Annual productivity in the tropical marsupials that have been studied to date is similar to that of tropical rodents and ranges from 7.5 to 12 (Table 12.6). These values are probably somewhat higher than net

Table 12.6. *Productivity in tropical marsupials*

Region and habitat	Species	No. of litters per year	Average litter size	Annual produc-tivity	Source
Panama: dry and moist tropical forests	*Didelphis marsupialis*	2	6.0	12.0	Fleming (1973*b*)
	Marmosa robinsoni	1	10.0	10.0	Fleming (1973*b*)
	Philander opossum	2	4.6	9.2	Fleming (1973*b*)
Queensland: moist sub-tropical forest	*Antechinus stuartii*	1	7.5	7.5	Wood (1970)

productivity (the number of young actually weaned) because litter mortality can sometimes be high in marsupials (Fleming, 1973*b*).

Mortality rates

As in the case of annual productivity, there is considerable variation in the mortality rates and consequently the average lifespan and population turnover rates of tropical mammals. Data presented in Table 12.4 indicate that the average annual probability of survival in 21 rodent species is about 0.21 (range, < 0.05–0.45). The average annual turnover rate of these populations, therefore, is 0.79, which results in an average population turnover time of 1.26 years (see Petrusewicz & MacFadyen, 1970). Turnover times range from 1.8 years in *Rattus exulans* on Kure atoll to 1.1 years or less in those species whose annual probability of survival is 0.05 or less.

Whenever population levels remain relatively stable, as they apparently do in many tropical species, birth and death rates must be equal. Hence, we should expect to find an inverse relationship between productivity and annual probability of survival, and the data in Table

12.4 support this expectation. Annual probability of survival is inversely correlated with productivity ($r = -0.65$, $P < 0.01$), as shown in Fig. 12.3.

Although longevities of one year or more have been reported for some tropical rodents, average lifespans in most species are usually less than one year. Lifespans of more than one year, however, are apparently not

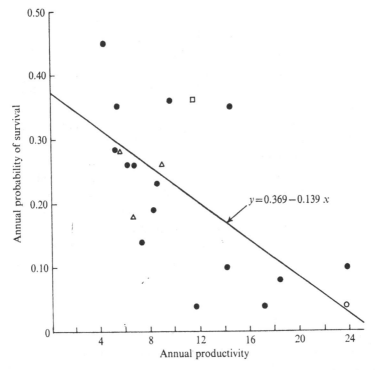

Fig. 12.3. Relationship between productivity and survivorship in 20 species of tropical rodents. Symbols as in Fig. 12.2. Sources of data are listed in Table 12.4.

uncommon in the Neotropical species *Liomys adspersus, L. salvini, Heteromys desmarestianus, Proechimys semispinosus,* and perhaps feral *Rattus rattus* (Gomez, 1960; Fleming, 1971, 1973*b*; in the African species *Tatera afra* and *Saccostomus campestris* (Allanson, 1958; Hanney, 1965) and, by inference, in the Australasian species *Rattus rattus jalorensis, R. r. argiventer, R. exulans* (on Ponape and Kure), *R. mülleri, R. rajah, R. bowersi, R. canus, R. fuscipes, Chiropodomys gliroides, Melomys cervinipes* and possibly others (Harrison, 1956; Jackson & Barbehenn, 1962; Wood, 1971; Wirtz, 1972).

Small mammals

On the other hand, particularly rapid turnover rates occur in the following species: in the Neotropics, *Oryzomys capito, Zygodontomys microtinus, Sigmodon hispidus, Akodon azarae* in temperate Argentina and *Rattus rattus* in El Verde forest, Puerto Rico (Pearson, 1967; Fleming, 1970, 1971; Weinbren *et al.*, 1970); in Africa, *Lophuromys flavopunctatus, Praomys jacksoni, Mus triton, Acomys cahirinus* (Hanney, 1965; Delany, 1971) and several Ugandan savanna species (Neal (1968), cited in Delany, 1971); and in Asia or the Pacific islands, *Rattus rattus* on Oahu, *R. rattus diardii*, and *R. exulans* on Oahu and in Malaya (Harrison, 1956; Tamarin & Malecha, 1971).

Seasonal variations in survivorship have been reported in several tropical species. In *Rattus rattus* in Venezuela and *Proechimys semispinosus* in Panama, rats born in the dry season survive better than those born in the wet season (Gomez, 1960; Fleming, 1971). Seasonal changes in food availability probably play an important part in seasonal survival differences in these two species. March-born individuals of *Lophuromys flavopunctatus* in Uganda have a maximum lifespan of 12 months whereas those born in September live only 8 months or less (Delany, 1971). Finally, individuals of *Rattus exulans* on Kure atoll first captured in late winter or summer when food levels were highest, survived better than those first captured in early winter (Wirtz, 1972).

When more data are available it may turn out that the mortality rates of tropical rodents, like their rates of productivity, will be more-or-less habitat-specific, as suggested by Harrison (1956). Species living in grassland, scrub, or in and around man's dwellings probably have faster population turnover rates than those living in forests because of their generally higher natality rates. The implication here – one that sorely needs testing – is that grassland species are so-called '*r*-strategists' in the sense of MacArthur & Wilson (1967) whereas forest species are more likely to be '*K*-strategists'. In addition to verifying whether in fact these two distinctions apply to rodents, the causal factors behind these differences are in need of study.

Conclusions

It is undoubtedly too early to reach strong conclusions about the role of small mammals in tropical ecosystems. This is particularly true because the trophic and other functional roles of bats, which usually comprise a large proportion of the species in tropical small mammal communities, are just beginning to be investigated. Despite the general lack of detailed information about the ecological impact of small mammals on tropical ecosystems, a number of tentative conclusions, many of which will probably need to be revised as further studies are conducted, have emerged from this review as follows:

(1) Many tropical small mammal communities are richer in number of species and probably number of individuals than those in temperate regions.

(2) Although taxonomic composition varies zoogeographically, the non-chiropteran portions of small mammal communities are trophically similar in all tropical areas. However, there are important trophic differences in the bat faunas and communities of the different zoogeographic regions.

(3) There appear to be consistent differences between habitats regarding species richness: grasslands usually contain fewer species than forests. Fully two-thirds of the species in some tropical small mammal communities are bats.

(4) Relative abundances and also biomass are inequitably distributed among species: one to three species are usually numerically dominant and biomass is concentrated in a few (not necessarily abundant) species.

(5) Small mammals are distributed three-dimensionally in spatially heterogeneous habitats, and there appear to be consistent zoogeographic differences in trophic and spatial distributions of species. African and Asian rodent, and probably bat, communities resemble each other more than they do Neotropical communities.

(6) Little is known at present about patterns of energy flow in tropical mammal communities. If mammal communities resemble tropical bird communities, then the energetic requirements, at least of the non-chiropteran portion of temperate and tropical mammal communities, may be similar. Because of the addition of new food habits (fruit, nectar, other vertebrates, etc.) in tropical bats, there undoubtedly are significant differences in patterns of energy flow in temperate and tropical bats.

(7) In lowland forest ecosystems, the most important functional roles played by tropical mammals are probably those of seed predators and/or seed dispersal agents and, in bats, pollinating agents.

(8) Population densities of tropical rodents appear to be higher in grasslands than in forests. Forest species appear to have relatively stable population levels from year to year.

(9) Wet–dry seasonality has a strong effect on the reproductive patterns of tropical mammals, many of which are seasonal breeders. Annual productivity per adult female in a number of rodent species is

low, averaging eleven offspring, and within-species productivity is inversely related to annual survivorship. Island-dwelling forms appear to have lower productivity and turnover rates than continental relatives. Rodent species associated with man, his crops, or habitats created by his disturbance, are characterized by particularly high productivity and turnover rates. Replacement of native forests and other vegetation by cultivation will favor these species and their potential economic impact will probably increase in the future.

Finally, it is necessary to stress that more work urgently needs to be done if the ecological and evolutionary importance of tropical mammals is to be fully understood and appreciated. In no other region of the world is man disturbing and disrupting highly complex, coevolved plant–animal associations faster than in the tropics. Although Delany (1971) noted that recent man-induced habitat modifications in the Mayanja Forest in Uganda increased, rather than reduced, the number of sympatric rodent species, such results cannot always be expected to occur. On the contrary, an ultimate decrease in the species richness of tropical habitats should be anticipated because of the increased probabilities of population extinction that accompany reductions in habitat size (Terborgh, 1973; Willis, 1973). For this reason it is imperative that research efforts directed towards gaining a better understanding of the ecological role of small mammals in tropical ecosystems be increased.

I thank M. J. Delany and J. Gliwicz for their comments on this paper. The Smithsonian Institution and US National Science Foundation have generously supported my research in tropical America.

13. The role of rodents in ecosystems of the northern deserts of Eurasia

N. P. NAUMOV

The objective of the IBP studies on biological productivity is not only to obtain a deeper insight into the principles and mechanisms of the productivity of different types of ecosystems, but also to find ways and means of their control for the sake of man. Detailed and extended investigations of basic processes of productivity and the linkage between these and other processes are necessary in order to detect those which can be controlled.

It is useful to distinguish in the complex ecosystem (biogeocoenosis) relatively separate and independent units of species interrelated through trophic relations in food chains or cycles. Two such units can be distinguished in desert ecosystems:

(a) Communities of wild and domestic ungulates, usually feeding on grasses and dwarf semishrubs (for instance, *Artemisia*). Their numbers and distribution depend not only on food abundance, but also on the presence of water-holes, particularly in the summer period. A group of large and medium-sized predators, birds and tetrapoda, and species of ecto- and endoparasites are associated with the ungulates. In addition, the condition of the plant cover subjected to the activity of troops of ungulate is influenced by the occurrence of a number of mammals, particularly Lagomorpha, birds, reptiles and invertebrates. The numbers and distribution of all these species largely depend on the number of ungulates and their impact on the pasture.

(b) Communities of small mammals, mainly rodents, which only partly utilize grasses and *Artemisia* and feed mainly on vegetative parts, fruit and seeds of many halophytes, shrubs and semishrubs, bulbs and other below-ground parts of perennial plants. Small mammals have more predators and parasites than ungulates. Predominate groups of predators include birds, mammals and also some reptiles. Endo- and ectoparasites are represented by many diversified and numerous groups such as arthropods (*Ixodes*, other ticks, fleas, lice), helminths, protozoans, bacteria, rickettsiae and viruses. Many of them carry and transmit dangerous diseases of man. In each region with rodent colonies several hundred species of arthropods and many dozen species of worms, protozoans and bacteria are associated with small mammals (Pavlovskii, 1964; Naumov *et al.*, 1972).

The intensive burrowing activity of the majority of desert rodent species has more pronounced effects on the desert soil and vegetation than grazing by ungulates. Due to the presence of a number of underground shelters, and sometimes very complex and deep burrows with stable microclimates, many vertebrate and invertebrate animals that need a relatively stable, high humidity can penetrate and inhabit deserts. For example, the majority of arthropod parasites can live only in the rodent burrows. Also, *Bufo viridis* occurring far into deserts and semideserts find their shelters in the burrows. A number of reptiles and birds construct their nests or lay eggs there and insectivorous mammals that frequently have no shelter of their own also live in the burrows. In this way, the group of plant and animal species associated with small mammals in desert ecosystems is more extensive than that associated with the ungulates.

Both these animal communities are interrelated through food resources, consuming a number of common plant species. But each utilizes the plants in different ways and interactions between these groups of animals within one biocoenosis are considerably less pronounced than those among the species within an animal community. Each of them preserves a large degree of autonomy.

Certainly, the group of small mammals is of particular importance in their community. In addition, one or more dominant or 'key' species, which determine the general pattern of the community, can usually be distinguished in any region of the desert zone. These species usually have high and relatively stable numbers, complex and persistent shelters, burrows and high burrowing activity. In semidesert areas *Citellus* frequently predominates in northern and southern deserts along with *Meriones, Rhombomys, Tatera* and less frequently *Spermophilopsis*. Jerboas (family Dipodidae), hamsters (subfamily Cricetinae) and other small mammals such as mice, voles, hedgehogs and shrews, rarely occur in large groups and do not dig much, preferring to utilize the retreats of other species. They are thus secondary members of the community having little importance.

In the semideserts of Kazakhstan and Ciscaucasia (Precaucasia), *Citellus pygmaeus* is the key or dominant species. Its burrows are up to 180 cm deep and become complicated because of successive reconstructions. They are distributed in groups of 10 to 15 burrows located around small hills of earth, formed of the excavated earth. These hills can be 60 cm high and 8 m in diameter. Breaking the soil and the excavation of salt earth changes the conditions of growth for plants and increases the degree of mosaic, species diversity and, frequently, the plant cover. In the dense populations of *C. pygmaeus* an average of about 30 such small hills per hectare can occur. The diet of *C. pygmaeus* contains about 60 species of wild plants. Thus, the role of these animals in the productivity of plant cover may be considerable (Formozov & Voronov, 1939; Sludskii, 1969) and this role is still greater due to the wide distribution of the species. In deserts, this species rarely occurs and does not form dense populations. It is frequently replaced there by another species, *Citellus fulvus* (Khodashova, 1953; Belyaev, 1954; Sludskii, 1969). The number of *Citellus* is relatively stable due not only to the stable food resources and the burrows it inhabits, but also to a winter hibernation of 5 to 7 months. Their burrows are used by many small mammals and sometimes by birds as temporary shelters. Seventy species of invertebrate inhabitants have also been found in *Citellus* burrows, including transmitters of a number of dangerous diseases (Nel'zina & Medvedev, 1962).

In the desert zone, species of *Meriones* are the most important group of small mammals. In river fens and fixed sands covered with vegetation, *Meriones tamariscinus* predominates, but it is sporadically distributed and only in a few areas are its numbers high. Thus, its role in desert ecosystems is rather insignificant and it cannot be regarded as a dominant or key species (Kim, 1960; Khrustselevskii *et al.*, 1963; Nurgel'dyev, 1969).

A similar role in semidesert ecosystems of Syria, Jordan, Israel, Turkey, western Iran and Transcaucasia is played by *Meriones blackleri* and in Iran and Afghanistan by *Meriones persicus*.

Sandy steppes of central, southern and northeastern Mongolia, and northern and northeastern China are inhabited by *Meriones unguiculatus*. The numbers of this species can be very high. In Transbaikalia 32 burrows with 760 outlets can be found per hectare (Leont'ev, 1954), and in Mongolia 504 outlets per hectare on sandy areas and up to 2 800 along roadsides with 70 to 176 animals per hectare have been recorded (Lavrinenko & Tarasov, 1968). In different years, the density fluctuates by dozens and even hundreds of times. The lifetime of these species does not exceed 2 years and averages 3 to 4 months. In practice the population is fully renewed during one year. The settled animals do not

Small mammals

undertake distant migrations, and generally travel no further than 700 m (i.e., they do not pass beyond the range of the population), but a case is known of a migration of 40 km made by a young animal along the side of a road (Leont'ev, 1962).

Feeding on a large number of plant species (with preference for the vegetative parts in summer and for seeds and fruit in winter) and also storing up to 20 kg of seeds per burrow, indicates that the role of *M. unguiculatus* in the life of semidesert and desert ecosystems of Central Asia is highly significant (Lavrinenko & Tarasov, 1968).

In sandy deserts of the Volga–Ural basin, Kazakhstan, Middle Asia, central and southern regions of Mongolia, northwestern and central China, northwestern Afghanistan and northeastern Iran, the most important species among small mammals is *Meriones meridianus*. This species prefers unstable sands and in fixed sands covered with vegetation, *Rhombomys opimus* dominates (Ismagilov, 1961; Khrustselevkii *et al.*, 1963; Nurgel'dyev, 1969; Reimov, 1972). The number of animals in favorable habitats (biotopes) averages 5 to 10 *Rhombomys*/ha, and fluctuates depending on the yields of basic food items, such as seeds of grasses and shrubs, but only in very unfavorable biotopes does it drop to a very low level. The burrows of these animals are very complex; their depth ranges from 2 to 4 m and the length is up to 4 m. They are characterized by a complicated system of galleries with two to four chambers where seeds are stored. The burrows of *Meriones meridianus* are less complex and resistant than those of *R. optimus*. The burrowing activity of *M. meridianus* affects the development of vegetation on sand. (Rall, 1938–9; Pavlov, 1959.)

The role of *M. meridianus* in sandy biotopes of deserts is a result not only of their high density, which is stable in optimum biotopes, but also of their high metabolic level. In *M. tamariscinus*, which often lives in the neighborhood of *M. meridianus*, there is not a large seasonal change in oxygen requirements (124–126 per cent difference), while the oxygen consumption in *M. meridianus* increases by 139 to 175 per cent. The metabolic rate in winter is 6 to 13 per cent higher than in summer for *M. meridianus*. *M. meridianus* is also more resistant to overheating, their critical temperature being 30 to 35 °C compared to 25 to 30 °C for *M. tamariscinus* (Kalabukhov, 1969). Also, in the northern regions, *M. meridianus* can breed throughout the year, and because of the rapid growth and maturity of the young they can reproduce at the age of 2 to 3 months. The role of *M. meridianus* in the ecosystem consists not so much in the direct effect on plant cover as an effect on higher trophic levels. This species is a host for a number of ecto- and endoparasites, including those transmitting and inducing many dangerous diseases of man and animals. Its burrows are inhabited by a number of invertebrate animals, while *Meriones meridianus* themselves are a food of almost all species of predatory birds, mammals and reptiles.

An important role in semidesert and desert ecosystems of Asia Minor and Central Asia, northwestern China and Transcaucasia is played by a subspecies of *Meriones libycus*, which has been named *M. erythrourus*. It occurs in ephemeral deserts with relatively thick loess and loess–sandy soils, fixed and well-fixed sands, and talwegs of dry valleys and ravines; it can be frequently observed in nonflooded parts of the fens of the desert rivers and also favors oases, orchards, alfalfa, cotton and grain crops, sides of ditches, waste lands, old-fields and dwelling houses (Rudenchik, 1959; Ismagilov, 1961; Davydov, 1962, 1964; Alekperov, 1966; Nurgel'dyev, 1969). The complexity of their burrows is similar to that of *R. opimus*; each of them occupies an area of several dozen square metres and has 10 to 60 outlets. When the number of animals increases, the burrows unite with one another and a continuous settlement is formed over large areas. These animals pass the winter in groups of 20 to 30 individuals which are gathered in one burrow from several adjacent ones. They feed mainly on seeds but in summer they consume large quantities of the vegetative parts of plants and in dry years also below-ground parts; nevertheless, in summer their diet is 80 per cent fruit and seeds. They store food for winter in their burrows and the amount of stored food can reach 10 kg per burrow.

Potentially, this species can reproduce throughout the year, but breeding is usually stopped during the dry period in summer (Vasil'ev *et al.*, 1963; Alekperov *et al.*, 1967). Reproduction rate in *M. libycus* can vary within large limits in relation to weather conditions and food supply in the environment. Thus, the numbers of these animals are more variable than for other species of this genus. In Turkmenya, very high densities of *M. libycus* occur every 7 to 8 years and at the peak density the population can contain up to 150 to 200 individuals/hectare (Fenyuk & Radchenko, 1957; Nurgel'dyev, 1969). Such increases are usually associated with considerable changes in the ecosystem. Intensive burrowing activity and grazing of plants can result in elimination of many plant species or considerable reduction in the abundance of others. In spring, these animals damage up to 20 per cent of the pasture vegetation, and they also damage the seedlings of fruit trees and crops.

303

Small mammals

In semideserts and deserts of Kazakhstan, Central Asia, northern Iran and Afghanistan, northwestern China, southern Mongolia and partly in Asiatic subtropical steppes, the most important small mammal species is certainly *Rhombomys opimus*. This species should be called a key species in that it largely determines the character of the ecosystem it occupies. Its morpho-physiological, ecological and ethological characters include dense fur in winter, feeding mainly on vegetative parts of shrub and woody halophytes (succulents and *Artemisia*), intensive storage of food in summer, high burrowing activity, construction of extremely persistent and complex retreats, use of burrows with an almost stable microclimate, diurnal activity, high degree of sociability and the development of complex acoustic, chemical and visual signaling. These characters indicate an excellent adaptation to the severe conditions of Asiatic deserts and secure its position in desert ecosystems of this region. This position explains its important economic role as a pest of pastures, saksaul forests and earth constructions and, at the same time, as a carrier–transmitter of a number of dangerous diseases of man, such as bubonic plague, rickettsiosis, and many others. For these reasons this species has been intensively studied from various points of view and now is one of the best known species of rodents. In the Soviet Union alone many hundreds of publications concerning this species have been issued.

The adaptation of *Rhombomys opimus* to the life in deserts consists of the reduction of life processes during unfavorable periods (without hibernation) and in maximum profit from favorable periods. This species has a strong body constitution and high physical ability and mobility, although its total metabolism is not high. Basal metabolism, for instance, is 1.5 times lower than in the rodents of the same size living in temperate zones and is lower than that of other gerbils (Shcheglova, 1962). Its adaptation to dry conditions are pronounced. Oxygen consumption at 20 °C is lower in *R. opimus* (2100 ml/kg) than in the laboratory rat of the same weight; accordingly, the pulse is lower (220–320 against 340–440/min) (Slonim & Shcheglova, 1963). The tissues of *R. opimus* are resistant to dehydration. Modifying its behavior and utilizing burrows under unfavorable weather conditions, this species uses its energy very economically (Kalabukhov, 1969).

The distribution of *R. opimus* depends largely on soil–ground conditions. These animals do not have effective burrowing abilities and avoid ground that is too hard, whereas *Ellobius* and jerboas can burrow without difficulty. *R. opimus* looks for ground of suitable mechanical composition, i.e., where there is a thick layer with stable moisture conditions between the surface water layer and the ground water level), and digs elaborate, deep burrows in which it spends most of its time, breeds and hibernates. These stable retreats are frequently restored and reconstructed and are utilized by the animals for a very long time. A 25-year study revealed that within that period the basic burrows did not essentially change their character. This is because a group of animals occupies several burrows, moving successively from one to another, restoring and reconstructing them, and transferring from generation to generation a ready system of persistent retreats. Some of the burrows can occupy an area of more than 1000 m² and have hundreds of outlets, systems of galleries at different levels and a number of large chambers for nests and stored food (Naumov, 1954; Rotshil'd, 1957; Balbabas *et al.*, 1965; Leont'eva, 1966; Naumov *et al.*, 1972).

The burrows of *R. opimus* are permanently or temporarily utilized by a number of invertebrate and vertebrate species which form a complex and stable community. Even in southern regions the inhabitants generally contain several hundred species (300 to 500) including worms, molluscs, myriapods, insects, ticks and spiders, as well as a number of species of reptiles, amphibians (*Bufo viridis*), some birds (*Saxicola, Upupa, Tadorna tadorna, T. ferruginea, Athene* and others), small mammals (shrews, hedgehogs, gerbils, jerboas, voles and *Citellus*), and also small and middle-size predators. All of them find permanent or temporary retreats in the burrows of *R. opimus* and a number of them can live only in them. In the burrows of *R. opimus* of Khe Kazakh desert, 219 species of invertebrate animals were recorded, while beyond them only 38 species were caught (Dubinin, 1946, 1954). Of 12 species of mosquitoes, 9 were observed in the burrows of *R. opimus*.

Mortality or extermination of *R. opimus* populations by man by burying outlets results in a decrease in numbers and even elimination of a number of invertebrate and vertebrate species. Moreover, cohabitants suffer even more than the owners of the burrows (Naumov *et al.*, 1970).

The life and stability of *R. optimus* populations are closely related to the possibility of constructing and supporting the burrows. Optimum environmental conditions exist in the alluvial grounds, where the burrows are the most solid, resistant, complex and deep, and they are particularly suitable for breeding and passing the winter. In such biotopes, which can be named 'centers of survival' or refuges, the most stable populations continuously exist (Naumov, 1954; Naumov *et al.*, 1972). Our studies indicate that such territories occupy no more than

305

Small mammals

5 to 10 per cent of the total area inhabited by *R. opimus*. They are surrounded by less suitable habitats where the animals can breed and pass the winter, but only in favorable years. Temporary populations are formed there which die and are formed again by animals coming from the refuges. Long-term observations of marked animals indicate, however, that the total number of animals emigrating from refuges during the period of several years is lower than the immigration of animals that settle in temporary habitats (Fig. 13.1).

The populations occurring in refuges are characterized by a high density of burrows per unit area and are mainly composed of settled and organized animals, while the surrounding biotopes are occupied by sparse populations of unstabilized and mobile animals, i.e., by less organized populations.

General stability of the population containing settlements of different types, increases with increasing area of refuges. Poorer territories are inhabited by the animals only temporarily, Such organization of the population, which contains temporary and permanent components with different degrees of stability and intensive exchange of animals between them, contributes to the dynamic utilization of the territory and effective response of the species to the changes of environmental conditions. Limiting the control of *R. opimus* only to the refuges enables one to economically and effectively regulate their numbers and make efforts to eradicate natural loci of bubonic plague, the permanent part of which (elementary loci) is related, as it could be expected, to the refuges of *R. opimus* (Naumov *et al.*, 1972).

R. opimus populations have a complex social organization. An elementary group is seldom composed of a one-pair family except early in the spring. Higher mortality of male rodents during the breeding season generally results in joining several families into a more complex group. They are called 'big families' (Eibl-Eibesfeldt, 1950; Anderson, 1965), 'parcel groups' or 'population parcels' (Naumov, 1967, 1971). These are groups of 15 to 120 animals, generally containing several adult males, more adult females and also their offspring. All animals of a group are in more or less stable and close contact, i.e., they are 'personally known to each other' and the relations between them are determined by less or more uniform behavior of the total group. The population parcel utilizes the whole system of nest and supplementary burrows, which are occupied either by single animals (more often males), or females with offspring or groups of young animals. Within a parcel, the inhabitants often move from one burrow to another. The territory of a parcel ranges from 4 to 20 ha and is defended against intruders by all members of the parcel. Unity of the group and uniform habitat of its members is realized by a complex system of acoustic, optical and chemical signals. The first kind of signal is generally used to warn of danger and the two remaining are used for intraspecies communication.

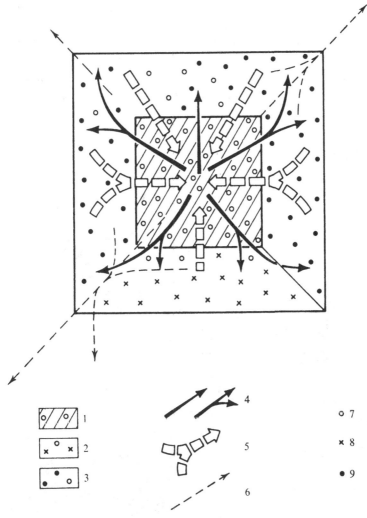

Fig. 13.1. The outline of interactions among elementary populations of a refuge of *Rhombomys opimus* (Naumov *et al.*, 1972). 1, stable population in a 'center of survival'; 2, temporary settlement in hilly sands; 3, the same in clay-sandy lowlands; 4, emigration of animals from the centers of survival; 5, re-emigration from temporary settlements; 6, long-distance emigration of young animals; 7, persistent, deep burrows; 8, deep but less persistent burrows; 9, shallow, not persistent burrows.

Small mammals

The parcels located in refuges are larger than the temporary ones and their internal organization is highly developed. They are the most stable groups and are composed chiefly of adult settled animals, the major part of which remain in the same place until the end of their life. Only young individuals move beyond the parcel, with a small number of them remaining in the parcel replacing dead or emigrating adult individuals.

Parcels located in less suitable biotopes contain smaller numbers of individuals; young individuals predominate in them and the parcels themselves are less stable and persistent. Accordingly, their internal organization is weaker. Their important feature, a high exchange of individuals among parcels, results from a generally high mobility of the animals. In such temporary populations animals of different origin meet

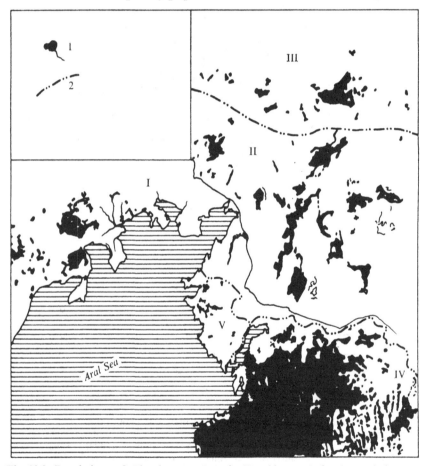

Fig. 13.2. Populations of *Rhombomys opimus* in Kazakhstan. 1, local populations; 2, geographic populations. I–V represent types of geographic populations.

308

with one another and thus exchange of genes occurs among them, i.e., natural selection is particularly intensive. The refuges, in turn, are the places of reproduction and accumulation of mutations in the population.

Study of the structure of *R. opimus* populations requires in practice the mapping of populations and distinguishing between different types of settlement (Naumov, 1954, 1967; Smirin, 1963; Rotshil'd, 1967; Naumov *et al.*, 1972). Such a map is shown in Fig. 13.2.

The burrowing activity of *R. opimus* and its feeding on vegetative and generative parts of many desert plants has a considerable and complex influence on plant cover. From the point of view of man's interests, this influence is ambiguous. The animals almost completely destroy vegetation on the surface of their burrows, where only barren patches are left. But loosening the soil, fertilizing with feces and plant remains, as well as partial grazing of the plants, stimulates plant growth and increases the productivity of plant cover. At the same time these impacts result in a number of new species appearing in the areas transformed by these animals. Among these new species are a number of annual grasses and Cruciferae that supply valuable food for domestic animals. Such zones of improved growth of basic plants and the introduction of new species occurs at the periphery of the burrow surface and within a large belt, 10 to 30 m wide, around it.

When the number of *R. opimus* naturally decreases, or if this species is exterminated by man, these valuable introduced species are eliminated after several years and the growth rate of the basic species of food plants of desert pastures decreases. The quality of the pasture deteriorates rather than improves (Rotshil'd, 1957, 1968). For that reason control of *R. opimus* cannot be total but should be undertaken with reference to the complex of local conditions.

The damages made by *R. opimus* to woody-shrub desert plants, particularly to saksaul, are significant. Drying of the ground by the burrows of *R. opimus* often results in death of this valuable tree. Moreover, these animals continuously feed on fresh shoots, which results in their deformation and also causes death of the saksaul. At the same time, well-developing, strong trees live at the periphery of the area occupied by the burrows of *R. opimus*, especially when a depression is formed on the burrow by the wind in which snow accumulates in winter. In these locations the soil is moister and more fertilized. These examples indicate that the role of *R. opimus* as the key species of the Kazackh desert is very important and the character of its effects on desert ecosystems is complex and often contradictory.

14. Applied research on small mammals

14.1. *Control of field rodents*

A. MYLLYMÄKI

With an increasing human population needing more food and animal feed, and striving for a higher standard of living, it has become necessary to channel all possible production to man. Vertebrate pests, predominantly small rodents, are one of the main competitors of man for agricultural products and, in certain conditions, for production of timber and wood fiber. Therefore, intensified rodent control is needed. However, they must be controlled without undue side effects from chemical pesticides and the resulting environmental deterioration.

Some of the most prominent rodent problems, such as the periodic outbreaks of microtines, have been known since the time of Aristotle and Pliny (Kemper, 1968) and are generally considered to be the result of the creation of large monocultures (Altum, 1876; Herold, 1954; van Wijngarden, 1957; Frank, 1956, 1957a). In addition to these classical outbreaks, new niches for field rodents have been created or old habitat extended, as in irrigation areas in the USA and elsewhere (Wolf, 1966; Howard, 1967b; White, 1967; Clark, 1972; Ryan & Jones, 1972). Although controversial, many new problems (or enlargements of old ones) are a consequence of the rationalization of agricultural or forest practices. For example, reforestation of old fields, clear-cutting of forests with fertile soil, and 'packeting' of cultivated land to prevent agricultural overproduction, has extended the favorite habitats of the field vole (*Microtus agrestis*) in Finland by one million hectares in the course of a few years (Myllymäki, 1970). Plantations of forest trees on these outbreak habitats are in immediate danger of being damaged by voles.

This paper reviews the impact of field rodents on agriculture and forestry, and considers the possibilities of controlling the injury. Even if the approach is global in principle, a great majority of examples are from the temperate region. This apparent restriction is due mainly to the shortage of exact information from the tropical areas and is certainly not due to the absence of problems. Also, the impact of mammals on

311

unmanaged tropical vegetation is discussed in Chapter 12 and on temperate vegetation in Chapter 10. For the same reason, the public health problems connected with rodent outbreaks are outside the range of the present review and are briefly considered in Chapter 14.2 and 14.3. However, principles and means of rodent control are, for the most part, independent of whether the problem is approached from the agricultural or medical viewpoint.

The types of damage and small mammal groups concerned

Problems in European agriculture, horticulture and forestry

According to the returns of a recent EPPO (European and Mediterranean Plant Protection Organization) questionnaire, damage by the common vole, *Microtus arvalis*, on cereals, alfalfa, sugar beets, pastures, etc., was recorded as the most common and economically most important type of rodent damage in European countries. Only the southern and middle parts of Italy, the Iberian peninsula, the British Isles, Iceland and the Scandinavian peninsula are outside the range of this pest. Detailed reviews on outbreaks of *M. arvalis* have been made in several countries, e.g. France (Harranger, 1967), Germany (Maercks, 1954), the Netherlands (van Wijngaarden, 1957), and Poland (Migula *et al.*, 1970). Some of these reviews include estimates of the economic value of the damage, but the basis for estimating the losses is rather subjective. In addition, much of the present information is written in languages which make it practically unavailable outside the particular countries' borders.

In the southeastern USSR the common vole is replaced by another species, *Microtus socialis*, and in the Middle East by the Levant vole, *Microtus guentheri*. Outbreaks of the Levant vole have been described by Bodenheimer (1949, 1958) and Wolf (1966).

In the Scandinavian countries (Norway, Sweden and, for a great part, Finland), the field vole, *Microtus agrestis* has replaced the common vole as the main agricultural rodent pest. However, outbreaks in field crops are restricted to special occasions such as the cases of hay destruction in northern Sweden and Finland referred to by Kanervo & Myllymäki (1970). *M. agrestis* mainly causes damage to fruit trees, woody-stemmed ornamental plants and, especially, forest trees. Quite new subjects for damage have been created recently; for example, grafts in the seed orchards of forest trees (Myllymäki, 1970). The damage is essentially the barking of tree stems during winter under the snow cover. In the 1920s and 1930s, problems of this kind were common in Great Britain (Elton, 1942) and after World War II in Germany (Frank, 1952; Schindler, 1956, 1972), but today the problem is crucial mainly in the Scandinavian countries. A summary of field vole outbreaks, damage and estimates of its economic value in these countries prior to 1969 was given by Kanervo & Myllymäki (1970).

In Central Europe, the bulk of fruit tree barking is done by *M. arvalis*, while part of the damage on forest trees is done by *Clethrionomys glareolus* (Turcek, 1960). There is little evidence of damage made by *C. glareolus* on forest trees in Scandinavia. Buchalczyk *et al.* (1970) reported that *Microtus oeconomus* barked trees in Poland.

The third common type of rodent-born damage in Europe is that caused by the subterranean rodents, including the seasonally subterranean forms of *Arvicola*. The most widely distributed species (or species group) is *Arvicola terrestris*, which is reported as an agricultural or forest pest in most of the European countries. The suberranean form occurs, at least, in France, Belgium, Germany, Austria and Switzerland (Meylan & Morel, 1970). Seasonal changes of habitat are characteristic of a part of North and East European *Arvicola* (Myllymäki, 1964; Panteleev, 1968), while some West European forms regularly live along the banks of rivers and other watersheds (van Wijngaarden, 1954; Davis, 1970). A thorough summary of the occurrence and damage by *Arvicola* in Germany is given by Klemm (1958). This small mammal nearly always damages the roots of trees and other cultivated plants. In Denmark (P. Bang, personal communication) and Finland, the populations of *Arvicola* have shown a recent upward trend and, consequently, increasing damage can be expected in the future.

Other subterranean, or partly subterranean, rodents considered as pests on cultivated plants are at least two species of the genus *Pitymys*, two species of *Spalax*, the sousliks, *Citellus* spp. and the hamster, *Cricetus cricetus*. Sousliks and hamsters are polyphagous and often eat the above-ground parts of crops. They are considered major pests in the countries on the lower Danube, and in the steppe regions of the USSR.

The generally synanthropic rats, *Rattus norvegicus*, *Rattus rattus*, and the house mouse, *Mus musculus* (especially *M. m. spicilecus* in the Balkans), also occur as pests on cultivated plants. Several species of *Apodemus* are also reported as important pests. Major problems from murids are common in the Balkan countries and the Southern USSR. Recently an unusual outbreak of *M. musculus* was reported in Australia (Ryan & Jones, 1972).

Small mammals

Arboreal rodents, *Sciurus carolinensis*, *Sciurus vulgaris* or *Glis glis*, are considered to be forest pests in some European countries (Platt & Rowe, 1964; Taylor, 1970). The first sign of squirrel problems has appeared in the oldest seed orchards of pine in Finland (Myllymäki, unpublished results).

Rodent problems in North America

The genus *Microtus* is common in North America and various types of damage by *Microtus* are also well-known and not negligible. However, it is impossible to compare European and New World economic losses, due to a lack of information on New World damage. Bird (1930) and Hanson & Whitman (1938) concluded that the main consumer of herbs and grasses on northern prairie grasslands, was not bison or any other big herbivore, but *Microtus drummondii*, a subspecies of *M. pennsylvanicus*. It is uncertain whether this and other species of *Microtus* may be pests within the framework of range management. Recently, there have been outbreaks of *Microtus* in irrigated fields with newly introduced alfalfa in the western USA (Anon. 1957–8; Howard, 1967a; White, 1967).

The meadow vole, *Microtus pennsylvanicus* (Ord.) is a pest in orchards and forests in the states surrounding the Great Lakes and in central Canada (Littlefield *et al.*, 1946; Jokela & Lorenz, 1959; Cayford & Haig, 1961; Hayne & Thompson, 1965; Buckner, 1972).

More attention has been paid in the USA to the destruction of conifer seed on reforestation areas by voles than on the debarking of trees (Radwan, 1963, 1970; Radvanyi, 1972). This is because in the USA direct seeding is a more important forest renewal method than in Europe. The white-footed mice, *Peromyscus* spp., are the most important seed eaters, but ground squirrels (*Citellus* spp.), chipmunks (*Tamias* and *Eutamias* spp.), and shrews (*Blarina*, *Sorex*) are of significance (Radwan, 1970). Buckner (1972) also lists *Microtus pennsylvanicus* as an important seed predator. The efficiency with which small mammals find conifer seeds could result in a failure of seeding programs in the absence of preventive measures. Despite relatively little attention to the seed depredation problem in Europe, this role of small mammals has recently been demonstrated by Myllymäki & Paasikallio (1972 and unpublished results).

American rodents occupying approximately the same niches as the European *Arvicola* are called pocket gophers (mainly *Thomomys* spp.).

314

Pocket gophers range from irrigated fields with alfalfa (Howard & Childs, 1959) to rangelands (Keith *et al.*, 1959; Hansen & Ward, 1966) and to the forests in all the western states of the USA (Hermann & Thomas, 1963). Howard & Childs (1959) considered pocket gophers as the most important rodent pest in Californian agriculture, and Canutt (1970) ranks pocket gophers second of the vertebrate forest pests in Oregon.

Ground squirrels (*Citellus* spp., *Spermophilus* spp.) and chipmunks (*Tamias, Eutamias*) (Alsager & Yaremko, 1972; White, 1972) are a variable group in regard to the number of species as well as to the range of habitats occupied, and the type of injury done. These animals are the counterparts of the European susliks. At least locally, many other rodent species such as cotton rats (*Sigmodon*) (Clark, 1972), and squirrels (*Sciurus* spp.) may cause damage to crops and forests.

Tropical rodent problems

There are few data on rodent losses in the tropics; one of the few exceptions, however, is the rat problem in Hawaiian sugarcane. Sugarcane grows through a two year cycle without any rotation of other plants. Three species of rats, the little rat (*Rattus exulans*), the roof rat (*R. rattus*), and the Norway rat (*R. norvegicus*), cause losses averaging 4.5 million dollars per year where 30 per cent of the stalks may be injured (Hood *et al.*, 1970; Teshima, 1970). Considerable damage to sugarcane has been reported from many other tropical areas, species of rats (*Rattus*) being the most common causative agents.

Another major problem in Pacific areas and also reported elsewhere in the tropics is coconut damage caused by *R. exulans* and *R. rattus* (Smith, 1967; Papers in Asia–Pacific Interchange Proceedings, 1968; Smith, 1968; Wodzicki, 1969*a, b*). Other tropical agricultures commonly suffering rodent damage are cocoa plantations in West Africa (Everard, 1964), and oil palm groves in the Malaysian–Polynesian region (Lever, 1962). Attacks on gum, tea gardens and various kinds of nuts have also been reported.

In India *Rattus rattus, R. norvegicus, Bandicota bengalensis, B. indica*, and the Indian gerbil *Tatera indica* cause considerable economic damage. Most of these species are field and commensal pests, and do most of their damage to cereals. Vague estimates on numbers of rodents and percentage damage to cereal crops are available. Pingale (1966, quoting Deoras), gives figures ranging from 10 to 35 per cent. Even if the more critical estimates by Singh (1967) are nearer the truth, the total losses to rodents are close to the annual import of grain to India. In addition, major damage is done to sugarcane, cotton, sesame, peanuts and watermelon.

Taylor (1968) reported a 34 per cent loss of wheat and 23 per cent loss of barley during a rodent outbreak in Kenya in 1962. Eight species of rodents were regarded as pests, the most important being *Rattus natalensis, Arvicanthis niloticus* and *Rhabdomys pumilio*. According to R. Chiomba (personal communication) a variety of rodent species play a detrimental role in Tanzanian agriculture. The relatively scanty literature concerning field rodent problems in tropical Africa and South America is mainly due to the lack of investigations.

Methods of field rodent control

Ecological control through habitat manipulation

All ecologically-minded scientists working on rodent control would agree that only through environmental manipulation of rodent habitats will the destructive outbreaks of field rodents be permanently removed. However, there are only a few articles in the literature reporting actual progress and factual solutions using this approach. Instead, principles of ecological control have been applied unconsciously as a by-product of normal agricultural practices.

Influence of cultivation methods on outbreaks of Microtus arvalis. Van Wijngaarden (1957) has documented the close relationship between the intensity of cultivation and the rise and disappearance of outbreaks of *Microtus arvalis* in the polder areas of Zuiderzee. New polders are characteristic plague zones during the transitional grassland phase after the water is pumped out and before the settlers have moved in. Later, intensive agriculture with grazing by sheep on dams and verges, totally prevents vole populations from increasing to outbreak numbers. Frank (1956) has, partly on the basis of the Dutch experience, presented an ecological control program for the plague zones of the common vole in northwestern Germany. There are no data to measure to what extent his ideas have been realized.

The conditions necessary for *M. arvalis* plagues in the Netherlands and Germany are not the same as outbreak conditions for the same species in Hungary. My experience during a recent visit to the Hungarian plague area was that permanent grasslands of extensively grazed pastures are often non-existent. The propagation of voles leading to outbreak numbers happens *in situ*, in the same fields of alfalfa or cereals, which are subject to damage later on. Ecological solutions must be more complicated here; consequently, in practice the rodents are controlled predominantly by chemical means. On the basis of the literature (Spitz, 1967, 1968), the situation is probably rather similar in France (Vendee).

Present views on the ecological control of attack on trees by Microtus agrestis. Ecological control of tree barking by *Microtus agrestis* (or *M. arvalis*) varies greatly according to circumstances. Total removal of food and cover by means of mechanical surface treatment may result in complete protection of apple trees in Danish orchards (P. Bang, personal communication), but cannot be considered to be of much value in Finland (Myllymäki, 1967, 1970). Here the snow cover shelters invading voles, and once invasion occurs the fruit trees are eaten as other food is depleted. The herb layer vegetation is usually luxuriant in seed orchards and the bulk of the damage to the trees is done by voles living in the orchard at the onset of winter.

Present trends in Finnish agriculture and forestry, as in Sweden and to a lesser extent Norway, create new habitats for the field vole (Myllymäki, 1970). Mechanical harvesting of timber produces large cut-off areas which are rapidly covered by grass. Seedlings on these clearings and on afforested arable land are in danger as soon as they are planted. Large-scale application of herbicides to remove food and cover from these areas would be prevented due to economic or environmental considerations. Feeding with preferable food might be suggested as a useful tool in preventing damage on fruit trees. Reliance on this method has, however, resulted in dangerous situations as exemplified by Kanervo & Myllymäki (1970).

Rodent-resistant varieties of fruit and forest trees could also be selected as tools of ecological control. Indeed, some local varieties of the Finnish spruce are clearly less palatable for the field vole than the Central European strains (Myllymäki, 1967).

Reduction of numbers of pocket gophers by means of herbicide application. Pocket gophers (*Thomomys talpoides*) were reduced in number by taking advantage of a weak point in their ecology and applying a selective herbicide (2,4-D) in Colorado rangelands (Keith *et al.*, 1959; Tietjen *et al.*, 1967). A reduction of 80 to 90 per cent compared to the pre-treatment population was not due to the direct toxicity or repellency of

317

the chemical, but to a subsequent change in the proportion of succulent forbs and grasses. Pocket gophers are dependent on the availability of certain forbs and a change in their diet from 82 per cent forbs and 18 per cent grass to 50 per cent forbs and 50 per cent grass was probably the explanation for the remarkable reduction in their numbers.

The effect of fire on seed eaters. The great interest in conifer seed loss in the USA has led several investigators to survey changes in rodent populations in connection with the burning of forest ground (Tevis, 1956; Arata, 1959; Cook, 1959). Results show that fire does not influence the number of seed-eating species, but changes the habitat of herbivorous species like *Microtus*. Thus, burning of slash cannot be considered an effective ecological method of preventing seed depredation.

Protection by means of mechanical barriers or chemical repellents

Mechanical prevention of barking on trees. Mechanical barriers preventing barking of stems are recommended both for their effectiveness and for being ecologically sound. The cost of the potential damage outweighs the costs of material and labor, at least in orchards and seed orchards. Appropriate materials are mesh fencing, aluminium foil, hard plastic, etc. Usually every tree is provided with a protective collar, but whole seed orchards have been fenced in Sweden (Hadders, 1968).

The least expensive material, aluminium foil, is widely used on fruit trees in Finland (Myllymäki, 1967, 1970). In seed orchards low aeration inside the alumium collar causes damage to young pine grafts and, consequently, the foil must be removed in the spring and put on in the fall or replaced by expensive mesh fencing.

Protective collars around the stems of fruit trees are also in common use in North America (Koval *et al.*, 1970). A related application of mechanical barriers is their widespread use in protecting coconuts from rats (papers in Asia-Pacific Interchange Proceedings, 1968).

Chemical repellents against rodent depredation. Because olfaction plays an important role in the feeding behavior of rodents, research toward averting these pests from cultivated plants, power cables and packages through application of chemical repellents has been rather intensive (Armour, 1963; Welch, 1967). Nearly 300 substances are tested yearly for mammal and bird repellency at the Denver Laboratory (Welch, 1967). Interest in rodent repellents is also high in Scandinavia, especially in Sweden, due to a lack of appropriate poisons and the public aversion to poisoning of vertebrates. Unfortunately, rodents can gnaw into their favorite food through the repellent-treated surface. The anatomy of the rodent mouth makes it possible to discard unpalatable or poisonous layers of a treated tree stem without harm (Myllymäki, 1970), and then devour the cambium beneath. Among the numerous repellents tested since the 1950s in Finland, only two have shown any marked effect; both of them were, however, phytotoxical.

The idea of making the grafts or seedlings of forest trees unpalatable for rodents by use of an excess potassium as suggested by some practical foresters in Finland, has not yet been verified experimentally. Indeed, excess potassium could become unfavorable to the growth rate of the trees. Another related, and theoretically possible, approach would be the use of systemic poisons applied in the same way (Myllymäki, 1970).

Seed coating repellents. Coating of conifer seed with repellents has been the main method of avoiding seed depredation in direct seeding programs in the USA and Canada (Cone, 1967; Radwan, 1970; Radvanyi, 1972). Aluminium powder–endrin–arasan–latex has been the most widely accepted coating mixture but recently Radvanyi (1972) found a new chemical, called R-55, to be superior to the earlier combination. It also improved the germination rate. When applied to tree stems in Finland, however, the same repellent was not able to protect the trees against attack by the field vole and probably the seed coating technique is not completely effective either as indicated by Cone (1967).

Reduction of numbers of rodents by means of rodenticide application

Advance prevention of rodent outbreaks is seldom achieved, and once the rodents are present, they usually are subjected to rodenticide treatment. However, large-scale application of rodenticides against field rodents has recently been criticized on several grounds including side-effects on game animals and other non-target wildlife, environmental pollution, and the sometimes questionable efficiency of the method. The following review shows that this criticism is partly justified.

Chemicals used for destruction of field rodents. Rodenticides (Gratz, 1966; Hermann, 1969) are divided into two main groups: (*a*) promptly acting *acute* rodenticides and (*b*) *chronic* rodenticides, usually anticoagulants. The first group of chemicals is by far the most important in

319

field rodent control. This is due both to economic reasons and to effectiveness. Some of the most important acute rodenticides in present use are described below.

Sodiumfluoroacetate, or '1080' and fluoroacetamide, '1081', are both extremely hazardous substances. Despite this fact and the lack of any antidote, these toxicants are, or have recently been, used against field rodents in several countries, including the USA and tropical areas. Good acceptance of baits treated with these compounds is one of the most important justifications for their continued use.

Strychnine is highly toxic and has the drawback of being poorly accepted by rodents (Howard, Palmateer & Nachman, 1968). It has been used recently for rodent control in the USA and France (e.g., Giban, 1967; Hood, 1972).

Thallium sulfate shares most of the drawbacks and advantages of sodium-fluoroacetate and fluoroacetamide. It also is still used in several countries. In connection with its present restricted use against *Arvicola* in Finland, cases of secondary poisoning have been detected.

Crimidine ('Castrix') is used in some European countries (Denmark, Norway, Germany, France) against *Microtus* and *Clethrionomys*. Its high toxicity is balanced by an effective antidote, sodium pentobarbital, and the fact that it is rather unpalatable. This same property is, in turn, a serious drawback in its use against the target species.

Gophacide, a highly toxic organophosphorous compound, has been one of the few new rodenticides developed during recent years. With known antidotes and a certain degree of specificity (pocket gophers, *Microtus*, *Pitymys* and *Dipodomys* being the most sensitive rodents), it was considered as one of the most promising rodent poisons (Richens, 1967). However, due to a high cutaneous toxicity, any further experiments were cancelled by the manufacturer.

Zinc phosphide is the most commonly used acute rodenticide today 'due to its fairly good safety record, low cost, and reasonably high effectiveness' (Gratz, 1966). Zinc phosphide disintegrates in the presence of alkalis or acids to zinc oxide and phosphine (PH_3) which is the gaseous substance killing the target animal. Neither of these decomposition products could be considered as serious environmental contaminants. The potential for secondary poisoning exists, especially for cats and dogs, but in practice accidents have been found to be rather minimal (Hood, 1972).

The weathering of zinc phosphide bait is a problem. In contrast to the general belief that zinc phosphide is well accepted by target species

(Gratz, 1966; Hood, 1972), *Microtus agrestis* can discriminate between a 4 per cent bait and a 2 per cent bait of zinc phosphide, and distinguish a 2 per cent bait from a non-poisonous bait (Myllymäki, unpublished results). Low acceptance, probably depending on the active ingredient, has been noted elsewhere (G. Nechay and A. Stenmark, personal communications), and in the USSR zinc phosphide is used as a seed coating repellent (V. A. Bykovski, personal communication). Still zinc phosphide is used against various rodent pests all over the world, especially in the developing countries.

Endrin, toxaphen and endosulfan, chlorinated hydrocarbon insecticides, were introduced for the control of microtine rodents in several European countries during the latter half of the 1950s (Schindler, 1955). Endrin treatments were found extremely effective against *Microtus agrestis* and *M. arvalis* (Frank, 1965; Myllymäki, 1970; Lund, 1972a). Treatments with toxaphen were slightly less effective, while the present endosulphane treatments, e.g., in Hungary, are questionable. As a consequence of a recommendation by EPPO (in the Report of the International Conference on Rodents and Rodenticides, 1967), the use of endrin and toxaphen was prohibited or restricted all over Europe in the 1960s. During the past five years two new chemicals, one organo-phosphate and one carbamate compound, were tested with promising results against *Microtus agrestis* in Finland. Both were later cancelled by the parent companies.

The use of chronic anticoagulant rodenticides is negligible on field rodents. However, chlorophacinone baits are recommended for the control of *Microtus arvalis* in France and some other European countries despite the fact that often only 70 per cent of target animals are killed (Giban, 1970; Grolleau, 1971; Nikodemusz, 1973). Chlorophacinone has also been tried in muskrat control but with variable results (Giban, 1968, 1972; Moens, 1968). In England, anticoagulant bait appears promising against *Sciurus carolinensis* (Taylor *et al.*, 1968). In tropical areas, where murid species are the most important field pests, anticoagulants have been used, but often their use is economically unfeasible.

Gaseous poisons, such as phosphine, carbon monoxide etc., are often used to control subterranean rodents (Van den Bruel & Bollaerts, 1960; Bollaerts & Tahon, 1968). Hamar & Sutova (1970) report excellent results with phosphine (aluminium phosphide) against hamsters (*Cricetus cricetus*), while the same treatment against *Spalax leucodon* was unsuccessful (Hamar *et al.*, 1970).

321

Modes of application of rodenticides. The simplest poison application method is surface spraying. In this treatment, all the food of a local population of herbivorous rodents is poisoned and the problems of bait acceptance circumvented. Seen from the point of view of efficiency, the short duration of the 'endrin era' was by far the most advanced period in field rodent control in Europe.

Baiting technique is by far the most common form of application of rodenticides. Here the rodenticide is incorporated with a foodstuff that inevitably competes with the natural food items of the target rodent. The weak point of the technique concerns problems with acceptance and bait shyness due to the initial warning signs from sublethal doses of poison. In pen tests Howard, Marsh & Cole (1968) could confirm that certain 'lures' like safflower oil and lecitin mineral oil improved both the detection and palatability of four types of grains by deer mice (*Peromyscus*). On the contrary, I was able to evoke only an orientation response by *Microtus* to baits provided with artificial fruit essences, and the bait was not consumed in the presence of a favorite natural food. Further, there was little success in preparing a bait pellet of 'green meal', made of young alfalfa or clover and dried apple, that could successfully compete with natural food, even though the same constituents would be preferred by *Microtus* if fresh (Myllymäki, unpublished results). Consequently, one must make a distinction between olfactory 'lures' initiating the rodent's response, and the final palatability of the bait.

While it is extremely difficult to find a bait base which is competitive with naturally occurring food items and is marketable and easily distributed, the acceptability of the bait is lowered further by the addition of the unpalatable poison substance. To my knowledge, only one rodenticide substance, the anticoagulant coumatetryl, has improved the palatability of the bait (Lund, 1972*b*).

The situation of poor palatability of baits is exaggerated in dealing with strictly herbivorous rodents such as *Microtus* and *Arvicola* in Europe. It is generally known that control campaigns on *Microtus arvalis* with grain baits (e.g., zinc phosphide as the active ingredient), are mostly unsuccessful. Even less success could be expected by the application of zinc phosphide, crimidine grain, or artificial commercial baits, to the control of *M. agrestis* or *Arvicola*, which are less fond of cereals than *M. arvalis*. However, baits with zero acceptability in critical field tests are generally sold in several European countries: this could be called purely commercial rodent control. The only type of bait readily accepted by these species and, e.g., the muskrat (*Ondatra zibethicus*), are fresh vegetable baits, like apples or carrots.

One more drawback of the baiting technique in its conventional form is the cost of distributing the bait, especially when manual preparation and placing of fresh vegetable baits is concerned. Only mechanical bait distribution is economically competitive with surface spraying (Frank, 1965), but unfortunately it also limits the types of baits that can be used. Large-scale aeromechanical broadcasting of poison baits has been used against several rodent pests in the USA (White, 1967; Marsh, 1968) and the USSR (Panteleev, 1968). Artificial burrow-builders/bait applicators (Ward & Hansen, 1962) have been used for pocket gopher control in the USA.

Fumigation of rodent burrows has, in some instances, proved to be a useful tool against subterranean species. The application of fumigants, mostly phosphine-producing granules or tablets of aluminium phosphide, is usually done manually. Van den Bruel & Bollaerts (1960) discussed a special mode of application planned to minimize personal risks, and Kemper & Kock (1969) used a burrow-builder for distribution of phosphine-producing granulate in orchards infested by *Arvicola*.

Biological control methods

Predators. The role of predators as a regulating factor on populations of rodents and as a potential agent in preventing or depressing rodent outbreaks, is a controversial question (Howard, 1967a). Much of our present understanding of the prey–predator relationship is based on the 'principle of compensation', first clearly documented by Errington (1946). Losses by predation, as many other losses in a rodent population, are compensated for through increased fecundity and life expectancy of the survivors. Howard (1967a) has stressed the beneficial effect of predation on rodent populations. This theory agrees well with the concepts of optimum harvesting (e.g., application of beneficial artificial predation) of game or fish populations. An illustrative case history of the responses of a population of *Arvicola terrestris* to artificial predation is given by Myllymäki (1974a).

New information about predation and about reproductive potentials of predators and their rodent prey, does not support the idea of controlling rodents by protecting naturally occurring predators or introducing new ones. However, there are some examples in the literature of temporary successes after the introduction of new predators, such as mustelids (Hiraiwa *et al.*, 1959; van Wijngaarden & Mörzer Bruijns,

1961), although the introduced predators have often also caused some harm (Howard, 1967*b*).

One predator which, perhaps, could keep rodent populations permanently at a low level in certain conditions is the domestic cat. Elton (1953) recognized that a sufficient number of cats could keep the surroundings of a farmhouse rat-free if provided only with supplementary food. Frank (1956) emphasized the advantages of scattered settlement in the country as it provides an even distribution of cat-zones around the farmhouses, which would function as 'service stations' for the vole-preying cats. Ryszkowski *et al.* (1973) recently tried to quantify the impact of domestic cat predation on a local population of *M. arvalis*. Supplementary feeding makes the domestic cat an exceptional predator which does not suffer from the periodic lows in rodent populations, which often decimate populations of natural predators.

Pathogens. Experimentation with several strains of *Salmonella*-type pathogens during the early decades of our century (Elton, 1942), provided rather pessimistic views on the usefulness of bacterial agents in rodent control. These early experiments were stopped for two reasons: first, the epidemic did not spread among the natural populations of field rodents as was anticipated, and secondly, objections based on verified or suspected health hazards for man and domestic animals were voiced, leading finally to a total ban of bacterial rodenticides in several European countries. However, specific strains of *Salmonella enteritidis*-type bacterial preparations are employed in large-scale field rodent control in the USSR (Kandybin, 1971). The percentage of kill has varied between 65 and 100 per cent, the microtine rodents being the most susceptible.

Evidence of bacterial (or other microbial) epidemics fatal to natural populations of field rodents is fragmentary. Elton *et al.* (1935), found an epidemic caused by *Toxoplasma* in a declining population of *Microtus agrestis*, and were able to transmit it with fatal outcome to healthy animals in the laboratory. A corresponding situation was found on the same species in Finland in the spring of 1962, *Erysipelothrix rhusiopathiae* being the causative agent this time (Myllymäki, unpublished results). Pearson (1966) hypothesized that *Pasteurella tularensis* was the controlling factor causing periodic die-offs of Scandinavian lemming (and vole) populations. In fact, the decline phase of unusually severe vole outbreaks and a widespread epidemic of tularemia in man coincided in northern Sweden and Finland in 1967.

324

These few examples show the potential of bacterial (or viral) diseases in declines of natural rodent outbreaks, and point out the reservoir of possible organisms for biological rodent control. According to Herman (1964, cited by Howard, 1967a), the following rules must be fully considered before any introduction to rodent control is attempted:

(1) The pathogen must be highly lethal to the subject species.
(2) Long-term changes in the survival of the subject species, including possible resistance, must be anticipated.
(3) The pathogen must be host specific.
(4) The pathogen must be self-perpetuating in the natural environment.
(5) It should be possible to monitor the control program both to insure the progress and to avoid unanticipated side-effects.

The discovery of a praxis-applicable pathogen is possible, as exemplified by the well-known case of myxoma virus in rabbits. However, the search for another effective pathogen may take a long time.

Chemosterilants. A chemosterilant is defined by Marsh & Howard (1970) as 'a chemical that can cause permanent or temporary sterility in either or both sexes or, through some other physiological aspect, reduce the number of offspring or alter the fecundity of the offspring produced'. The physiological means of interference could be variable: damage to developing or mature gametes, prohibition of implantation, abortion, neonatal interference with sexual development in the offspring, reduced lactation, etc. The sterile male technique, predominantly applied in economic entomology, is considered less successful in regard to the polygamous and perpetually reproducing rodents than female sterilization.

The main interest in the search for antifertility agents has centered on steroid hormones. Two highly potent synthetic steroids, mestranol and diethylstilbestrol, have been in the focus of recent interest (Howard & Marsh, 1969; Marsh & Howard, 1970; Alsager & Yaremko, 1972). Howard & Marsh (1969) experimented with mestranol on rats (*Rattus norvegicus*) and voles (*Microtus californicus, M. montanus*) and found that this chemical, consumed by lactating voles, was effectively passed on to the offspring resulting in irreversible sterility in most of the offspring. An apparent drawback was initially limited acceptance, leading to increased aversion to successive feedings, which may prove critical in field conditions. Of the male sterilants, 'U-5897' (3-chloro-

1, 2-propenedrol), a single feeding antisterility agent, seems to be most promising (Marsh & Howard, 1970).

One of the weak points of application of chemosterilants is that they must be exposed in bait form. Besides the disadvantages of poor acceptance and conditioned bait shyness, the costs of application would not be economical because at least several feedings are needed with most of the present compounds.

While the theory and knowledge of the physiological action of the chemosterilants is rather advanced, the ecological application of antisterility compounds in practical rodent control is still in its initial stage. At present, the most sensible application would be to prevent population recovery after conventional poisoning with rodenticides in garbage dumps, croplands and recreational areas. Marsh & Howard (1970) discussed the possible use of chemosterilants in suppressing diseased populations of ground squirrels (*Spermophilus*), and Alsager & Yaremko (1972) have initiated experiments including chemosterilants as a part of an integrated control program on Richardson's ground squirrel (*S. richardsonii*) in Alberta.

Elements of integrated rodent control programs

The basic starting point for any integrated rodent control program should be a proper knowledge of the species to be controlled, their potentials for causing damage if not controlled, actual and predictable levels of population size and limitations and hazards of the control method to be applied.

Assessment of rodent-born damage

Determination of the pest species. The initial step of dealing with the rodent-born damage is the identification of the pest species concerned. Material for identification is usually collected by means of trapping (the trapping procedures are dealt with in the next chapter). General cues for handling the catch and acquiring of expert advice is briefly discussed by Taylor (1971).

Attributing a certain type of damage to a given species cannot usually be done on the mere basis of its presence and abundance. Feeding experiments, microscopical diet analysis (e.g., Hansson, 1970) or tagging of the damaged item by radioactive tracers (Myllymäki & Paasikallio, 1972; Radvanyi, 1972; etc.) are examples of methods usable for pest identification.

Unit counts. The least ambiguous type of damage is tree barking by *Microtus* or de-rooting by *Arvicola*. A girdled tree will die, and its value is totally lost. In the case of valuable items, as fruit trees or grafts in seed orchards, total unit counts could be performed, while in forest plantations an adequate number of samples should be taken. As pointed out by Taylor (1971), a completely random sampling scheme and small sampling plots leads to a predominance of zero counts and should be avoided. Postponing counting of the damaged items towards the end of the growing season is recommended; then the dying seedlings are easily discovered by the color of the needles or leaves. According to Taylor (1971) unit counts could be applied in crops like coconut or cacao, where the nuts and pods are the units. Maize or sugarcane, however, are more complicated because the estimates of the final value of the damage on cane must include several kinds of secondary losses, including discontinued growth of the stalks and attacks by insects and fungi as well as lowered sugar content of the harvested cane (Hood *et al.*, 1970; Teshima, 1970).

Damage to field crops. Judenko· (1967) provided an example of the complexity and pitfalls of estimation of rodent damage in cereals. He studied various consequences of rat (*Rattus norvegicus*) destruction on maize. Initial destruction of 28.0 per cent of the stalks at an early stage was followed by compensatory growth of the remaining plants, so no more than 18.4 per cent was lost in total green weight, 15.1 per cent was lost in the total weight of all cobs and only 8.5 per cent was lost in the current value of marketable cobs.

Ryszkowski *et al.* (1973) in Poland have developed an 'exact' procedure for estimating the impact of *Microtus arvalis* on field crops in an agricultural ecosystem consisting of four main components: alfalfa, rye, potato fields and shelter-belts. Their approach consists of the following steps: (*a*) determination of the density of the rodent population, (*b*) indication of food consumption by the voles, (*c*) estimation of the ratio consumption/destroyed crop, and (*d*) consideration of the long-term effect from year to year. However, this procedure is too laborious to be accepted as a general procedure in primitive conditions. At least two more simple, but consequently less precise, approaches can be considered. These are the use of isolated plots which the rodents cannot enter or, enclosures from which they cannot escape, and the determination of damage at given threshold densities in controlled conditions, and consequent generalization of these findings to large-scale

327

density determinations. All three methods are more or less dependent on the estimation of density, which is discussed in detail in Chapter 2.

Myllymäki (unpublished results) has used the isolated plot technique in experimental field studies on *Microtus agrestis* by studying the grazing by *Microtus* in an outdoor enclosure planted with three species of grasses and red clover. Selective feeding on clover at an early stage was highly effective and resulted in the disappearance of this vegetation

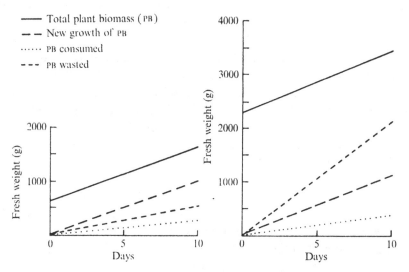

Fig. 14.1. Green plant biomass consumed by a female *Microtus agrestis* in relation to the material wasted and the new growth during two consecutive 10 day feeding trials in outdoor enclosures.

component. The rodents cut off and wasted an exponentially increasing amount of grass, but the wasted material was compensated by the regrowth induced and accelerated by the depredation. Another experiment with individually fenced *Microtus* on a mixture of two grasses and red clover (Fig. 14.1) demonstrated a rapid increase in the proportion of wasted material from cut-off green forage when the vegetation grew older. One of the main contributions of these findings might be the concept that threshold density is not independent of the growth stage of the vegetation.

The preceding experiments concentrated on finding useful methods to predict anticipated damage. Measurements of the final yield at harvest time are usually too late for that purpose, but could be useful in elucidating the economic importance of rodent problems and in demonstrating the benefits of control measures.

Surveillance of rodent populations and prediction of damage

Surveillance based on visual observation. The simplest method of estimating rodent abundance is the 'reading of signs', e.g., grass cuttings, droppings, trails in the grass, mounds, earth plugs, etc. Hayne & Thompson (1965) developed an appraisal system of estimating the *Microtus* hazard to orchards in Michigan and Wisconsin. As crtiteria for abundance and, consequently, hazard they used a graded scale from 0 to 10, based on the frequency as well as the 'heaviness' of signs on the 25 random inspection points in each plot. By means of this simple method Hayne was able to survey 150 to 200 observation plots twice a year. The success of the method is largely dependent on the personal skill and tenacity of the observer, and is surely most applicable when all the observations are made by a single person, as Hayne did.

In Europe, a comparable method, the counting of filled and reopened galleries, has been widely used for surveillance of the abundance of *Microtus arvalis* on cultivated fields (Bernard, 1959). Recently, most of the scientific workers on this species have been critical at least of the predictive value of these counts, and are stressing the necessity of more advanced methods. However, the counting of mounds and earth plugs was found to be related to the population counts by intensive trapping in the case of the mountain pocket gopher (*Thomomys talpoides*) in Colorado (Reid *et al.*, 1966).

Surveillance based on sampling by traps. Several kinds of trapping procedures can be considered as basic means of surveillance of rodent populations today. At least for immediate response and short-term prediction, simple index methods are most suitable. Generally they are based on snap trapping on lines, as the well-known NACSM-trap lines in the USA, the index lines recommended by Linn (1954), Spitz (1965) and Hansson (1967) or the 'trap row' of Schindler (1959). Recently, some apparent drawbacks of the line arrangement of traps led the author to experiment and propose (Myllymäki *et al.*, 1971), an alternative index method, known as the Small Quadrat Method (SQM). The main advantages of the SQM are that (*a*) the sampling unit could more easily be placed to represent a given, definable type of habitat than the trap line, and that (*b*) the SQM catches could probably be transformed to density values better than the line catches (cf., however, Brant 1962). Hansson (1972) concluded that 'at present no better method is available for forecasting studies'.

Small mammals

Elaborate trapping procedures, designed for determination of 'absolute density' values, such as exhaustive snap trappings, in the method by Stein & Reichsten (1957) or the Standard Minimum for IBP (Grodziński *et al.*, 1966) or various modifications of CMR (catch-mark–release) methods, are too time-consuming for general surveillance of agricultural and forest pests. A critical review on trapping methods is presented by Smith *et al.* in Chapter 2.

Short-term forecasts. Predictions of rodent outbreaks are usually based on simple sign-count indices or on numbers of trapped animals. The work is generally carried out by plant protection services or similar organizations. Sign counts have recently been used for the prediction of *Microtus arvalis* in Central and Eastern Europe, while trap indices are commonly used to predict population trends of *Microtus agrestis* in Northern Europe, including Germany (Schindler, 1959; Myllymäki *et al.*, 1971; Hansson, 1972; Myllymäki, 1974*b*). More advanced methods are being developed simultaneously in several countries.

Critical seasons approach. In the USSR an organized warning service concerning all important pests and diseases of cultivated plants has existed for nearly half a century (Polyakov, 1958). According to Polyakov (1959) the methods are generally based on mapping techniques and the concept of a 'critical season', e.g., the spatial distribution, numbers and physiological condition of animals during the most unfavorable season. The production at the end of the favorable season is also measured. Two surveys are therefore conducted each year, but, depending on local conditions, even three or four could be needed. The critical periods are separately determined for each species and geographical region. For example, Polyakov (1959) showed that, for *Microtus socialis*, the critical season is winter in the steppe region of Crimea and summer in Azerbaijan.

Klemm (1960), apparently influenced by the Russian work, has applied the concept of a critical season to German populations of *Arvicola terrestris*. On the basis of a ten year mapping of the relative indices of the species' occurrence, he believed he was able to predict the population development six months in advance, on the basis of precipitation in February and March.

Forecasts based on population parameters. Spitz (1970) bases his 'medium-term' forecasts (about two months) on density indices and determinations of population structure, e.g., reproduction parameters and age structure, and uses maps of the distribution of infested fields in autumn as a criterion for predicting population levels next spring. Buckner (1972), in turn, found that key factor analysis (Morris, 1959) based on juvenile survival was a reliable indicator of population trends in *Microtus pennsylvanicus* in Saskatchewan and Manitoba. Unfortunately, he did not describe the primary method of estimating the juvenile survival.

The Scandinavian approach to predicting outbreaks of *Microtus* (and *Clethrionomys*) involves large-scale surveys based on SQM which are run every year. At the moment only short-term warnings have appeared, but records on environmental and population parameters are being collected according to a common scheme. It is hoped that a key method for the further processing of these data will be found in the near future. Among others the potentials of systems analysis (cf. French, 1969; Bunnell, 1972) are studied.

The discussion above concentrates solely on the problems and their solutions in temperate regions. So far as the tropics are concerned, the situation may be summarized by quoting Taylor (1968), who states that 'lack of knowledge about the natural regulation of rat numbers prevents outbreaks being forecast with any certainty'.

Efficacy and selection of control methods

This section presents a brief discussion of control measures that can be applied when a rodent population has been detected and its potential to cause damage has been assessed. In such circumstances it is often too late to apply ecological control methods (which are usually designed to prevent rather than to control an outbreak) and the remaining alternatives are destruction of the rodent population and the protection of the objects of potential damage.

Efficacy of rodent control agents. Independent of the theoretical classification of rodent control agents into poisonous or biological (pathogenic, sterilizing, etc.) substances, their biological efficacy should be tested prior to large-scale application in field conditions. In most countries, such tests are required by law. With the additional consideration that they must be safe to use, only rodenticides that perform well in these tests are, or should be, officially approved and registered for general use. The main tests that a potential rodenticide substance must undergo are:

Small mammals

(1) Screening tests designed to evaluate the toxicological properties (e.g., LD_{50}) of the substance on the target animal.

(2) Additional laboratory tests on substances showing promising performance in screening tests. Tests for acceptance by animals are essential when the potential rodenticide is to be applied in bait.

(3) Field tests, or at least tests in outdoor pens, are necessary for any substance intended to be used against field rodents. The field tests should be conducted in conditions such that comparisons can be made between population numbers before and after the treatment, as well as between treated and untreated plots.

It is not possible to give general priorities for any single control method since its efficacy often depends on the situation in which it is applied. However, the three examples discussed below may be of interest since they illustrate different problems and solutions.

Organized warning service and obligatory control of the common vole in Hungary. Twice a year each of 19 local plant protection stations estimates the abundance of *Microtus arvalis* on a number of plots. The method used now is the 'reopened holes method', but a more advanced system is under development (G. Nechay, personal communication). If the results of a survey indicate that an outbreak is imminent, estates are obligated to institute control procedures.

Two control methods are currently used: surface spraying with endosulphane and zinc phosphide baits, of which the former now has priority. The treatments are not considered to be very effective. In order to minimize hazards, endosulphane is used in suboptimal doses, and zinc phosphide baits are not readily accepted by the voles. From this point of view the system could be criticized, as well as for the probability that the compensatory responses in the vole population create a continuous need for control. However, through this feedback mechanism serious outbreaks have been prevented since 1965. In principle, the Hungarian system probably provides an example where chemical rodent control, involving adequate rodenticides without environmental hazard, could effectively be used alone. It has also been demonstrated that the wheat yield on a treated field was 4.070 kg/ha compared to 2.480 kg/ha on an untreated field. The experiment was done in strictly comparable conditions during an outbreak year of *Microtus arvalis* (K. Grozdics, personal communication).

Control of the field vole in seed orchards in Finland. Cooperation between the Institute of Pest Investigation, the State Forest Board and the central organization of private forestry ('Tapio'), has made it possible to develop a centralized system integrating information on the abundance of *Microtus agrestis* and practical control measures (Mylly-mäki, 1974*b*). Prior to the cooperation (in the 1960s) the control of voles was based on incidental surface spraying with endrin, and the number of lost trees was considerable.

Since 1969, yearly surveys by means of small quadrat trappings have been conducted in nearly all seed orchards, now amounting to about 3000 ha (Fig. 14.2). Recommendations for control have been based on catch indices and local conditions. As Fig. 14.2 shows, the tendency has been towards non-poisonous control (e.g., aluminium collars and guards of wire net on the trees) even though investments in these materials have accounted for up to 60 per cent of all control costs. In addition to restricting the treatments with endrin to the minimum and to areas where the effect of a treatment is not likely to be invalidated by invasions, our supervision has produced up to a 60 per cent saving in the total running costs (mostly labor).

Experiments on ground squirrels in Canada. Alsager & Yaremko (1972) have provided an interesting example of a search for an economic, effective and safe solution (even though it is still not a final solution) to the control of ground squirrels (*Spermophilus richardsonii*). The authors were experimenting with methods of bait application and found mechanical bait applicators and portable bait stations unsuitable, while the best method was found to be the use of disposable tubes, and aerial application. The authors suggest that numbers of ground squirrels could be reduced by means of poisoned baits and kept at a low level by subsequent aerial distribution of chemosterilants in baits, i.e., a large-scale realization of the ideas put forth by Marsh & Howard (1970). Still another method, the use of amphetamine to reduce body weight and increase winter mortality during hibernation, is discussed.

The three examples above show that there can be a variety of solutions to rodent control problems, depending on the species to be controlled and the local conditions. Two of these case histories also serve as examples of the integration of surveillance and control measures.

Safe use of rodenticides. Due to the relatively close physiological relationships between rodents, other wildlife and man, effective rodent killers often have undesirable side-effects. Unfortunately, the ideal specific rodenticide has not yet been found.

Hazards connected with the use of traditional acute rodenticides are already well-known and predominantly consist of direct poisoning of

333

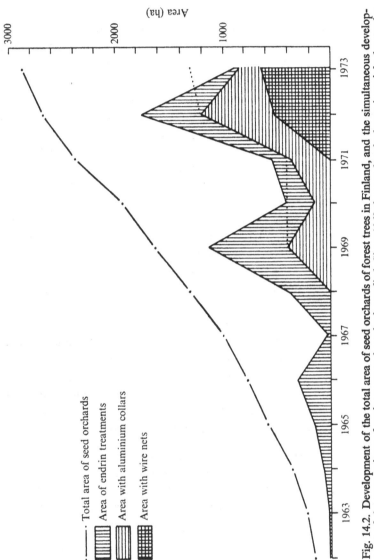

Fig. 14.2. Development of the total area of seed orchards of forest trees in Finland, and the simultaneous development of the proportions of various control methods applied. Since 1969 the non-hatched area is mainly due to the absence of unnecessary treatments.

Legend:
- .... Total area of seed orchards
- Area of endrin treatments
- Area with aluminium collars
- Area with wire nets

Area (ha)

3000
2000
1000

1963 1965 1967 1969 1971 1973

pest controllers themselves, children, non-target animals, or, as in the case of thallium sulfate, mostly of secondary poisoning of predatory animals. In the recent report by the WHO Expert Committee on Insecticides (WHO Technical Report 513, 1973), the acute rodenticides are placed in three groups from the point of view of safe use.

(1) Acute rodenticides requiring ordinary care. Except for two typical raticides, 'scillirosid' and 'norbormide', the most important poison against field rodents, zinc phosphide, is placed in this category. The Committee thus endorses its use.

(2) Acute rodenticides requiring maximal precautions. These rodenticides are recommended only for restricted use by educated professional pest controllers. The group includes sodium fluoroacetate and fluoroacetamide, and strychnine (the last mentioned, however, is predominantly used for the control of vertebrate pests other than rodents).

(3) Acute rodenticides too dangerous for use. These rodenticides, recommended to be totally banned, are arsenic trioxide, phosphorus, thallium sulfate, alphanaphthylthiourea, and gophacide. Of these, thallium sulfate is the only one generally used in rodent control.

The Committee's recommendations include an endorsement of the use of anticoagulants, when appropriate in field rodent control. The group of rodenticide substances not dealt with in the Committee's recommendations mentioned above, the chlorinated hydrocarbons, are not in general use since the ban on endrin. Hurter *et al.* (1966), and Rautapää *et al.* (1972), found clearly recognizable amounts of endrin in the top soil 1 to 3 years after the (last) treatment. However, Morris (1970, 1972) could not detect any detrimental effect of endrin residues of 1 to 2 ppm in grass on the recovery of populations of *Microtus* subjected to experimental treatment with endrin. On the basis of this finding and the relatively rapid excretion of sublethal doses of endrin reported by Klein *et al.* (1967) it could be reasoned that a very restricted use of endrin in controlled circumstances, as is practiced in Finland today (Myllymäki, 1974*b*) does not necessarily involve any serious environmental risk. This statement should not be interpreted as any endorsement of large-scale endrin, or endosulphane, treatments on cultivated fields.

General discussion and conclusions

The destruction of vertebrate animals, however serious a pest problem, is often more offensive to human feelings than is insect control or other forms of plant protection. Much of the topical discussion on methods applied in vertebrate pest control is greatly influenced by ideological reasoning instead of rational thinking. A sound basis for all control programs is a proper knowledge of the behavior and population ecology of the target species, which, however, does not imply that only ecological methods, or 'biocontrol' (Howard, 1967a) could, or should, be envisaged. The seriousness of rodent problems demands that all possible approaches towards new solutions should be considered.

As pointed out by Howard (1967a) 'any artificial manipulation of habitats will alter the entire ecosystem more drastically than if members of specific species of animals were removed by some selective control method'. To be effective and to avoid the creation of new problems, application of strictly ecological tools must be accompanied by a proper knowledge of the functioning of the whole ecosystem concerned. Especially in the tropical areas, lack of this basic information often blocks any advance by this means at its very beginning (R. Brown, 1970), while 'economic realities', short-sighted or well-founded, are most often the counteracting forces in the developed countries. So-called 'biological' control agents, e.g., utilization of predators, diseases or sterilizing chemicals, are of restricted use in field rodent control today. Only exceptionally can predators, such as the domestic cat, be used as effective control agents. The same is generally true with regard to bacterial and viral diseases: a really successful solution, such as myxomatosis, may be found once in a lifetime. The use of chemosterilants, the most promising 'biological' approach today, is still in an experimental stage.

Little progress on the usability of chemical repellents can be recorded at the moment, with the possible exception of the seed coating technique. Also, the future development of effective repellents used on the conventional surface-treatment basis must be considered with a degree of scepticism. Systemic chemicals (repellents, poisons or artificial fertilizers) could theoretically work better, but only little published information exists on this subject. Instead, mechanical barriers have proved to be effective in certain conditions.

The potentialities of chemical control of field rodents have not been sufficiently developed, the most obvious gap being the lack of effective chemicals against microtines and other herbivorous rodent pests. I share the opinion of Lund (1972a) that new selective poisons for surface spraying are badly needed, especially because long-term solutions based on habitat manipulation are either unacceptable or difficult to carry out. Simultaneously, all approaches towards a deeper understanding of factors influencing bait acceptance such as studies on olfaction, pheromones (Christiansen & Døving, 1973), artificial lures, etc., should be encouraged.

Recent restrictions in the use of the most poisonous chemicals, such as endrin, have affected the efficiency of the control of certain field rodents. However, there are other, strictly ecological factors, which affect the efficacy of rodenticide treatments, such as those discussed by Morris (1970, 1972). He studied the effect of endrin spraying on enclosed and field populations of *Microtus pennsylvanicus*. After a reduction in numbers due to the initial kill, the remaining (or invading) populations responded by higher growth and survival rates than was detected in the reference populations.

These findings, and the experience in Hungary cited above, are examples of the compensatory responses that would be expected in response to natural (Howard, 1967a) or artificial (Myllymäki, 1974a) predation. Too low a percentage kill, due to underdosing of poison, poor acceptance of baits, etc., would prove to be beneficial to the target population. The same applies if only parts of larger suitable habitat complexes are treated and sources of immediate invasion left untreated. At present, at least in Europe, much field rodent control suffers from defects in organizational efficiency.

Repeated but ineffectual poison applications might also create conditions favoring the development of resistance. Already, widespread resistance to anticoagulants (Lund, 1967; Jackson & Kaukeinen, 1972a) has a considerable influence on the control of field rodents. Only one case is known of resistance to acute poisons in field conditions and that is the resistance of the pine vole (*Pitymys pinetorum*) to endrin in California (Webb & Horsfall, 1967); a resistance that may be acquired (Drummond, 1970) or hereditary. However, resistance to acute poisons cannot be ruled out as a potential problem in the control of field rodents.

Methods for the objective estimation and prediction of rodent-born damage, especially to field crops, are still in their infancy. Such methods are urgently needed as a basis for integrated control programs, as are reliable, but not too complicated, methods for the surveillance of rodent populations. Procedures for medium- or long-term forecasts are under development in several countries.

The development of new methods for surveillance, the testing of the efficacy of rodent control agents and basic research for new solutions is both laborious and time-consuming, and, therefore, intensified international cooperation is urgently needed. The European and Mediterranean Plant Protection Organization (EPPO) is already sponsoring a Working Party on field rodents, with the biological basis of field rodent control as its main objective. Another small group of EPPO is working on outlines for standard tests for the evaluation of efficiency of various control agents, which, if once agreed upon and prudently used, would save a lot of effort now wasted in undue double work.

14.2. *Control of rodents in stored products and urban environments*

F. P. ROWE

Compared with the numbers of small mammals that conflict with man's interests in agriculture, horticulture and forestry, few species have developed a close association with human settlements. Although the shrew *Suncus murinus* inhabits buildings in southeast Asia, the principal pests in food storage premises and urban environments generally, are rodents. Locally important species of these rodents are the coucha rat, *Praomys natalensis*, of Africa, the little rat, *Rattus exulans*, of Asia and the Pacific area, the spiny mouse, *Acomys cahirinus*, of Egypt and its environs and the lesser bandicoot rat, *Bandicota bengalensis*, of India. Of greater world-wide significance are three rodents that have become particularly well adapted to a commensal existence. They are the Norway(or brown or common)rat, *Rattus norvegicus*, the roof(or ship or black) rat, *Rattus rattus* and the house mouse, *Mus musculus*. The remarkable success of these species can largely be attributed to their ability to live in a wide range of habitats, to their immense reproductive capacity and to their omnivorous feeding habits.

The Norway rat is particularly well-established in temperate zones where it occurs commonly in rural as well as in urban areas; it is also the main rodent found in sewers. In tropical regions, it is more confined to larger towns and port areas, although it is also found in some inland towns where there are no indigenous competitors. The roof rat, an arboreal species which climbs effectively, is generally more successful in tropical than in temperate zones where, in the main, it tends to inhabit port and riverside areas and rarely forms colonies living outdoors. Many tropical towns and villages are infested with roof rats and in certain parts of the world, notably southeast Asia, distinct varieties of this species have evolved. In some countries, roof rats seem unable to compete with naturally occurring species in the fields; for example, while they occur in many isolated villages and farms in Africa, they are rarely found living at any distance from human settlements. In contrast, where no natural competitors exist, such as in the Pacific, Atlantic and Caribbean islands, roof rats are common in fields, scrubland and forest. The house mouse is cosmopolitan being found in cities and towns throughout the world and infesting dwellings in remote villages and isolated settlements. In parts of Europe and North America, mice thrive in natural colonies outdoors and in parts of Australia, where local competition proved weak, the mouse has replaced some indigenous species over much of their former ranges. Occasionally, in each of these three regions, the field populations of mice increase to plague proportions. A fuller account of the distribution and the life history, biology and behavior of these three important commensal rodents has been given (Rowe & Taylor, 1970).

The urban rodent problem

The need to control commensal rodents is important at both human public health and economic levels. Both rats and mice are implicated as vectors in carrying diseases from natural endemic foci to human settlements and in transmitting to humans such diseases as plague, leptospirosis, murine typhus, food poisoning and lymphocytic choriomeningitis. Therefore, on this ground alone, effective measures against them are required. In both rural and urban environments, the commensals are responsible for economic losses, mainly as a result of their depredations among many kinds of foodstuffs. The losses are undoubtedly severe and undesirable in view of the demands on food supplies made by ever-increasing human populations. Although numerous estimates of losses due to rodent attack have been made, the majority are unreliable for few critical assessments have been made. Many of the estimates have been calculated from the average daily rate of consumption of cereal or other foods by rats or mice together with rough assessments of rodent numbers and, thus, have no valid quantitative basis. The 'invisible' losses may in fact assume greater significance than the amount of food eaten. For example, although an adult mouse eats only about 3 g of cereal food per day, losses of cereals in food stores based on this figure, even if the numbers of mice were known, would be unrealistic. This is because the mouse is an exceptionally wasteful feeder and through its habit of discarding partially eaten food, it spoils more food than it actually consumes. Southern & Laurie (1946) found that a high proportion of the grain in heavily mouse-infested cereal stacks was damaged and rendered useless for milling purposes and that this loss far exceeded the calculated amount of grain eaten.

Apart from the food they eat or destroy, commensal rodents are responsible for other largely unrecorded losses. Bagged foodstuffs under long-term storage can collapse due to heavy rodent attack – in part due to the search for nesting material – and again the costs involved in re-bagging and cleaning can be considerable. Food supplies contaminated by rodent droppings and hairs can also cause economic loss; the contaminants are difficult to remove at an economical cost and processed foods exported to countries with high sanitation standards that are found to contain rodent hair fragments have been condemned. Further losses attributable to commensal rodents can occur in factories, warehouses and dwellings and arise as a result of damage to manufactured goods and the fabric of buildings. Rats and mice may also gnaw insulating materials causing electrical installations to be put out of action and creating fire hazards.

The lack of accurate information on the magnitude of losses incurred by commensal rodents, as with many other pests, has been attributed to the difficulty of measuring the losses, the lack of suitable methods for assessing them and past neglect on the part of protection scientists (Chiarappa *et al.*, 1972). The need to procure such data for policy decision purposes and to help determine the development of and trends in control programs is now realized at both national and at international levels (Shuyler, 1972).

Control measures

Some of the diverse aspects not associated with commensal rodent behavior, but that are fundamental in the initiation, organization and conduct of urban rodent control in different countries, are appropriate legislation, responsibility for control, regulations pertaining to public health and safety, support for research work at both laboratory and field levels, choice, cost and availability of rodenticides, bait and equipment, and the numbers of suitably trained operational staff.

While the most intensive studies into the biology, behavior and control of commensal rodents have been made by countries in temperate zones, the more serious problems exist in tropical regions. However, even allowing for the different environmental conditions of the tropics and elsewhere, the basic problems in relation to control are similar. Thus, providing that existing control techniques are suitably adapted to meet local conditions, their application is universal.

The most obvious and effective method of commensal rodent control is to deny rodents access to food and harborage. Although this need is continually stressed by authorities concerned in urban rodent control, the practical difficulties are often immense. In the case of stored food preservation, however, where economic losses can be greatest, this approach to the problem is by far the most appropriate, and it can also avoid the use of toxic chemicals. The basic principles involved in protecting stored food from attack by the use of suitably designed and well-maintained premises have been described by Jenson (1965) and Drummond & Taylor (1970). It is emphasized here that newly constructed food storage premises should be made rodent-proof from the outset, using the most appropriate local materials and in consultation with architects and builders; furthermore, food premises should be sited, if possible, where infestation risk is low. The vegetation should be destroyed in the immediate vicinity, and the storage premises maintained in a good state of repair.

Constant high standards of hygiene are required both inside and outside food storage premises. Commensal rodents can be attracted to a store in the first place by waste food or materials left lying outside it and it is essential to prevent such accumulation. It is equally important to keep the inside of the store clean by cleaning up and then protecting any spilled food and waste packaging material. To facilitate the inspection of stacks of bagged foodstuffs for signs of rodents and to assist in the carrying out of possible control measures, the stacks should not be built in corners or be more than about 10 m wide. Should infestation occur, it is important to detect it early before the rodents have had the opportunity to breed and disperse; for populations allowed to become well-established and dense, apart from the damage they inflict, often prove difficult to eradicate without a great deal of effort. One of the best ways of detecting the penetration of a food store by rodents is to lay smoothed patches of an inert dust at strategic points inside the store and then to examine the patches regularly for rodent footprints.

From the rodent control point of view, the increased storage of loose cereals in bulk rather than in bags is advantageous in reducing the amount of cover for rodents. Large bulk storage containers can be maintained rodent-proof without difficulty and, at a smaller level, Taylor (1972) found that village stores in Africa, carefully made with such rudimentary materials as sticks and sun-baked mud, proved to be effective rat-proof storage units.

Unfortunately, in many parts of the world, large quantities of food continue to be stored in unsuitable premises and are easily vulnerable to rodent attack. Current measures to control commensal rodents in stores and in urban areas generally depend on the application of poisons (or rodenticides) incorporated in either bait, powder or drinking water. These are usually classified either as acute (single dose, quick-acting) or as chronic (multiple dose, slow-acting) poisons. General accounts of compounds of both kinds that are currently used to varying extent in different countries for the control of commensal rodents have been given recently by Gutteridge (1972) and Greaves (1971) and an extensive review of the chronic anticoagulant rodenticides has been made by Bentley (1972). The previous chapter (Chapter 14.1) also discusses these topics in considerable detail. Those of the acute type include zinc phosphide, arsenious oxide, sodium fluoroacetate, fluoroacetamide, thallium sulfate, alphachloralose, norbormide, crimidine, antu, strychnine, red squill, gophacide and silatrane. Of these compounds, only one, norbormide, possesses a highly specific toxicity to the genus *Rattus* and particularly to *R. norvegicus*, but its performance in field trials proved to be disappointing due to development of tolerance and acceptance problems (Greaves *et al.*, 1968; Rennison, *et al.*, 1968). Of the remainder, zinc phosphide has probably had the most extensive use. Alphachloralose has been introduced recently as a mouse poison but its efficacy diminishes with increasing environmental temperature. Fluoroacetamide has been shown to be an effective poison against *R. norvegicus* living in sewer systems (Bentley *et al.*, 1961). However, each of the acute rodenticides now at hand has some disadvantage or another either in relation to toxicity, acceptability, safe usage or secondary poisoning hazards. Also, although the acute poisons differ widely in their physiological effects, they are commonly quick-acting and the onset of poison symptoms in affected animals is therefore fairly rapid. Initially, rats tend to feed slowly at baiting points and if acute poison bait is laid directly some individuals may well take a sublethal dose of poison and refuse to feed further on the poison bait – a phenomenon commonly referred to as bait or poison shyness. In order to induce rapid feeding and obtain the most effective results from acute poison treatments, it is necessary, therefore, to lay well-distributed attractive 'pre-bait' for several days before the poison is included. Even so, in practice, it is found often that two or more treatments, using a different bait and poison each time, are required before complete control is achieved. Acute rodenticides still have a part to play in commensal rodent control but nowadays they are principally and most gainfully employed in situations demanding a rapid reduction of infestations.

Not surprisingly, in view of the various difficulties involved in both the use and efficiency of acute rodenticides, the chronic anticoagulant

343

Small mammals

compounds are now regarded as first choice rodenticides against commensal rodents in most countries. This choice has largely stemmed from their rather unique combination of efficacy and relative safety. They are readily accepted by commensal rodents when they are included in bait at low concentrations and their effect is cumulative over a period of days. Vitamin K_1 is a strong antidote. The most widely used anticoagulant has been warfarin, but since its advent in 1950 other important coumarins and indandiones have been discovered and made available, notably coumatetralyl, chlorophacinone, pival and diphacinone. The anticoagulants can be obtained as concentrates for mixing with any suitable bait or in ready-mixed bait. To enable changes of bait to be made, in case baiting problems arise, it is important to always have some of the concentrate at hand. When treating commensal rodents infesting food stores and urban habitats with anticoagulant-treated bait it is essential to maintain surplus bait in place until feeding ceases. Experience in western Europe and elsewhere has shown that well-conducted treatments, carried out after thorough surveys of infested areas, are capable of eradicating rat populations living in varied habitats and thus they continue to be widely used. In the last decade, however, resistance to warfarin has been encountered in all three commensal species in the United Kingdom and, as far as *R. norvegicus* and *M. musculus* are concered, to other commonly available anticoagulants also (Rowe & Redfern, 1965; Drummond, 1966; Greaves, Rennison & Redfern, 1973). There is disquieting evidence that this phenomenon is becoming more widespread (Lund, 1964; Ophof & Langeveld, 1969; Jackson & Kaukeinen, 1972b).

The application of poison dusts, solutions and foams, and fumigation, are minor, but in some situations important, control methods against commensal rodents. Rodenticidal dusts depend for their effect on rats and mice picking them up on their bodies when travelling through them and then ingesting the poison during grooming. The effectiveness of rodenticidal dusts is therefore independent of idiosyncrasies in rodent feeding behavior and they can be particularly useful where difficulties are encountered in drawing commensal rodents to baits. Since a relatively high concentration of poison has to be employed in a dust formulation in order to obtain good control, the utmost care must be exercised when using dusts to avoid contaminating foods, humans or domestic animals. Thus, dusts should not be placed on unprotected foods, between bags, or in places where they can be dispersed easily by air currents. In the main, rodenticidal dusts are used to supplement other control methods

344

by laying them in such places as roof spaces, the cavities between walls, beneath floorboards and in ducts and conduits. At one time, dust preparations containing high concentrations of DDT were extensively and effectively used against mice in some countries, but nowadays, although gamma-BHC is occasionally employed, anticoagulant dusts are preferred.

Another supplementary or alternative and relatively cheap method of controlling commensal rodents is to provide them with poisoned water. This method is particularly appropriate against rats and mice inhabiting warm dry environments where free water is normally absent and in cases where solid bait is expensive or scarce. Although water solutions of acute poisons such as sodium fluoroacetate and thallous sulfate have been used on occasion, the hazards involved have largely precluded their use and water soluble forms of anticoagulants, particularly the sodium salts of warfarin and pival, are most commonly employed. The combined application of an anticoagulant dust placed around anticoagulant-poisoned water has been found to be a useful technique against mice infesting food storage premises.

A quick and efficient, albeit expensive, means of destroying rodents, that either through neglect or inefficient control measures have been allowed to reach high population densities in large stocks of foodstuffs held in stores or in craft, is to fumigate. Often the best way of doing this in buildings is to cover the foods with gas-proof sheeting and to apply the fumigant (methyl bromide is often used in this manner) beneath the sealed sheets. Fumigations can only be undertaken by professional operators familiar with the use of fumigants and with the precautionary measures that need to be observed. Outdoor colonies of *R. norvegicus* that are living well away from buildings in burrows excavated in relatively damp consolidated soil can also be gassed effectively by using one of the proprietory powders that produces hydrogen cyanide gas or aluminium phosphide pellets producing phosphine.

Traps, both of the snap and live-catch types, are commonly employed in commensal rodent control but they have very limited value in dealing with populations that are either dense or widespread. They are most effective when used to eliminate small numbers of rats and mice that have become recently established in dwellings, stores, factories and other buildings. Traps are also useful in removing the survivors of poison treatments or where the use of rodenticides is considered undesirable. The effectiveness of trapping campaigns is largely determined by the sensitivity of the traps, their siting and abundance. Even in carefully

345

planned campaigns, however, it is sometimes difficult to remove some individuals that tend to be 'trap-shy' and for the best success frequent trappings conducted with many traps over a few days are preferable to continuous campaigns.

In rural areas, commensal and other rodents are naturally subject to attack by native hawks, owls and carnivores. Such predators can play a useful role in intercepting rats and mice moving near the vicinity of, or between, farm buildings. The domestic cat fulfils a similar function in rural and urban environments but its value in limiting well-established high rat populations in urban areas tends to be over-estimated (Jackson, 1951).

The use of bait containing cultures of the bacterium *Salmonella enteritidis* as a means of killing commensal rodents was first made about 70 years ago. Although such preparations are still used in some countries for this purpose, notably in the USSR, they have not found wide application because their overall efficacy was considered doubtful and also becuase of the possibility of human infection. The reported development of bacterial preparations lethal to rodents but non-pathogenic to man may lead, however, to wider consideration of this control method (Laird, 1966).

The future in urban rodent control

Advances in the field of urban rodent control have been made in the past 30 years, although existing knowledge has not yet been fully exploited in all parts of the world. However, outstanding problems remain. Some of these are of a biological nature, but others are more concerned with human psychology and economics.

As a result of highly organized and systematic work, some success has been achieved recently in western Europe in eliminating *R. norvegicus* over large urban areas (Telle, 1962; Drummond, 1969). The 'rat-free town' concept, however, is heavily dependent for its success on the deployment of chronic poisons and it could be undermined seriously (for a time, at any rate) should the problem of resistance to the anti-coagulants increase in commensal rodent populations. The development of this resistance has stimulated research in two main fields. First, it has provoked a study of resistant animals, from genetical, biochemical and ecological aspects, in order to elucidate the mode of inheritance, the physiological mechanisms involved and the likely spread of resistance in affected areas. Second, it has encouraged the search for alternative rodenticides. In several countries advantageous collaboration and liaison has been established between various laboratories, the World Health Organization and chemical and pharmaceutical manufacturers, in the examination of compounds with diverse modes of action. These compounds include anti-metabolites, anti-inflammatory drugs and immunosuppressive agents. In the United Kingdom, each compound is first screened in the laboratory and promising ones are then subjected to field evaluation (Rowe *et al.*, 1970). One outcome of this work has been the recent development of a new chronic poison that is effective against anticoagulant-resistant rats and mice (Ministry of Agriculture, Fisheries and Food, unpublished results).

Research work is in progress in other fields in an attempt to improve control measures against commensal and other rodent pests. Increasing attention is being given to the possible application of reproduction inhibitors or chemosterilants and investigations have been conducted with both steroidal and non-steroidal compounds. In a recent field trial, a dense *R. norvegicus* population inhabiting a refuse tip was effectively controlled after the application of a long-acting synthetic oestrogen (Kendle *et al.*, 1973); such a compound could also find a use in the control of rats in sewer systems. Although it is unlikely that chemosterilants will be employed as a method of first choice in most urban habitats (because of the demand for a more immediate method of control), they could find a use in an integrated control program. Other recent experimental studies with *R. norvegicus* and *M. musculus* have shown that both species respond adversely to high intensity ultra-sound (Greaves & Rowe, 1969). At present, the practical and economic difficulties need to be overcome in order to take full advantage of this finding in relation to preventing the movement of rats and mice into food stores, mills and bakeries.

Increased or renewed research is required in a number of other areas that are relevant to the improvement of control measures against commensal rodents. For example, there has been little recent advance in regard to either repellents or attractants. Investigations of this nature in the past have been mainly concerned with the use of synthetic additives to packaging or to bait. However, naturally occurring pheromones have now been shown to significantly affect various behavioral traits in rodents and the identification of the substances involved could lead to more fruitful lines of investigation in these two fields. Other aspects impinging directly or indirectly on commensal rodent control that require further attention include bait formulation and presentation techniques, the physiological mechanisms involved in the development of 'poison bait shyness', intra- and interspecific competition, seasonal and other movements, and the development of accurate census techniques. Little study has been made of the possible importance to commensal rodents of bait particles of different shape or size, although attempts have been made to increase the acceptance of poison baits by microencapsulation of toxicants (Greaves *et al.*, 1968; Cornwell, 1970).

There is little doubt that such research studies will lead to further improvements in commensal rodent control. It is encouraging that the problems posed to mankind by commensal rodents have become more universally appreciated and that, as a result, there has been increased consultation and exchange of information. The overall aim in commensal rodent control, however, is to obtain a permanent, not just temporary, reduction in numbers and progress in this direction has been slow. For one reason or another, poor planning and lack of coordination of effort is too common. Control programs generally are insufficiently evaluated and the benefit to be expected in relation to the cost of control is rarely defined. At operational level, there are frequently inadequate numbers of trained staff. Furthermore, the effective and long-lasting control of commensal rodents is clearly dependent on the active co-operation of the public and such cooperation, in view of different cultural backgrounds and economic and social pressures, can prove difficult to obtain. It is in these areas of organization, evaluation, training and education that the more significant contributions to improved commensal rodent control are to be found.

14.3. *The importance of small mammals in public health*

A. A. ARATA

The historical association of wild mammals, especially rodents, with certain human epidemic diseases is well documented, and it is evident that this relationship was understood in broad terms centuries before the scientific developments of the nineteenth century led to the definition of infectious cycles and the intricate roles played by reservoirs, vectors and pathogenic agents (Pollitzer, 1954; Meyer, 1963). Of these diseases, plague has had the greatest notoriety and its major pandemics may well have changed the course of human history as much as insect-born diseases such as malaria, yellow fever and typhus (Zinsser, 1967). Rodent pests that destroy field crops and stored food, even if not associated directly with the spread of disease, are also of considerable importance in public health. Although the figures are often unreliable, estimates of grain destruction in developing countries range from 10 to 25 per cent and in certain instances are probably higher (Pingale, 1966). The resultant human malnutrition and lowered resistance to common infectious and chronic diseases of man continue to take a heavy toll of human life as they have done for millennia.

Despite the importance of small mammals in public health, there has been a very slow development of specialists in this field as compared to medical entomology. Few institutions offer courses and there are no specialized scientific societies, journals, etc. Most people who have entered the field have done so because their studies on specific diseases (plague, murine typhus, tularemia, etc.) have required some expertise in this field. At the same time mammalian ecology was developing as an academic science with some association with agriculture and forestry, but the principles of habitat selectivity, population density and dynamics had, until recently, little effect on studies of zoonotic diseases transmitted by small mammals. The same can be said about rodent control as opposed to simply 'rat killing'. Only recently has this been based upon ecological principles and the most effective means of control evaluated by an understanding of reproductive potential and the importance of harborage and food in controlling rodent pests by regulating the carrying capacity of the environment (Elton, 1942; Davis, 1953).

349

Small mammals

The purpose of this brief report will be: (*a*) to outline some of the major historical and current public health problems related to small mammals by specific diseases and geographically; (*b*) to describe those aspects of mammalian ecology that should be most useful to public health workers in understanding the nature and extent of small mammal involvement in human disease, including certain problems typical of temperate and tropical environments. Such a short review cannot be comprehensive from either an epidemiological or ecological viewpoint, but the need to seek a common denominator is a reflection of the relationship between the two fields.

Rodent-born diseases

One should distinguish those reports which simply note that a small mammal can be or has been observed in nature to be infected with an agent pathogenic to man from those which show that the species truly harbors the agent and plays a role in transmission to man. The former are too abundant to summarize and even the latter are numerous. A partial list of rodent-born diseases is provided in Table 14.1: lists of

Table 14.1. *Selected list of representative rodent-born diseases*; *the list is not considered to be comprehensive*

A. Viral Diseases:
    Crimean haemorrhagic fever
    Omsk haemorrhagic fever
    Russian spring–summer encephalitis
    Western encephalitis
    California encephalitis
    Central European encephalitis
    Argentine haemorrhagic fever
    Bolivean haemorrhagic fever

B. Rickettsial Diseases:
    Murine typhus
    Q fever
    Rickettsial pox
    Scrub typhus (tsutsugamushi)
    Spotted fever group

C. Bacterial Diseases:
    Brucellosis
    Glanders
    Plague
    Listeriosis
    Salmonellosis
    Shigellosis
    Tularemia

D. Spirochetal Diseases:
    Endemic relapsing fevers
    Leptospirosis
    Rat-bite fever

E. Fungoid Diseases:
    Actinomycosis
    Adiaspiromycosis
    Histoplasmosis
    Ringworm

F. Protozoan Diseases:
    Cutaneous leishmaniasis
    Visceral leishmaniasis
    Toxoplasmosis
    American trypanosomiasis (Chagas' disease)

G. Helminth Diseases:
    Echimococcosis
    Asian schistosomiasis

H. Arthropod Diseases:
    Chigger mite dermatitis
    Rat mite dermatitis

350

Importance in public health

bat-born diseases are found elsewhere (Tesh & Arata, 1967; Constantine, 1970). The rodent-born diseases discussed below are presented only as examples of the nature and diversity of rodent involvement in human disease, emphasizing 'classical' and 'newly-emerging' rodent-born diseases.

Plague

Plague is an acute infectious disease of rodents caused by *Yersinia pestis* and maintained in many regions of the world in a continuous infection chain through rodent reservoirs and flea vectors. Man becomes infected by accidental intervention into this cycle (Baltazard *et al.*, 1963). Although the black rat, *Rattus rattus* and its fleas, especially *Xenopsylla cheopsis*, are most often implicated in human outbreaks, a large number of rodent and flea species maintain the infection in natural foci.

Among the rodents we find most commonly associated with plague are the Gerbillinae (*Meriones, Rhombomys, Tatera* and other genera) in Asia and Africa; Cricetinae (*Oryzomys, Neotoma* and others) in North and South America as well as *Cricetus* in Central Asia; Microtinae (*Microtus, Ellobius*) in North America and Central and Northern Asia; a large assortment of Murinae (*Rattus, Acomys, Arvicanthis, Bandicota, Mus, Mastomys*, etc.) throughout Africa and Southeast Asia; Sciurinae (*Citellus, Marmota, Funambulus, Cynomys* and other ground squirrels) in Asia and North America and tree squirrels (*Xerus* and *Sciurus*) in Africa and South America. Other groups (*Otomyinae, Dendromyinae, Dipodidae, Heteromyidae, Caviidae* and *Echimyidae*) have been reported to be naturally infected or have been strongly incriminated through positive findings in their ectoparasites. Lagomorphs (*Lepus* and *Sylvilagus*) are also reservoirs in Africa, North and South America and parts of Central Asia (Pollitzer, 1954).

In all, some 200 or more species of rodents of all major families throughout the world can naturally harbor plague. They differ considerably in their etiological importance as some are more highly susceptible than others. One explanation of cyclic plague outbreaks, especially in the steppes of Central Asia, involves the interaction of two sympatric gerbil species (usually *Meriones*) of which one is somewhat resistant to infection and acts as the primary reservoir and the other is highly susceptible. It is the latter which serves to spread the infection when its populations are high. Following a rapid population crash the infection rests in the resistant species as a new cycle of population builds up in the susceptible species. Although suggested for the tropics, the hypothesis is not as convincing in this region since tropical rodent populations do not fluctuate as drastically from year to year and reproduction is more evenly distributed. Although the susceptibility and resistance phenomena doubtless play an important role even in the tropics, there are differences in nesting and movement patterns of rodents in tropical and steppe areas and the spread of the infection from natural foci to domestic *Rattus* (and hence to man), may be influenced by patterns of rainfall, agricultural practices and other human behavioral patterns. At the time of writing the World Health Organization has completed two years of a three year ecological study in Central Java, where plague is persistent, to determine more precisely the role of the rodents and fleas in maintaining the infection in nature and its spread.

The known distribution of world plague foci is shown in Fig. 14.3. The diversity of habitats from rain forest to desert and the numerous rodents involved indicate how small is our knowledge of the ecology of the rodent reservoirs. Fortunately, human plague is treated successfully with antibiotics at present, but the cost of surveillance and of medical supplies is severe in developing countries where programs for control of other diseases such as cholera and malaria also compete for limited financial resources. Thus, even though foci may be quiescent, certain public health problems still exist, and the severity of past plague outbreaks reinforces this attitude. In India, over 12 500 000 plague deaths occurred between 1898 and 1948, and 235 000 between 1939 and 1952 (Pollitzer, 1954). In recent years over 3000 cases have been noted in Vietnam.

Although ecological studies of mammals and fleas are certainly not the only aspects of plague that need study, they are very important as our increased knowledge concerning population densities and turnovers, movements and associated phenomena will aid us greatly in establishing whether rodent surveillance may be of use in predicting potential outbreaks of plague.

Tularemia

Like plague, tularemia is a natural disease of rodents and lagomorphs which may also infect man. The numerous strains of *Francisella tularensis* vary greatly in virulence to man, associated in part with the mode of transmission. Tularemia may be transmitted from rodents to man by blood-sucking arthropods (mites, ticks, flies, midges, fleas,

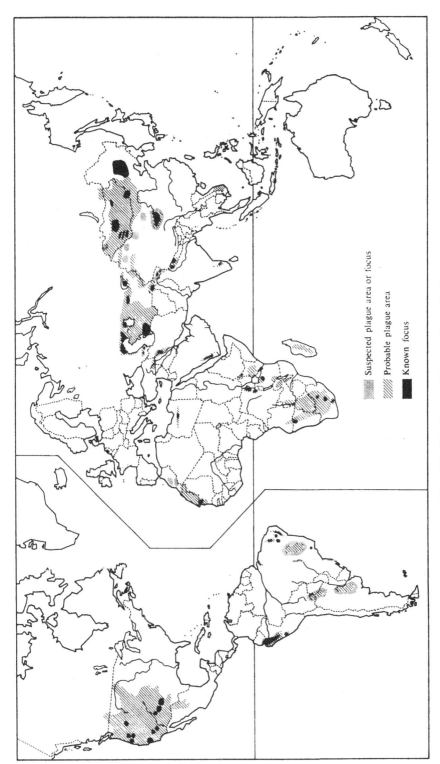

Suspected plague area or focus

Probable plague area

Known focus

Fig. 14.3. World map of known plague foci.

mosquitos), by contaminated water, aerogenically by inhalation of feces-contaminated dust, contact with infected live mammals, their carcasses and insufficiently cooked meat.

The infection in man is essentially cosmopolitan, although in North America hares are associated with most reported cases, whereas in Central Asia and Eastern Europe *Microtus* and *Arvicola* are the predominant reservoirs. No attempt is made here to summarize the vast literature on this disease and its association with man and small mammals, but it will be noted how this disease has been closely linked to human activities that influence mammal populations. Clearing of forests in Eastern Europe and Central Asia for pasture and meadowlands has increased the available habitat for *Microtus* in these areas and the dynamics of population numbers, themselves major determinants of epizootics, are closely connected with agricultural activities (Maksimov, 1965).

Over and above being a disease of man, tularemia can be a major factor in regulating rodent populations themselves. As the distribution of the reservoir rodents is so closely linked to agricultural practice, especially in Europe and Asia, there can be few better examples of the intimate association between field rodents, human activities and zoonotic diseases. As a result of this link tularemia in man has often been associated with opening new areas for agricultural purposes. Although tularemia has been known for a long time, the public health problems described are new, but apparently follow the same pattern of human–rodent contact.

Viral diseases

Lassa fever. This infection is caused by a virus first isolated from human cases in West Africa during recent years. It is antigenically related to Machupo virus (causative agent of Bolivean haemorrhagic fever) and other viruses known to have associations with rodents. In a recent report (WHO, 1973) an outbreak was noted in Sierra Leone where the overall attack rate was 2.2 per thousand in the human population with a case-fatality of 38 per cent among hospitalized patients. From a survey of 480 rodents, 110 bats and 23 insectivores, isolates were recovered from one species, *Mastomys natalensis* (*Rattus natalensis* = *Mastomys concha*). This rat, the multimammate or coucha rat, is well known as a plague reservoir in arid areas of southern and eastern Africa, but its appearance as the reservoir of an apparently newly-emerging disease in forested West Africa is surprising. We have little information in this area on the ecology, population patterns, seasonality, etc., of this rodent now shown to be a reservoir.

Bolivean haemorrhagic fever. As noted above, the causative agent of this infection is related antigenically to Lassa virus. Similarly, a South American cricetine rodent, *Calomys callosus*, was found to be chronically infected and when it becomes commensal with man, virus excreted by the rodents in urine is transmitted to man, probably by the direct contamination of food. It has been speculated that such a 'new' disease (as may also be the case with Lassa fever) may be the result of extension of the range by the reservoir rodent or a modification of its spatial or behavioral relationship with man, perhaps caused by human activities (Mackenzie, 1972).

Korean haemorrhagic fever. This infection was first noted in man in the 1930s and reached epidemic proportions during the early 1950s in the troops of various armies. Despite repeated attempts to isolate a virus from human patients, field rodents and suspect arthropod vectors, no success has been obtained. During 1973 and 1974, a World Health Organization research team was continuing to work on this problem. However, unless a viral isolation is made and subsequent antigenic material prepared, it will be impossible to determine the role of rodents in this disease, which is probably widespread in northern and eastern Asia.

Rabies. Although classically a disease spread by wild and domestic carnivores, the role that bats may play in transmission has become more obvious in recent years, especially in the Americas. This is very well documented elsewhere (Constantine, 1970) and will not be discussed in this paper, which deals largely with the role of rodents in public health. However, in the past few years isolation of a number of strains of rodent variants of rabies virus has been made in Europe from *Apodemus flavicollis*, *Clethrionomys glareolus* and *Microtus arvalis*. There is no evidence to link these observations to rabies in foxes (the principle vector and reservoir) or any other animal species. However, the possible existence of a natural focus of variants of rabies virus in rodents cannot be completely ignored and further research on all aspects of these viruses should be encouraged (WHO, 1972).

Small mammals

Other rodent-born diseases. The few examples given above characterize some of the classical and some of the newly-emerging (or newly-recognized) rodent-born diseases. Similar examples could be given for rickettsial infections (the spotted-fever group, murine and scrub typhus), other viral infections (tick-born encephalitis, W. Nile virus, RSSE, etc.), parasitic diseases (Asian schistosomiasis and Chagas' disease), and those of great importance to veterinary public health (leptospirosis and salmonellosis).

Ecological studies

There are several basic criticisms that can be made of many studies on rodents in association with zoonotic diseases. The first is that proper taxonomical determinations are too often lacking. Ecological studies must assume proper identifications, yet the rodent faunas of Africa, Southeast Asia, and South America are complex and the results of numerous studies have become invalidated due to misidentifications, confusion of nomenclature and the failure to retain reference material for subsequent re-examination. Second, in a human population one would not expect individuals in the age classes 1 to 5, 11 to 15, and 41 to 45 years to have the same antibody levels to a common local infection (e.g., measles); however, in serological studies of rodents, age distinctions between populations, or even within populations are seldom made, even though an individual population may consist of 30 per cent adults at one season and 70 per cent adults at a different time of the year. As a result, comparisons of the prevalence of infections made within or between populations differing in space or time are misleading. Thirdly, in studies on infections in wild animals a common denominator is too often lacking. Does a collection of 100 or 1000 specimens represent 10 per cent or 50 per cent of the population sampled? Although reliable density estimates are often difficult to obtain in surveys, certain crude indices (relative abundance of species/trap night, or of individuals of a single species/trap night) can be calculated from trapping records and are of considerable value – especially when these data must be linked to abundance of ectoparasites (e.g., to determine flea indices in studies of rodent plague in natural foci). Finally, although studies of home range, movements and dispersal require repeated observations of individuals in some instances, these data are often necessary to determine potential contact rates between reservoirs, interchange of ectoparasite vectors, and the potential spread of pathogenic agents.

356

The observations mentioned above as being all too often lacking in studies of rodent-born diseases, are fundamental for the construction of quantitative models of disease transmission which may ultimately allow the development of predictive models rather than a series of disjointed descriptive studies.

Example of a national study

Between 1969 and 1971 WHO in conjunction with the Institute of Public Health Research, University of Teheran, conducted a survey of small-mammal-born diseases in Iran. Mammals were collected during 1969 and 1970 at 47 localities in the country and examined for a large number of infections. Table 14.2 shows a broad listing of mammalian groups and the infections observed. Fig. 14.4 is an example of the type

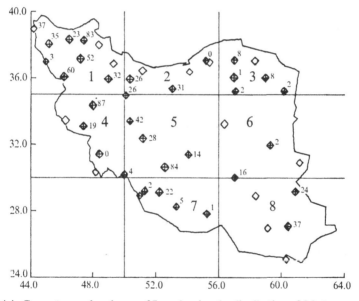

Fig. 14.4. Computer produced map of Iran showing the distribution of *Meriones persicus*, the Persian jird. Hatched diamonds represent collection sites where the species was collected and what percentage of the total number of mammals it constituted; open diamonds show the collection sites where the species was not collected.

of computer maps that were produced to show the distribution and relative abandance (as a percentage of all mammals taken at that point, and not 'density') of individual species. By means of such mapping based only on longitude and latitude, new data on distribution and

357

Small mammals

relative abundance can be added sequentially. Distribution of infections and habitat data (not shown on this map) can be entered on the same computer program.

Table 14.2. *Results of studies of small-mammal-born diseases in Iran, 1969–1971 (+, examined and found positive; −, examined and found negative; blank, not examined)*

Infections	Mammalian order			
	Rodentia	Lagomorpha	Chiroptera	Insectivora
Bacterial				
Pasteurellosis	−	−	−	−
Plague	+			
Tularemia				+
Salmonellosis	+		+	
Erysipeloid	−	−	−	−
Listeriosis	−	−	−	−
Leptospirosis	+			
Rickettsial				
R. prowazeki	+	+		
R. sibiricus	+	+	+	+
R. tsutsugamushi	+			+
R. canada	+	+		+
R. mooseri	+			+
C. burneti	+	+	+	+
Viral				
Rabies	−	−	−	−
West Nile virus	+	+	+	+
Sindbis	+	+	+	+
Crimean haemorrhagic fever	+	+		
Sicilian spotted fever		+		
Bhanja	+			+
Russian spring–summer encephalitis	+		+	+
Blood and tissue parasites				
Leishmaniasis	+			
Toxoplasmosis				
Hoemogregarinidae	+			+
Trypanosoma	+		+	
Spirochete	+			
Microfilaria	+		+	
Piroplasmida	+			+

Urbanization

In most developed countries urban rodents are pests but only rarely disease carriers. In the developing countries the situation is different. Outbreaks of plague in certain cities of Southeast Asia are not inconceivable; the generally poor sanitation in many tropical cities provides not only excellent rodent harborage, but also ideal circumstances for transmission of salmonellosis and leptospirosis both directly to man and to his domestic and food animals and thence to man. It is often argued that this is an economic and not a biological problem, but this is not so. Rodent control programs in many developed countries have failed because they were not based on ecological principles. In developing countries, the economic and ecological situations are compounded.

In Southeast Asia, *Bandicota* spp. are reported to be becoming more urbanized, and other examples can be given of the displacement of one rodent species by another in urban and peri-urban conditions. In our lack of knowledge we are not certain if such changes will bring about increased public health problems.

Conclusions

These can be summarized briefly as follows:

1. Traditional and newly-emerging rodent-born diseases continue to present important problems in public health. Man-made changes in the environment, especially through urbanization and agriculture practice, have contributed to the spread and augmentation of these problems.

2. Although many of the rodent-born diseases are responsive to treatment in large areas of the world, the facilities and supplies necessary to provide such treatment are not easily available. It is not possible at present to truly assess the economic and public health impact of rodents.

3. Studies on rodents of importance in public health have not to date taken proper advantage of the developments in mammalian ecology, especially concerning population structure, density, turnover, movements etc. In general, there has been poor contact between those concerned with rodents in relation to public health, including their control, and those interested in basic mammalian ecology.

List of mammal names cited in the text†

The following list contains the names used by the authors of the fore-going chapters. Orders and families are arranged according to Simpson (1945); genera and species are arranged alphabetically; common names are mostly from Chasem (1940), Ellerman & Morrison-Scott (1951), Ellerman *et al.* (1953) and Hall & Kelson (1959).

Order Marsupialia

Family Didelphidae

Caluromys Allen	Woolly opossum
Didelphis marsupialis Linnaeus	Opossum
Marmosa robinsoni Bangs	Central American crouse-opossum
Metachirus Burmeister	Brown-masked opossum
Philander opossum Linnaeus	Four-eyed opossum

Family Dasyuridae

Antechinus stuartii Macleay	Broad-footed marsupial-mouse

Order Insectivora

Family Erinaceidae

Erinaceus europaeus Linnaeus	European hedgehog

Family Soricidae

Blarina brevicauda Say	Short-tailed shrew
Cryptotis parva Say	Least shrew
Microsorex hoyi Baird	Pygmy shrew
Neomys anomalus Cabrere	Mediterranean water-shrew
Neomys fodiens Pennant	European water-shrew
Sorex alpinus Schinz	Alpine shrew
Sorex araneus Linnaeus	Common shrew
Sorex arcticus Kerr	Arctic shrew
Sorex cinereus Kerr	Masked shrew
Sorex longirostris Bachman	Southeastern shrew
Sorex minutus Linnaeus	Lesser shrew

† We are grateful to Dr Charles Handley, Museum of Natural History, The Smithsonian Institution, for checking the names in the list and providing complete citations for many species outside of North America and Europe.

List of mammal names

Sorex murinus = Suncus murinus
Sorex tscherskii Ognev   Pygmy shrew
Sorex vagrans Baird   Vagrant shrew
Suncus etruscus Savi   Savi's pygmy shrew
Suncus murinus Linneaus   House shrew
Suncus tscherskii = Sorex tscherskii

Family Talpidae
Talpa caucasica Satunin   Caucasian mole
Talpa europaea Linnaeus   Common mole

Order Chiroptera

Family Rhinolophidae
Rhinolophus hipposideros   Lesser horseshoe bat
Bechstein

Family Phyllostomatidae
Artibeus jamaicensis Leach   Jamaican fruit-eating bat
Artibeus lituratus Olfers   Big fruit-eating bat
Carollia perspicillata Linnaeus   Greater short-tailed bat
Carollia subrufa Hahn   Lesser short-tailed bat
Phyllostomatus discolor Wagner   Pale-nosed bat
Phyllostomatus hastatus (Pallas)   Spear-nosed bat
Sturnira lilium E. Geoffroy   Yellow-shouldered bat
St.-Hilaire

Family Desmodontidae
Desmodus rotundus E. Geoffroy   Vampire bat
St.-Hilaire

Family Vespertilionidae
Lasiurus cinereus Beauvois   Hoary bat
Myotis myotis Borkhausen   Large mouse-eared bat
Nyctalus noctula Schreber   Common noctule

Family Molossidae
Tadarida brasiliensis E. Geoffroy   Brazilian free-tailed bat
St.-Hilaire

Order Lagomorpha

Family Leporidae
Lepus arcticus Ross   Arctic hare
Lepus europaeus Pallas   European hare
Lepus timidus Linnaeus   Mountain hare
Sylvilagus floridanus J. A. Allen   Eastern cottontail rabbit

Order Rodentia
Family Ochotonidae
    Ochotona princeps (Richardson)   Pika
Family Aplodontidae
    Aplodontia rufa Rafinesque   Mountain beaver
Family Sciuridea

Ammospermophilus leucurus Merriam	White-tailed antelope squirrel
Callosciurus Gray	Tricolored squirrel
Citellus = Spermophilus	
Cynomys leucurus Merriam	White-tailed prairie dog
Cynomys ludovicianus Ord	Black-tailed prairie dog
Eutamias amoenus J. A. Allen	Yellow-pine chipmunk
Eutamias minimus Backman	Least chipmunk
Eutamias panamintinus Merriam	Panamint chipmunk
Eutamias sibiricus Laxmann	Siberian chipmunk
Funambulus pennanti Wroughton	Northern palm squirrel
Funisciurus anerythrus Thomas	Redless striped squirrel
Heliosciurus Trouessart	Sun squirrel
Marmota baibacina Brandt	Asiatic alpine marmot
Marmota caligata Eschscholtz	Hoary marmot
Rhinosciurus Gray	Long-nosed squirrel
Sciurus carolinensis Gmelin	Gray squirrel
Sciurus granatensis Humboldt	Tropical red squirrel
Sciurus niger Linnaeus	Fox squirrel
Sciurus variegatoides Ogilby	Variegated squirrel
Sciurus vulgaris Linnaeus	European red squirrel
Spermophilopsis Blasius	Long-clawed ground squirrel
Spermophilopsis leptodactulus Lichtenstein	Long-clawed ground squirrel
Spermophilus armatus Kennicott	Unita ground squirrel
Spermophilus beecheyi Richardson	California ground squirrel
Spermophilus citellus Linnaeus	European souslik
Spermophilus franklinii Sabine	Franklin's ground squirrel
Spermophilus fulvus Lichtenstein	Large-toothed souslik
Spermophilus lateralis (Say)	Golden-mantled ground squirrel
Spermophilus leucurus Merriam = *Ammospermophilus leucurus* Merriam	
Spermophilus parryii Richardson = *Spermophilus undulatus* Pallas	
Spermophilus pygmaeus Pallas	Little souslik
Spermophilus relictus Kashkarov	Red-cheeked souslik
Spermophilus richardsonii Sabine	Richardson's ground squirrel
Spermophilus suslicus Güldenstaedt	Spotted souslik
Spermophilus tridecemlineatus Mitchell	Thirteen-lined ground squirrel

363

List of mammal names

Spermophilus undulatus Pallas — Arctic ground squirrel
Sundasciurus (Geoffroy) — Lowe's or horsetailed squirrel
Tamias striatus Linnaeus — Eastern chipmunk
Tamiasciurus hudsonicus Erxleben — Red squirrel
Tamiscus emini Stuhlmann — African striped squirrel
Xerus sp. — Ground squirrel

Family Geomyidae

Thomomys bottae Eydoux and Gervais — Southern pocket gopher
Thomomys talpoides Richardson — Northern pocket gopher

Family Heteromyidae

Dipodomys agilis Gambel — Agile kangaroo rat
Dipodomys merriami Mearns — Merriam's kangaroo rat
Dipodomys microps Merriam — Chisel-toothed kangaroo rat
Dipodomys nitratoides Merriam — Fresno kangaroo rat
Dipodomys ordii Woodhouse — Ord's kangaroo rat
Heteromys desmarestianus Gray — Desmarest's spiny pocket mouse
Heteromys goldmani Merriam — Goldman's spring pocket mouse
Liomys adspersus Peters — Panama spring pocket mouse
Liomys pictus Thomas — Painted spiny pocket mouse
Liomys salvini Thomas — Salvin's spring pocket mouse
Perognathus californicus Merriam — California pocket mouse
Perognathus fallax Merriam — San Diego pocket mouse
Perognathus fasciatus Wied — Olive-backed pocket mouse
Perognathus formosus Merriam — Long-tailed pocket mouse
Perognathus longimembris Coues — Little pocket mouse

Family Castoridae

Castor canadensis Kuhl — American beaver

Family Cricetidae

Subfamily Cricetinae

Akodon azarae Fischer — Azara's field mouse
Calomys callosus Rengger — Greater vesper mouse
Cricetus cricetus Linnaeus — Common hamster
Cricetus raddei = *Mesocricetus auratus* Waterhouse — Golden hamster
Mesocricetus raddei Nehring — Caucasian golden hamster
Myospalax fontaniere Milne-Edwards — Common Chinese zokor
Nectomys Saussure — Neotropical water rat
Neotoma cinerea (Ord) — Bushy-tailed wood rat
Neotoma floridana Ord — Eastern wood rat
Neotoma fuscipes Baird — Dusky-footed wood rat
Neotoma lepida Thomas — Desert wood rat
Neotoma micropus Baird — Southern plains wood rat

364

Nyctomys Peters	Vesper rat
Ochrotomys nuttalli Harlan	Golden mouse
Onychomys leucogaster Wied-Neuwied	Northern grasshopper mouse
Onychomys torridus Coues	Southern grasshopper mouse
Oryzomys capito Olfers	Forest rice rat
Oryzomys concolor Wagner	(No common name)
Oryzomys palustris Harlan	Marsh rice rat
Peromyscus boylii Baird	Brush mouse
Peromyscus californicus Gambel	California mouse
Peromyscus crinitus Merriam	Canyon mouse
Peromyscus eremicus Baird	Cactus mouse
Peromyscus gossypinus LeConte	Cotton mouse
Peromyscus leucopus Rafinesque	White-footed mouse
Peromyscus maniculatus Wagner	Deer mouse
Peromyscus nuttalli = *Ochrotomys nuttalli*	
Peromyscus polionotus Wagner	Old-field mouse
Peromyscus truei Shufeldt	Piñon mouse
Reithrodontomys fulvescens J. A. Allen	Fulvous harvest mouse
Reithrodontomys halicoetes = *R. raviventris halicoetes* Dixon	
Reithrodontomys humulis Audubon and Bachman	Eastern harvest mouse
Reithrodontomys megalotis Baird	Western harvest mouse
Reithrodontomys raviventris Dixon	Salt-marsh harvest mouse
Sigmodon hispidus Say and Ord	Hispid cotton rat
Tylomys Peters	Climbing rat
Zygodontomys microtinus Thomas	Cane rat

Subfamily Microtinae

Alticola lemminus Miller	Lemming vole
Alticola roylei Gray	Royle's high-mountain vole
Alticola strelzowi Kastschenko	Flat-skulled vole
Arvicola terrestris Linnaeus	Water vole
Clethrionomys gapperi Vigors	Southern red-backed vole
Clethrionomys glareolus Schreber	Common red-backed vole
Clethrionomys rufocanus Sundevall	Large-toothed red-backed vole
Clethrionomys rutilus Pallas	Northern red-backed vole
Dicrostonyx groenlandicus Traill	Greenland collared lemming
Dicrostonyx hudsonius Pallas	Labrador collared lemming
Dicrostonyx torquatus Pallas	Arctic lemming
Ellobius talpinis Pallas	Northern mole-vole
Lagurus lagurus Pallas	Steppe lemming
Lemmus lemmus Linnaeus	Norway lemming
Lemmus obensis Brants = *Lemmus sibiricus* Kerr	

List of mammal names

Lemmus sibiricus Kerr	Siberian lemming
Lemmus trimucronatus Richardson	Brown lemming
Microtus agrestis Linnaeus	Field vole
Microtus arvalis Pallas	Common vole
Microtus californicus Peale	California vole
Microtus drummondii = *M. pennsylvanicus drummondii* Audubon and Bachman	
Microtus gregalis Pallas	Narrow-skulled vole
Microtus guentheri Danford and Alston	Günther's vole
Microtus juldaschi Severtzov	Pamir vole
Microtus middendorffi Poliakov	Middendorff's vole
Microtus miurus Osgood	Singing vole
Microtus montanus Peale	Montane vole
Microtus nivalis Martins	Snow vole
Microtus ochrogaster Wagner	Prairie vole
Microtus oeconomus Pallas	Tundra vole
Microtus oregoni Bachman	Creeping vole
Microtus pennsylvanicus Ord	Meadow vole
Microtus pinetorum LeConte	Pine vole
Microtus savii de Sélys Longchamps	Savi's pine vole
Microtus socialis Pallas	Social vole
Microtus subterraneus de Sélys Longchamps	European pine vole
Myopus schisticolor Lilljeborg	Wood lemming
Ondatra zibethicus Linnaeus	Muskrat
Pitymys = *Microtus*	

Subfamily Gerbillinae

Gerbillus dasyurus Wagner	Wagner's gerbil
Gerbillus pyramidum I. Geoffróy St. Hilaire	Large Egyptian gerbil
Meriones blackleri Thomas	Turkish jird
Meriones erythrourus Gray = *Meriones libycus*	
Meriones hurrianae Jerdon	Indian desert gerbil
Meriones libycus Lichtenstein	Libyan jird
Meriones meridianus Pallas	Midday gerbil
Meriones persicus Blanford	Persian jird
Meriones shawi Duvernoy	Shaw's jird
Meriones tamariscinus Pallas	Tamarisk gerbil
Meriones tristrami Thomas = *Meriones shawi* Duvernoy	
Meriones unguiculatus Milne-Edwards	Clawed jird
Meriones vinogradovi Heptner	Vinogradov's jird
Rhombomys opimus Lichtenstein	Great gerbil
Tatera afra Gray	Cape gerbil

Tatera brantsi A. Smith	Highveld gerbil
Tatera indica Hardwicke	Indian gerbil
Tatera leucogaster Peters	Bushveld gerbil
Taterillus emini Thomas	Naked-soled gerbil

Taterillus pygargus F. Cuvier = *Gerbillus pyramidum* I. Geoffroy St.-Hilaire

Family Spalacidae

Spalax leucodon Nordmann	Lesser mole rat
Spalax microphthalmus Güldenstaedt	Russian mole rat

Family Muridae

Subfamily Murinae

Acomys cahirinus Desmarest	Cairo spiny mouse
Aethomys chrysophilus de Winton	Red veld rat
Aethomys kaiseri Noack	The Kaiser's rat
Apodemus agrarius Pallas	Striped field mouse

Apodemus argenteus Temminck = *Apodemus sylvaticus* Linnaeus

Apodemus flavicollis Melchoir	Yellow-necked field mouse
Apodemus sylvaticus Linnaeus	Common field mouse
Arvicanthis niloticus Desmarest	Nile rat
Bandicota bengalensis Gray and Hardwicke	Lesser bandicoot rat
Bandicota indica Bechstein	Large bandicoot rat
Beamys major Dollman	Larger long-tailed pouched rat
Conilurus penicillatus Gould	Brush-tailed rabbit-rat
Cricetomys Waterhouse	Giant pouched rat

Grammomys dolichurus = *Thamnomys dolichurus* Smuts

Hybomys univittatus Peters	One-striped mouse
Hylomyscus stella (Thomas)	Stella wood mouse

Lophuromys aquilus True = *Lophuromys flavopunctatus* Thomas

Lophuromys flavopunctatus Thomas	Speckled harsh-furred rat
Malacomys longipes Milne-Edwards	Milne-Edward's swamp rat

Mastomys = *Rattus*

Micromys minutus Pallas	Harvest mouse
Melomys cervinipes Gould	Fawn-footed melomys
Mus musculus Linnaeus	House mouse
Mus triton Thomas	Larger pygmy mouse
Oenomys hypoxanthus Pucheran	Rufous-nosed rat

Praomys jacksoni de Winton = *Praomys morio* Trouessart

Praomys morio Trouessart	Soft-furred rat

Praomys natalensis = *Rattus natalensis*

Rattus argentiventer Robinson and Kloss	Ricefield rat

List of mammal names

Rattus blanfordi Thomas — Blanford's rat
Rattus bowersi Anderson — Bower's rat
Rattus canus Miller — Grey tree rat
Rattus cremoriventer Miller — Dark-tailed rat
Rattus diardii = *Rattus rattus diardii* Jentink — Malaysian house rat
Rattus exulans Peale — Little rat
Rattus fuscipes Waterhouse — Western swamp rat
Rattus mülleri Jentink — Muller's rat
Rattus natalensis Smith — Coucha or multimammate rat
Rattus norvegicus Berkenhout — Norway, brown or common rat
Rattus rajah Thomas — Brown spiny rat
Rattus rattus Linnaeus — Black, ship or roof rat
Rattus sabanus Thomas — Noisy rat
Rattus surifer = *Rattus rajah surifer* Miller
Rattus tiomanicus Miller — Malaysian wood rat
Rattus villosissimus Waite — Long-haired rat
Rattus whiteheadi Thomas — Little spiny rat
Rattus wroughtoni Hinton = *Rattus rattus* Linnaeus
Rhabdomys pumilio Sparrman — Four-striped rat
Saccostomus campestris Peters — Cape pouched mouse
Thamnomys cometes Thomas and Wroughton — Mozambique forest mouse
Thamnomys dolichurus Smuts — Cape forest mouse
Thamnomys rutilans Peters — Shining thicket rat

Subfamily Dendromurinae

Dendromus insignis Shortridge and Carter = *D. melanotis* A. Smith
Dendromus melanotis A. Smith — Gray pygmy tree-mouse

Subfamily Otomyinae

Subfamily Phloeomyinae
Chiropodomys gliroides Blyth — Pencil-tailed tree mouse

Family Gliridae
Glis glis Linnaeus — Fat dormouse
Graphiurus Smuts — African dormice

Family Zapodidae
Napaeozapus insignis Miller — Woodland jumping mouse
Zapus hudsonicus Zimmermann — Meadow jumping mouse
Zapus princeps J. A. Allen — Western jumping mouse

Family Dipodidae
Allactaga elater Lichtenstein — Small five-toed jerboa
Jaculus jaculus Linnaeus — Lesser Egyptian jerboa

Family Hystricidae
Atherurus Cuvier — Brush-tailed porcupine

368

Family Erethizontidae
    Coëndou Lacépède   Prehensile-tailed porcupine

Family Dasyproctidae
    Agouti paca Linnaeus   Paca
    Dasyprocta punctata Gray   Agouti

Family Capromyidae
    Geocapromys ingrahami   Ingraham's hutia
        J. A. Allen
    Myocaster coypus Molina   Nutria

Family Octodontidae
    Spalacopus cyanus Molina   Coruro
    Thryonomys Fitzinger   Cane rat

Family Ctenomyidae
    Ctenomys talarum Thomas   Tuco-tuco

Family Echimyidae
    Diplomys Thomas   Arboreal spiny rat
    Proechimys dimidiatus Günther   Rio spiny rat
    Proechimys iheringi Thomas   Ihering's spiny rat
    Proechimys semispinosus Tomes   Central American spiny rat

Order Cetacea
    Suborder Mysticeti   Baleen whales

Order Carnivora
Family Canidae
    Alopex lagopus Linnaeus   Arctic fox

Family Procyonidae
    Bassaricyon gabbii J. A. Allen   Olingo
    Nasua narica Linnaeus   Coati
    Potos flavus Schreber   Kinkajou
    Procyon lotor Linnaeus   Raccoon

Family Mustelidae
    Eira barbara Linnaeus   Tayra
    Mustela erminea Linnaeus   Stoat or ermine
    Mustela frenata Lichtenstein   Long-tailed weasel
    Mustela nivalis Linnaeus   Least weasel
    Mustela rixosa Bangs = *Mustela nivalis* Linnaeus

Family Felidae
    Lynx rufus (Schreber)   Bobcat

Family Phocidae
    Phoca groenlandica Erxleben   Harp seal

List of mammal names

Order Artiodactyla
Family Suidae
      Sus scrofa Linnaeus   Wild boar

Family Cervidae
      Capreolus capreolus Linnaeus   Roe deer

AUTHORITIES USED

Allen, G. M. (1939). A checklist of African mammals. *Bulletin of The Museum of Comparative Zoology, Harvard*, **83.**

Chasen, F. N. (1940). *A Handlist of Malaysian mammals.* Bulletin of the Raffles Museum, Singapore, No. 15.

Ellerman, J. R. & Morrison-Scott, T. C. S. (1951). *Checklist of palaearctic and Indian mammals, 1758 to 1946.* British Museum (Natural History), London.

Ellerman, J. R., Morrison-Scott, T. C. S. & Hayman, R. W. (1953). *Southern African mammals, 1758 to 1951: a reclassification.* British Museum (Natural History), London.

Hall, E. R. & Kelson, K. R. (1959). *The mammals of North America.* Ronald Press, New York.

Rosevear, D. R. (1969). *The rodents of West Africa.* British Museum (Natural History), London.

Simpson, G. G. (1945). *The principles of classification and a classification of mammals.* Bulletin of The American Museum of Natural History, Vol. 45.

Walker, E. P. (1968). *Mammals of the world*, 2nd edition, 2 vols. John Hopkins Press, Baltimore.

References

Abaturov, B. D. (1972). The role of burrowing animals in the transport of mineral substances in the soil. *Pedobiologia*, **12**, 261–6.

Abaturov, B. D., Devyatykh, V. A. & Zubkova, L. V. (1969). Rol royushchei deyatelnosti suslikov (*Citellus pygmaeus* Pall) v peremeshchenii mineralnykh veshchestv v polupustynnykh pochvakh Zavolzhya. *Pochvovedenie*, **12**, 93–9.

Abaturov, B. D. & Zubkova, L. V. (1969). Vliyanie malykh suslikov (*Citellus pygmaeus* Pall) na vodno-fizycheskie svoistva soloncovykh pochv polupustynii Zavolzhya. *Pochvovedenie*, **10**, 59–69.

Abbott, H. G. (1961). White pine seed consumption in small mammals. *Journal of Forestry*, **59**, 197–201.

Acton, F. S. (1966). *Analysis of straight-line data.* Dover Publications, New York.

Adams, L. (1959). Analysis of a population of snowshoe hare in Northwestern Montana. *Ecological Monographs*, **29**, 141–70.

Adamczewska, K. A. (1961). Intensity of reproduction of *Apodemus flavicollis* (Melchior 1834) during the period 1954–1959. *Acta Theriologica*, **5**, 1–21.

Adamczewska-Andrzejewska, K. A. (1966). Variations in the hardness of the teeth of *Sorex araneus* Linnaeus, 1758. *Acta Theriologica*, **11**, 55–69.

Adamczewska-Andrzejewska, K. A. (1967). Age reference model for *Apodemus flavicollis* (Melchior, 1834). *Ekologia Polska*, Ser. A, **15**, 787–90.

Adamczewska-Andrzejewska, K. A. (1971). Methods of age determination in *Apodemus agrarius* (Pallas 1771). *Annales Zoologici Fennici*, **8**, 68–71.

Adamczewska-Andrzejewska, K. A. (1973a). The lens weight as indicator of age of the wild *Microtus arvalis* population. *Bulletin de L'Académie Polonaise des Sciences*, Cl. 2, **21**, 331–36.

Adamczewska-Andrzejewska, K. A. (1973b). Growth, variations and age criteria in *Apodemus agrarius* (Pallas, 1771). *Acta Theriologica*, **18**, 353–94.

Adamczyk, K. & Ryszkowski, L. (1968). Estimation of the density of a rodent population using stained bait. *Acta Theriologica*, **13**, 295–311.

Adamczyk, K. & Walkowa, W. (1971). Compensation of numbers and production in a *Mus musculus* population as a result of partial removal. *Annales Zoologici Fennici*, **8**, 145–53.

Adams, L. & Watkins, S. G. (1967). Annuli in tooth cementum indicate age in California ground squirrels. *Journal of Wildlife Management*, **31**, 836–9.

Aho, J. & Kalela, O. (1966). The spring migration of 1961 in the Norwegian lemming, *Lemmus lemmus* (L.), at Kilpisjarvi, Finnish Lapland. *Annales Zoologici Fennici*, **3**, 53–65.

Alekperov, A. M. (1966). *Mlekopitayushchie yugo-zapadnogo Azerbaïdzhana*, pp. 1–178. Baku.

Alekperov, A. M., Eïgelis, Yu. K., Poltavtsev, N. N. & Akhverzov, N. M. (1967). Dinamika chislennosti i razmnozheniya krasnokhvostoï peschanki

References

(*Meriones erythrourus* Gray) v Azervaĭdzhanskoĭ SSR. *Izvestiya Akademii Nauk Azerbaidzhanskoi SSR, Seriya Biologicheskikh Nauk*, **1**, 70–6.

Aleksiuk, M. (1964). Observations of birds and mammals in the Perry River region, N.W.T., Canada. *Arctic*, **17**, 263–8.

Allanson, M. (1958). Growth and reproduction in the males of two species of gerbils. *Proceedings of the Zoological Society of London*, **130**, 373–96.

Allen, K. R. (1950). The computation of production in fish population. *New Zealand Science Review*, **8**, 89–101.

Alsager, D. E. & Yaremko, R. (1972). Experimental population suppression of Richardson's ground squirrels (*Spermophilus richardsonii*) in Alberta. Proceedings of the IV Vertebrate Pest Conference, pp. 93–100.

Altum, B. (1876). *Forstzoologie. I. Säugetiere*. Berlin.

Ambrose, H. W., III (1972). Effect of habitat familiarity and toe-clipping on rate of owl predation in *Microtus pennsylvanicus*. *Journal of Mammalogy*, **53**, 909–12.

Ambrose, H. W., III (1973). An experimental study of some factors affecting the spatial and temporal activity of *Microtus pennsylvanicus*. *Journal of Mammalogy*, **54**, 79–110.

Anderson, P. K. (1961). Density, social structure, and nonsocial environment in house-mouse populations and the implications for regulation of numbers. *Transactions of the New York Academy of Science*, Ser. II, **23**, 447–51.

Anderson, P. K. (1965). The role of breeding structure in evolutionary processes of *Mus musculus* population. Mendel Symposium, Prague, pp. 17–21.

Anderson, P. K. (1970). Ecological structure and gene flow in small mammals. *Symposia of the Zoological Society of London*, **26**, 299–325.

Anderson, S. (1960). The baculum in microtine rodents. *University of Kansas Publications Museum of Natural History, Lawrence*, **12**, 181–216.

Anderson, S. & Jones, J. K., Jr. (1967). *Recent Mammals of the World: A Synopsis of Families*. Ronald Press, New York.

Andersson, A. & Hansson, L. (1966). Smadäggdjursundersökningar i norra Norge 1965 (Small mammal investigations in northern Norway 1965). *Fauna and Flora, Upsala*, **61**, 49–72.

Andrewartha, H. G. & Birch, L. C. (1954). *The distribution and abundance of animals*. University of Chicago Press, Chicago.

Andrzejewski, R. (1962). Settling by small rodents a terrain in which catching out had been performed. *Acta Theriologica*, **6**, 257–74.

Andrzejewski, R., Bujalska, G., Ryszkowski, L. & Ustyniuk, J. (1966). On a relation between the number of traps in a point of catch and trappability of small rodents. *Acta Theriologica*, **11**, 343–9.

Andrzejewski, R., Fejgin, H. & Liro, A. (1971). Trappability of trap-prone and trap-shy bank voles. *Acta Theriologica*, **16**, 401–5.

Andrzejewski, R. & Gliwicz, J. (1969). Standard method of density estimation of *Microtus arvalis* (Pall.) for the investigation of its productivity. *Small Mammals Newsletter*, **3**, 45–53.

Andrzejewski, R., Kajak, A. & Pieczyńska, E. (1963). Efekty migracji. *Ekologia Polska*, Ser. B, **9**, 161–72.

References

Andrzejewski, R., Petrusewicz, K. & Walkova, W. (1963). Absorption of newcomers by a population of white mice. *Ekologia Polska*, Ser. A, **11**, 223–40.

Andrzejewski, R. & Rajska, E. (1972). Trappability of bank vole in pitfalls and live traps. *Acta Theriologica*, **17**, 41–56.

Andrzejewski, R. & Wrocławek, H. (1961). Mortality of small rodents in traps as an indication of the diminished resistance of the migrating part of a population. *Bulletin de l'Académie Polonaise des Sciences*, Cl. 2, **9**, 491–2.

Andrzejewski, R. & Wrocławek, H. (1963). Metal cylinder as a live trap with a bait. *Acta Theriologica*, **6**, 297–300.

Anonymous (1957–8). *The Oregon meadow mouse irruption of 1957–1958*. Bulletin of the Federal Cooperative Extension Service, Oregon State College, Corvallis, Oregon.

Antonov, A. S. (1973). Éksperimental'noe obosnovanie nekotorȳkh kontseptsiĭ genosistematiki. PhD Thesis. Moscow State University.

Antonov, A. S. & Belozarskiĭ, A. N. (1972). *Stroenie DNA i polozhenie organizmov v sisteme.* Izdatel'stvo Moskovskogo gosudarstvennogo Universiteta, Moscow.

Arata, A. A. (1959). Effects of burning on vegetation and rodent population in a longleaf pine/turkey oak association in North Central Florida. *Quarterly Journal of the Florida Academy of Scientists*, **22**, 94–104.

Arata, A. A., Negus, N. C. & Downs, M. S. (1965). Histology, development, and individual variation of complex murid bacula. *Tulane Studies in Zoology*, **12**, 51–64.

Armour, C. J. (1963). The use of repellents for preventing mammal and bird damage to trees and seeds: a revision. *Forestry Abstracts*, **24**, 27–38.

Artimo, A. (1964). The baculum as a criterion for distinguishing sexually mature and immature bank voles, *Clethrionomys glareolus* Schr. *Annales Zoologici Fennici*, **1**, 1–6.

Artimo, A. (1969). The baculum in the wood lemming, *Myopus schisticolor* (Lilljeb.), in relation to sexual status, size and age. *Annales Zoologici Fennici*, **6**, 335–44.

Ashby, W. R. (1956). *An introduction to cybernetics.* Chapman & Hall, London.

Ashby, K. R. (1959). Prevention of regeneration of woodland by field mice and voles. *Quarterly Journal of Forestry*, **53**, 228–36.

Ashby, K. R. (1967). Studies on the ecology of field mice and voles (*Apodemus sylvaticus*, *Clethrionomys glareolus* and *Microtus agrestis*) in Houghall Wood, Durham. *Journal of Zoology, London*, **152**, 389–513.

Asia Pacific Interchange Proceedings (1968). Rodents as factors in disease and economic loss. *Asia Pacific Interchange Proceedings, Honolulu*, **513**, 1–56.

Askaner, T. & Hansson, L. (1967). The eye lens as an age indicator in small rodents. *Oikos*, **18**, 151–3.

Aulak, W. (1967). Estimation of small mammal density in three forest biotopes. *Ekologia Polska*, Ser. A, **15**, 755–78.

Aulak, W. (1973). Production and energy requirements in a population of the bank vole in a deciduous forest of *Circaeo–Alnetum* type. *Acta Theriologica*, **18**, 167–90.

References

Aumann, G. D. (1965). Microtine abundance and soil sodium levels. *Journal of Mammalogy*, **46**, 594–604.

Babinska, J. & Bock, E. (1969). The effect of prebaiting on captures of rodents. *Acta Theriologica*, **14**, 267–71.

Bailey, C. B., Kitts, W. D. & Wood, A. J. (1960). Changes in the gross chemical composition of the mouse during growth in relation to the assessment of physiological age. *Canadian Journal of Animal Science*, **40**, 143–55.

Bailey, N. T. J. (1951). On estimating the size of mobile populations from recapture data. *Biometrika*, **38**, 293–306.

Bailey, N. T. J. (1952). Improvements in the interpretation of recapture data. *Journal of Animal Ecology*, **21**, 120–7.

Baker, C. E. & Dunaway, P. B. (1969). Retention of ^{134}Cs as an index of metabolism in the cotton rat (*Sigmodon hispidus*). *Health Physics*, **16**, 227–30.

Baker, H. G. (1972). Evolutionary relationships between flowering plants and animals in American and African tropical forests. In: *Tropical forest ecosystems in Africa and South America: a comparative review* (ed. B. J. Meggers, E. S. Ayensu & W. D. Duckworth), pp. 145–50. Smithsonian Institution Press, Washington.

Baker, R. H. (1968). Habitats and distribution. In: *Biology of* Peromyscus (ed. J. A. King), pp. 98–126. American Society of Mammalogists, Special Publication 2.

Balabas, N. G., Kuzin, N. P. & Trofimenko, I. P. (1965). O stroenii nor bol'shikh peshchanok v severo-vostochnȳkh Muyukumakh. In: *Materialȳ IV nauchnoĭ konferentsii po prirodnoĭ ochagovosti i profilaktike chumȳ*, pp. 25–7. Alma Ata.

Balph, D. F. (1968). Behavioral responses of unconfined Uinta ground squirrels to trapping. *Journal of Wildlife Management*, **32**, 778–94.

Baltazard, M., Karimi, Y., Ektekhari, M., Chamsa, M. & Mollaret, H. (1963). La Conservation interepizootique de la peste en foyer invétéré. Hypotheses de Travail. *Bulletin de la Societé de Pathologie exotique*, **46**, 1230–41.

Barbehenn, K. R. (1969). Responses of rodents and shrews to patterns of removal trapping. *Indian Rodent Symposium* (convener, D. W. Parrack), pp. 247–52. Calcutta.

Barbehenn, K. R. (1974). The use of stratified traps in estimating density. In press.

Barkalow, F. S., Jr, Hamilton, R. B. & Soots, R. F., Jr (1970). The vital statistics of an unexploited gray squirrel population. *Journal of Wildlife Management*, **34**, 489–500.

Barrett, G. (1969). Bioenergetics of a captive least shrew, *Cryptotis parva*. *Journal of Mammalogy*, **50**, 629–30.

Barrier, M. J. & Barkalow, F. S., Jr (1967). A rapid technique for ageing gray squirrels in winter pelage. *Journal of Wildlife Management*, **31**, 715–19.

Bartholomew, G. A. & Hudson, J. W. (1964). Terrestrial animals in dry heat: estivators. In: *Handbook of physiology. Section 4: Adaptation to the environment* (ed. D. B. Dill), Chapter 34. American Physiological Society, Washington.

Batzli, G. O. (1973). Population determination and nutrient flux through lemmings, US IBP Tundra Biome Data Report 73–19.

Batzli, G. O. (1974). Production, assimilation and accumulation of organic matter in ecosystems. *Journal of Theoretical Biology* (In press).

Batzli, G. O. & Pitelka, F. A. (1970). Influence of meadow mouse populations on California grassland. *Ecology*, **51**, 1027–39.

Batzli, G. O. & Pitelka, F. A. (1971). Condition and diet of cycling populations of the California vole, *Microtus californicus. Journal of Mammalogy*, **52**, 141–63.

Batzli, G. O., Stenseth, N. C. & Fitzgerald, B. M. (1974). Growth and survival of suckling brown lemmings, *Lemmus trimucronatus. Journal of Mammalogy*, **65**, 828–31.

Beale, D. M. (1962). Growth of the eye lens in relation to age in fox squirrels. *Journal of Wildlife Managment*, **26**, 208–11.

Beasom, S. L. (1970). Turkey productivity in two vegetative communities in South Texas. *Journal of Wildlife Management*, **34**, 166–75.

Beckman, W. A., Mitchell, J. W. & Porter, W. P. (1971). Thermal model for prediction of a desert iguana's daily and seasonal behavior. Transactions of the ASME, Winter meeting. pp. 1–7.

Bee, J. W. & Hall, E. R. (1956). Mammals of Northern Alaska on the Arctic slope. University of Kansas Miscellaneous Publication of the Museum of Natural History, No. 8.

Beer, J. R. (1964). Bait preferences of some small mammals. *Journal of Mammalogy*, **45**, 632–4.

Beer, J. R. & MacLeod, C. F. (1966). Seasonal population changes in prairie deer mouse. *American Midland Naturalist*, **76**, 277–89.

Bellier, L. (1967). Recherches écologiques dans la savane de Lamto (Cote D'Ivoire): densites et biomasses des petits mammifères. *Terre et Vie*, **114**, 319–29.

Belyaev, A. (1954). *Vrednŷe grŷzunŷ Kazakhstana i merŷ bor'bŷ s nimi*, pp.1–57. Alma-Ata.

Benedict, F. G. (1938). Vital energetics: a study in comparative basal metabolism. Carnegie Institute of Washington Publications, **503**, 1–215.

Bentley, E. W. (1972). A review of anticoagulants in current use. *Bulletin of the World Health Organisation*, **47**, 275–80.

Bentley, E. W., Hammond, I., E., Bathard, A. H. & Greaves, J. H. (1961). Sodium fluoracetate and fluoracetamide as 'direct' poisons for the control of rats in sewers. *Journal of Hygiene, Cambridge*, **59**, 413–17.

Bergerud, A. T. (1967). The distribution and abundance of arctic hares in Newfoundland. *Canadian Field Naturalist*, **81**, 242–8.

Bergstedt, B. (1965). Distribution, reproduction, growth and dynamics of the rodent species *Clethrionomys glareolus* (Schreber), *Apodemus flavicollis* (Melchoir) and *Apodemus sylvaticus* (Linné) in southern Sweden. *Oikos*, **16**, 132–60.

Bergström, U. (1967). Observations on Norwegian lemmings, *Lemmus lemmus* (L.) in the autumn of 1963 and spring of 1964. *Arkiv für Zoologi*, **20**, 321–65.

Bernard, J. (1959). Methods used in Belgium for estimating populations of

375

References

Microtus arvalis Pall. Report of the International Conference on Harmful Mammals and their Control (OEPP/EPPO), pp. 49, 51, 53. Paris.

Berry, R. J. (1968). The ecology of an island population of the house mouse. *Journal of Animal Ecology*, **37**, 445–70.

Berry, R. J. & Murphy, H. M. (1970). The biochemical genetics of an island population of the house mouse. *Proceedings of the Royal Society of London*, Ser. B, **176**, 87–103.

Berry, R. J. & Truslove, G. M. (1968). Age and eye lens weight in the house mouse. *Journal of Zoology, London*, **155**, 247–52.

Beyers, R. J., Smith, M. H., Gentry, J. B. & Ramsey, L. L. (1971). Standing crop of elements and atomic ratios in small mammal communities. *Acta Theriologica*, **14**, 203–11.

Bider, J. R. (1968). Animal activity in uncontrolled terrestrial communities as determined by a sand transect technique. *Ecological Monographs*, **38**, 269–308.

Bird, R. D. (1930). Biotic communities of the Aspen parkland of Central Canada. *Ecology*, **11**, 356–442.

Birdsell, J. B. (1957). Some population problems involving Pleistocene man. Population studies: animal ecology and demography, Cold Spring Harbor Symposia on Quantitative Biology, **22**, 47–69.

Birkebak, R. C. (1966). Heat transfer in biological systems. *International Review of General and Experimental Zoology*, **2**, 269–344.

Birkebak, R. C., Cremers, C. J. & LeFebvre, E. A. (1966). Thermal modeling applied to animal systems. *Journal of Heat Transfer, Transactions of the ASME*, Ser. C, **88**, 125–30.

Black, R. F. (1954). Permafrost – a review. *Bulletin of the Geological Society of America*, **65**, 839–56.

Blaedel, W. J. & Meloche, V. W. (1963). *Elementary quantitative analysis. Theory and practice*, 2nd edition. Harper & Row, New York.

Blair, W. F. (1941). Techniques for the study of mammal populations. *Journal of Mammalogy*, **22**, 148–57.

Blair, W. F. (1943*a*). Ecological distribution of mammals in the Tularosa Basin, New Mexico. *Contributions of Laboratory of Vertebrate Biology, University of Michigan*, **20**, 1–24.

Blair, W. F. (1943*b*). Populations of the deermouse and associated small mammals in the mesquite association of southern New Mexico. *Contributions of Laboratory of Vertebrate Biology, University of Michigan*, **21**, 1–40.

Blair, W. F. (1950). Ecological factors in speciation in *Peromyscus*. *Evolution*, **4**, 253–75.

Blair, W. F. (1951). Population structure, social behavior and environmental relations in a natural population of the beach mouse, *Peromyscus polionotus of leucocephalus*. *Contributions of Laboratory Vertebrate Biology, University of Michigan*, **48**, 1–47.

Bliss, L. C. (1962). Net primary productivity of tundra ecosystems. In: *Die Stoffproduktion der Pflanzendecke* (ed. H. Lieth), pp. 35–46. G. Fischer, Stuttgart.

References

Blus, L. (1966). Some aspects of golden mouse ecology in southern Illinois. *Transactions of the Illinois State Academy of Science*, **59**, 334–41.
Bobek, B. (1969). Survival, turnover and production of small rodents in a beech forest. *Acta Theriologica*, **15**, 191–210.
Bobek, B. (1971). Influence of population density upon rodent production in a deciduous forest. *Annales Zoologici Fennici*, **8**, 137–44.
Bobek, B. (1973). Net production of small rodents in a deciduous forest. *Acta Theriologica*, **18**, 403–34.
Bodenheimer, F. S. (1949). *Problems of vole populations in the Middle East*. Report of the population dynamics of the Levant vole (*Microtus guentheri* D. et A.). Government of Israel, Jerusalem.
Bodenheimer, F. S. (1958). *Animal ecology today*. W. Junk, The Hague.
Boguslavsky, G. W. (1956). Statistical estimation of the size of a small population. *Science*, **124**, 317–18.
Bollaerts, D. & Tahon, J. (1968). Utilisation de l'hydrogene phosphore' dans la lutte contre les petits mammifères souterrains. *Mededelingen Rijksfaculteit Landbouwwetenschappen Gent.*, **33**, 3.
Bondaf, E. P., Burdelov, A. S. & Shatalov, M. S. (1967). Mezhvidovȳe kontaktȳ pozvonochnȳkh v norakh bol'shikh peschanok. In: *Materialȳ V nauchnoĭ konferentsii protivochumnȳkh uchrezhdeniĭ Sredneĭ Azii i Kazakhstana*, pp. 100–2. Alma-Ata.
Borowski, S. & Dehnel, A. (1953). Augabei zur Biologie der Sorricidae. *Annales Universitatis Mariae Curie – Sklodowska*, Sectio C, **7**, 305–448.
Borutzky, E. V. (1939). Dynamics of *Chironomus plumosus* in the profundal of Lake Beloie. *Arb. Proc. Limnol. Sta. Kossino*, **22**, 156–95, 196–218. (In Russian with English summary.)
Boshell, M. J. & Rajogopalau, P. K. (1968). Small rodents and shrews in the Sagar-Sorab area, Mysore State India: Population studies 1961–1964. *Indian Journal of Medical Research*, **56**, 527–40.
Bowen, H. J. M. (1966). *Trace elements in biochemistry*. Academic Press, New York.
Boyd, C. E. & Walley, W. W. (1972). Production and chemical composition of *Saururus cernuus* L. at sites of different fertility. *Ecology*, **53**, 927–32.
Bradley, W. G. (1967). Home range, activity patterns, and ecology of the antelope ground squirrel in southern Nevada. *Southwestern Naturalist*, **12**, 231–51.
Bradley, W. G. & Mauer, R. A. (1971). Reproduction and food habits of Merriam's kangaroo rat, *Dipodomys merriami*. *Journal of Mammalogy*, **52**, 497–507.
Bradt, G. W. (1938). A study of beaver colonies in Michigan. *Journal of Mammalogy*, **19**, 139–62.
Brambell, F. W. R. & Davis, D. H. S. (1941). Reproduction in the multimammate mouse. *Proceedings of the Zoological Society of London*, **111**, 1–11.
Bramble, W. C. & Goddard, M. K. (1942). Effect of animal coaction and seedbed conditions on regeneration of pitch pine in the barrens of central Pennsylvania. *Ecology*, **23**, 330–5.

377

References

Brant, D. H. (1962). Measures of the movements and population densities of small rodents. *University of California Publications in Zoology*, **62**, 105–84.

Bree, P. J. H., Jensen, B. & Kleijn, L. J. K. (1966). Skull dimensions and the length/weight relation of the baculum as age indications in the common otter *Lutra lutra* (Linnaeus, 1758). *Danish Review of Game Biology*, **4**, 98–104.

Briese, L. A. (1973). Variations in elemental composition and cycling in the cotton rat, *Sigmodon hispidus*. MS Thesis. University of Georgia, Athens.

Briese, L. A. & Smith, M. H. (1974). Seasonal patterns of abundance and movement of nine species of small mammals. *Journal of Mammalogy*, **55**, 615–29.

Brisbin, I. L. (1966). Energy-utilization in a captive hoary bat. *Journal of Mammalogy*, **47**, 719–20.

Britton, M. E. (1957). Vegetation of the arctic tundra. In: *Arctic biology* (ed. H. P. Hansen), pp. 67–130. Oregon State University Press, Corvallis.

Brjuzgin, V. K. (1939). K metodike issledovanija vozrasta i rosta reptilij. *Doklady Akademii Nauk SSSR*, **23**.

Broadbrooks, H. E. (1970). Home ranges and territorial behavior of the yellow-pine chipmunk, *Eutamias amoenus*. *Journal of Mammalogy*, **51**, 310–26.

Brody, S. (1945). *Bioenergetics and growth*. Hafner Publishing Co. Inc., New York.

Broekuizen, S. (1973). Age determination and age composition of hare populations. Actes du Xe Congrès de l'Union Internationale des Biologistes du Gibier, pp. 477–89. Office National de la Chasse, Paris.

Bronson, F. H. & Marsden, H. M. (1964). Male-induced synchrony of estrus in deermice. *General and Comparative Endocrinology*, **4**, 634–7.

Brooks, J. E. (1972). An outbreak and decline of Norway rat populations in California rice fields. *California Vector Views*, **19**, 5–14.

Brooks, R. J. (1970). Ecology and acoustic behavior of the collared lemming, *Dicrostonyx groenlandicus* (Trail). PhD Thesis. University of Illinois.

Brown, J. & Bowen, S. (eds.) (1971). The structure and function of the tundra ecosystem. US IBP Tundra Biome 1971 Progress Report.

Brown, J. & Bowen, S. (eds.) (1972). Proceedings of the 1972 US IBP Tundra Biome Symposium.

Brown, J. & West, G. C. (eds.) (1970). Tundra biome research in Alaska. US IBP Tundra Biome Report 70–1.

Brown, J. H. (1968). Adaptation to environmental temperature in two species of woodrats, *Neotoma cinerea* and *N. albigula*. *Miscellaneous Publications of the Museum of Zoology, University of Michigan*, **135**, 1–48.

Brown, J. H. & Lasiewski, R. C. (1972). Metabolism of weasels: the cost of being long and thin. *Ecology*, **53**, 939–43.

Brown, L. E. (1966). Home range and movements of small mammals. *Symposia of the Zoological Society of London*, **18**, 111–42.

Brown, L. E. (1969). Field experiments on the movements of *Apodemus sylvaticus* using trapping and tracking techniques. *Oecologia*, **2**, 198–222.

Brown, L. N. (1967). Ecological distribution of six species of shrews and comparison of sampling methods in the central Rocky Mountains. *Journal of Mammalogy*, **48**, 617–23.

Brown, L. N. (1970). Population dynamics of the western jumping mouse (*Zapus princeps*) during a 4-year study. *Journal of Mammalogy*, **51**, 651–8.

Brown, R. (1970). Rodent control in developing countries. Proceedings of the Vth Vertebrate Pest Conference, pp. 140–3.

Brown, R. Z. (1953). Social behavior, reproduction, and population change in the house mouse (*Mus musculus*). *Ecological Monographs*, **23**, 217–40.

Bruce, H. M. (1959). An exteroceptive block to pregnancy in the mouse. *Nature, London*, **184**, 105.

Bruce, H. M. & Parrott, D. M. V. (1960). Role of olfactory sense in pregnancy block by strange males. *Science*, **131**, 1526.

Bruel, W. E. van den (1969). Le campagnol des champs *Microtus arvalis* Pallas; état du probleme en Belgique. *Parasitica*, **25**, 117–51.

Bruel, W. E. van den & Bollaerts, D. (1960). Mise au point du dispositif utilisé pour détruire au moyen de l'hydrogène phosphoré les mammifères dissimulés dans des terriers profonds. Proceedings of the IVth International Congress of Crop Protection Hamburg, 1957, pp. 1335–41.

Buchalczyk, A. (1970). Reproduction, mortality and longevity of the bank vole under laboratory condition. *Acta Theriologica*, **15**, 153–76.

Buchalczyk, T., Gebczyńska, Z. & Pucek, Z. (1970). Numbers of *Microtus oeconomus* (Pallas, 1776) and its noxiousness in forest plantations. *EPPO Publications*, Ser. A, **58**, 95–9.

Buchalczyk, T. & Pucek, Z. (1968). Estimation of the numbers of *Microtus oeconomus* using the Standard Minimum method. *Acta Theriologica*, **13**, 461–82.

Buckner, C. H. (1964). Metabolism, food capacity and feeding behavior in four species of shrews. *Canadian Journal of Zoology*, **42**, 259–79.

Buckner, C. H. (1966*a*). Populations and ecological relationships of shrews in tamarack bogs of Southeastern Manitoba. *Journal of Mammalogy*, **47**, 181–94.

Buckner, C. H. (1966*b*). The role of vertebrate predators in the biological control of forest insects. *Annual Review of Entomology*, **11**, 449–70.

Buckner, C. H. (1969). Some aspects of the population ecology of *Sorex araneus* near Oxford, England. *Journal of Mammalogy*, **50**, 326–32.

Buckner, C. H. (1972). The strategy for controlling rodent damage to pines in the Canadian Mid-West. Proceedings of the Vth Vertebrate Pest Conference, pp. 43–8.

Bujalska, G. (1970). Reproduction stabilizing elements in an island population of *Clethrionomys glareolus* (Schreber, 1780). *Acta Theriologica*, **15**, 381–412.

Bujalska, G. (1973). The role of spacing behaviour among females in the regulation of reproduction in the bank vole. *Journal of Reproduction and Fertility*, Supplement, **19**, 465–74.

Bujalska, G., Andrzejewski, R. & Petrusewicz, K. (1968). Productivity investigation of an island population of *Clethrionomys glareolus* (Schreber, 1780). II. Natality. *Acta Theriologica*, **13**, 415–25.

Bujalska, G., Cabon-Raczyńska, K. & Raczyński, J. (1965). Studies on the

References

European hare. VI. Comparison of different criteria of age. *Acta Theriologica*, **10**, 1–10.

Bujalska, G. & Gliwicz, J. (1968). Productivity investigations of an island population of *Clethrionomys glareolus* (Schreber, 1780). III. Individual growth curve. *Acta Theriologica*, **13**, 427–33.

Bujalska, G. & Ryszkowski, L. (1966). Estimation of the reproduction of the bank vole under field conditions. *Acta Theriologica*, **11**, 351–61.

Bunnell, F. L. (1972). Lemmings – models and the real world. Report of the US IBP Tundra Biome.

Bunnell, F. L. (1973). Computer simulation of nutrient and lemming cycles in an arctic tundra wet meadow ecosystem. PhD Thesis. University of California, Berkeley.

Burt, W. H. (1936). A study of the baculum in the genera *Perognathus* and *Dipodomys*. *Journal of Mammalogy*, **17**, 145–56.

Burt, W. H. & Grossenheider, R. H. (1964). *A field guide to the mammals.* Houghton Mifflin Co., Boston.

Caboń-Raczyńska, K. & Raczyński, J. (1972). Methods for determination of age of the European hare. *Acta Theriologica*, **17**, 75–86.

Caldwell, F. T., Hammel, H. T. & Dolan, F. (1966). A calorimeter for simultaneous determination of heat production and heat loss in the rat. *Journal of Applied Physiology*, **21**, 1665–71.

Caldwell, L. D. & Connell, C. E. (1968). A precis on energetics of the old-field mouse. *Ecology*, **49**, 542–8.

Calhoun, J. B. (1962). A 'behavioral sink'. In: *Roots of behavior* (ed. E. L. Bliss), pp. 295–315. Hafner Publishing Co. Inc., New York.

Calhoun, J. B. (1964). The social use of space. In: *Physiological mammalogy* (ed. W. Mayer & R. van Gelder), Vol. I. Academic Press, New York.

Calhoun, J. B. & Arata, A. A. (1950–7). Population dynamics of vertebrates, compilations of research data. In: Releases Nos. 3–9. Annual report, North American census of small mammals. US Department of Health, Education and Welfare, National Institute of Mental Health, Bethesda, Maryland.

Canutt, P. R. (1970). Pocket gopher problems and control practices on National forest lands in the Pacific northwest region. Proceedings of the IVth Vertebrate Pest Conference, pp. 120–5.

Carl, E. A. (1971). Population control in Arctic ground squirrels. *Ecology*, **52**, 395–413.

Carley, C. J. & Knowlton, F. F. (1968). Trapping wood rats: Effectiveness of several techniques and differential catch by sex and age. *Texas Journal of Science*, **19**, 248–51.

Carmon, J. L., Golley, F. B. & Williams, R. G. (1963). An analysis of the growth and variability in *Peromyscus polionotus*. *Growth*, **27**, 247–54.

Carson, J. D. (1961). Epiphyseal cartilage as an age indicator in fox and grey squirrels. *Journal of Wildlife Management*, **25**, 90–3.

Cayford, J. H. & Haig, R. A. (1961). Mouse damage to forest plantations in Southeastern Manitoba. *Journal of Forestry*, **59**, 124–5.

Chambers, R. E. (1962). In discussion of Lord, R. D. (1962). Ageing deer and determination of their nutritional status by the lens technique. Proceedings of the 1st National White-tailed Deer Disease Symposium, Athens, Georgia, pp. 89–93.

Chapman, B. M., Chapman, R. F. & Robertson, I. A. D. (1959). The growth and breeding of the multimammate rat, *Rattus* (*Mastomys*) *natalensis* (Smith), in Tanganyika Territory. *Proceedings of the Zoological Society of London*, **133**, 1–9.

Chapman, D. G. & Junge, C. O., Jr (1956). The estimation of the size of a stratified animal population. *Annals of Mathematical Statistics*, **27**, 375–89.

Chapskii, K. K. (1952*a*). Determination of age in some mammals by microstructure of bone. *Izvest. Estest. Nauk. Inst. P. F. Lesgafta Akad. Pedagog. Nauk ESFSR*, **25**, 47–66. (In Russian.)

Chapskii, K. K. (1952*b*). A method of determining age in mammals. Structure of claws as age indication in Greenland seal. *Izvest. Estest. Nauch. Inst. P. F. Lesgafta Akad. Pedagog. Nauk RSFSR*, **25**, 67–77. (In Russian.)

Chase, H. B. & Eaton, G. J. (1959). The growth of hair follicles in waves. *Annals of the New York Academy of Sciences*, **83**, 363–8.

Chelkowska, H. (1967). An attempt at comparing two methods of trapping small rodents (in pitfalls and live traps). *Ekologia Polska*, Ser. A, **15**, 779–85.

Chelkowska, H. & Ryszkowski, L. (1967). Causes of higher abundance estimates of small rodents at the edges of sampling areas in forest ecosystems. *Ekologia Polska*, Ser. A, **15**, 737–46.

Chen, Pao-Lai, Tsai, K. C., Chou, N.W. & Fung, K. N. (1966). On the population density of mole-rats in the upper Who-Tuo valley. *Acta Zoologica Sinica*, **18**, 21–7.

Chew, R. M. (1971). The excretion of ^{65}Zn and ^{54}Mn as indices of energy metabolism of *Peromyscus polionotus*. *Journal of Mammalogy*, **52**, 337–50.

Chew, R. M. & Butterworth, B. B. (1964). Ecology of rodents in Indian Cove (Mojave Desert), Joshua Tree National Monument, California. *Journal of Mammalogy*, **45**, 203–25.

Chew, R. M. & Chew, A. E. (1970). Energy relationships of the mammals of a desert shrub (*Larrea tridentata*) community. *Ecological Monographs*, **40**, 1–21.

Chiarappa, L., Chiang, H. C. & Smith, R. F. (1972). Plant pests and diseases: Assessment of crop losses. *Science*, **176**, 769–73.

Childs, H. E., Jr (1969). Birds and mammals of the Pitmegea River region Cape Sabine, Northwest Alaska. Biological Papers of the University of Alaska, No. 10.

Chipman, R. K. (1965). Age determination of the cotton rat (*Sigmodon hispidus*). *Tulane Studies in Zoology*, **12**, 19–38.

Chipman, R. K. & Fox, K. A. (1966*a*). Oestrous synchronization and pregnancy blocking in wild house mice (*Mus musculus*). *Journal of Reproduction and Fertility*, **12**, 233–6.

Chipman, R. K. & Fox, K. A. (1966*b*). Factors in pregnancy blocking: age and reproductive background of females: numbers of strange males. *Journal of Reproduction and Fertility*, **12**, 399–403.

References

Chipman, R. K., Holt, J. A. & Fox, K. A. (1966). Pregnancy failure in laboratory mice after multiple short-term exposure to strange males. *Nature, London,* **210,** 653.

Chitty, D. (1952). Mortality among voles (*Microtus agrestis*) at Lake Vyrnwy, Montgomeryshire in 1936–39. *Philosophical Transactions of the Royal Society of London,* Ser. B, **236,** 505–52.

Chitty, D. & Phipps, E. (1966). Seasonal changes in survival in mixed populations of two species of vole. *Journal of Animal Ecology,* **35,** 313–32.

Chitty, H. & Austin, C. R. (1957). Environmental modification of oestrus in the vole. *Nature, London,* **179,** 592–3.

Christian, J. J. (1971). Fighting, maturity and population density in *Microtus pennsylvanicus. Journal of Mammalogy,* **52,** 556–67.

Christiansen, E. & Døving, K. (1973). Kommunikasjon hos smågnagere viser nye veier til bekjempelse. *Tidsskrift før Skogsbruk,* **81,** 293–306.

Clark, D. O. (1972). The extending of cotton rat range in California – their life history and control. Proceedings of the Vth Vertebrate Pest Conference, pp. 7–14.

Clark, T. W. (1970). Richardson's ground squirrel (*Spermophilus richardsonii*) in the Laramie Basin, Wyoming. *Great Basin Naturalist,* **30,** 55–70.

Clark, T. W. (1971). Ecology of the western jumping mouse in Grand Teton National Park, Wyoming. *Northwest Science,* **45,** 229–38.

Claude, C. (1967). Morphologie und Altersstrukter von zwei schweizerischen Rötelmauspopulationen, *Clethrionomys glareolus* (Schreber, 1780). *Zeitschrift für Säugetierkunde,* **32,** 159–66.

Claude, C. (1970). Biometrie und Fortpflanzungsbiologie der Rötelmaus *Clethrionomys glareolus* (Schreber 1780) auf verschiedenen Höhenstufen der Schweiz. *Revue Suisse de Zoologie,* **77,** 435–80.

Clough, G. C. (1968). Social behavior and ecology of Norwegian lemmings during a population peak and crash. *Papers of the Norwegian State Game Research Institute,* Ser. 2, **28,** 1–50.

Clough, G. C. (1972). Biology of the Bahaman hutia, *Geocapromys ingrahami. Journal of Mammalogy,* **53,** 807–23.

Collett, R. (1878). On *Myodes lemmus* in Norway. *Journal of the Linnaean Society (Zoology),* **13,** 327–34.

Collett, R. (1895). *Myodes lemmus*: its habits and migrations in Norway. *Forhzandlinger Videnskabs-Selskabs Kristiania,* **3,** 1–62.

Collier, B. D., Stenseth, N. C., Osborn, R. & Barkley, S. (1974). *Models of the brown lemming at Barrow, Alaska.* US IBP Tundra Biome report.

Conaway, C. H. (1952). Life history of the water shrew *Sorex palustris navigator. American Midland Naturalist,* **48,** 219–48.

Cone, J. B. (1967). Rodent problems on private forest lands in northwestern California. Proceedings of the IIIrd Vertebrate Pest Conference, California, pp. 128–34.

Connolly, G. E., Dudziński, M. L. & Longhurst, W. M. (1969). The eye lens as an indicator of age in the black-tailed jack rabbit. *Journal of Wildlife Management,* **33,** 159–64.

Constantine, D. G. (1970). Bats in relation to the health, welfare and economy of man. In: *Biology of bats* (ed. W. A. Wimsatt), Vol. 2, pp. 320–449. Academic Press, New York.

Cook, S. E., Jr (1959). The effects of fire on a population of small rodents. *Ecology*, **40**, 102–8.

Cormack, R. M. (1968). The statistics of capture–recapture methods. In: *Oceanography and marine biology* (ed. H. Barnes), pp. 455–506. George Allen & Unwin Ltd., London.

Cormack, R. M. (1972). The logic of capture–recapture estimates. *Biometrics*, **28**, 337–43.

Cornwell, P. B. (1970). Studies in microencapsulation of rodenticides. Proceedings of the IVth Vertebrate Pest Conference, pp. 83–97. University of California, Davis.

Crawley, M. C. (1969). Movements and home-ranges of *Clethrionomys glareolus* Schreber and *Apodemus sylvaticus* L. in north-east England. *Oikos*, **20**, 310–19.

Crawley, M. C. (1970). Some population dynamics of the bank vole, *Clethrionomys glareolus*, and the wood mouse, *Apodemus sylvaticus*, in mixed woodland. *Journal of Zoology, London*, **160**, 71–89.

Crowcroft, P. & Rowe, F. P. (1957). The growth of confined colonies of the wild house-mouse (*Mus musculus* L.). *Proceedings of the Zoological Society of London*, **129**, 359–70.

Crowcroft, P. & Rowe, F. P. (1958). The growth of confined colonies of the wild house-mouse (*Mus musculus* L.): the effect of dispersal on female fecundity. *Proceedings of the Zoological Society of London*, **131**, 357–65.

Crowell, K. L. (1973). Experimental zoogeography: introductions of mice to small islands. *American Naturalist*, **107**, 535–58.

Curry-Lindahl, K. (1962). The irruption of the Norway lemming in Sweden during 1960. *Journal of Mammalogy*, **43**, 171–84.

Cygankov, D. S. (1955). Metodika opredelenija vozrasta i prodolžitel nost' žizni ondatry (*Fiber zibethicus* L.). *Zoologiceskii Zhurnal*, **34**, 640–51.

Daniel, M. (1964). Temperature and humidity in the nest of *Clethrionomys glareolus* observed in continuous experiment. *Acta Soc. Zool. Bohemoslov*, **28**, 278.

Dapson, R. W. (1968). Reproduction and age structure in a population of short-tailed shrews, *Blarina brevicauda*. *Journal of Mammalogy*, **49**, 205–14.

Dapson, R. W. & Irland, J. M. (1972). An accurate method of determining age in small mammals. *Journal of Mammalogy*, **53**, 100–6.

Dapson, R. W., Otero, J. G. & Holloway, W. R. (1968). Biochemical changes with age in the lenses of white mice. *Nature, London*, **218**, 573.

Davenport, L. B., Jr (1964). Structure of two *Peromyscus polionotus* populations in old-field ecosystems at the AEC Savannah River Plant. *Journal of Mammalogy*, **45**, 95–113.

Davis, D. E. (1945). The annual cycle of plants, mosquitoes, birds, and mammals in two Brazilian forests, *Ecological Monographs*, **15**, 244–95.

Davis, D. E. (1953). The characteristics of rat populations. *Quarterly Review of Biology*, **28**, 373–401.

Davis, D. E. & Christian, J. J. (1956). Changes in Norway rat populations

References

induced by introduction of rats. *Journal of Wildlife Management*, **20**, 378–83.

Davis, D. E. & Golley, F. B. (1963). *Principles in mammalogy*. Van Nostrand Reinhold, New York.

Davis, D. W. & Sealander, J. A. (1971). Sex ratio and age structure in two red squirrel populations in northern Saskatchewan. *Canadian Field Naturalist*, **85**, 303–8.

Davis, R. A. (1970). Control of voles in the United Kingdom. *EPPO Publications*, Ser. A, **58**, 87–94.

Davȳdov, G. S. (1962). K razprostraneniyu peschanok v Tadzhikistane. *Trudȳ Instituta Zoologii i Parazitologii*, **22**, 58–69.

Davȳdov, G. S. (1964). *Grȳzunȳ Severnogo Tadzhikistana*, pp. 1–271. Izdatel 'stvo Akademii Nauk Tadzhikskoĭ SSR, Dushanbe.

Dawson, N. J. (1970). Body composition of inbred mice (*Mus musculus*). *Comparative Biochemistry and Physiology*, **37**, 589–93.

Deanesly, R. & Allanson, M. (1967). The effects of external stimuli on reproduction. *Ciba Foundation Study Group*, **26**, 71–80.

Degn, H. J. (1973). Systematic position, age criteria and reproduction of Danish red squirrels (*Sciurus vulgaris* L.). *Danish Review of Game Biology*, **8**, 3–24.

Dehnel, A. (1949). Studies on the genus *Sorex* L. *Annales Universitatis Mariae Curie-Sklodowska*, Sectio C, **4**, 17–102.

Delany, M. J. (1964*a*). A study of the ecology and breeding of small mammals in Uganda. *Proceedings of the Zoological Society of London*, **142**, 347–70.

Delany, M. J. (1964*b*). An ecological study of the small mammals in the Queen Elizabeth Park, Uganda. *Revue of Zoology and Botany of Africa*, **70**, 129–47.

Delany, M. J. (1971). The biology of small rodents in Mayanja Forest, Uganda. *Journal of Zoology*, **165**, 85–129.

Delany, M. J. (1972). The ecology of small rodents in tropical Africa. *Mammal Review*, **2**, 1–42.

DeLong, K. T. (1967). Population ecology of feral house mice. *Ecology*, **48**, 611–34.

De Moor, P. P. (1969). Seasonal variation in local distribution, age classes and population density of the gerbil *Tatra brantsi* on the South African highveld. *Journal of Zoology, London*, **157**, 399–411.

Dennis, J. G. & Johnson, P. L. (1970). Shoot and rhizome-root standing crops of tundra vegetation at Barrow, Alaska. *Arctic and Alpine Research*, **2**, 253–66.

Dennis, J. G. & Tiezsen, L. (1972). Seasonal course of dry matter and chlorophyll by species at Barrow, Alaska. Proceedings of the 1972 US IBP Tundra Biome Symposium (ed. J. Brown & S. Bowen), pp. 16–21.

Deparma, N. K. (1954). The method of age determination in moles. *Bull. Soc. Nat. Moscou Biol.*, **59**, 11–25. (In Russian.)

Depocas, F. & Hart, J. Sanford (1957). Use of the Pauling oxygen analyzer for measurement of oxygen consumption of animals in open-circuit systems and in a short-lag, closed-circuit apparatus. *Journal of Applied Physiology*, **10**, 388–92.

References

Dice, L. R. (1968). Speciation. In: *Biology of* Peromyscus. American Society of Mammalogists, Special Publication 2 (ed. J. A. King), pp. 75–97.

Dieterlen, F. (1967a). Ökologische Populationsstudien an Muriden des Kivugebietes (Congo), Teil I. *Zoologische Jahrbuch Systematische*, **94**, 369–426.

Dieterlen, F. (1967b). Jahrezeiten und Fortpflanzungsperioden bei den Muriden des Kivusee-Gebietes (Congo). *Zeitschrift für Säugetierkunde*, **32**, 1–44.

Dieterlen, F. (1967c). Eine neue Methode für Lebendfang, Populationsstudien und Dichtebestimmungen an Kleinsäugern. *Acta Tropica*, **24**, 244–60.

Dill, D. B., Adolph, E. F. & Wilber, C. G. (eds.) (1964). *Handbook of Physiology.* Section 4: *Adaptation to the Environment.* Williams & Wilkins, Baltimore.

Dinesman, L. G. (1961). *Vliyanie dikikh mlekopitayushchikh na formirovanie drevostoev.* Izvestiya Akademii Nauk SSSR, Seriya Biologicheskaya, Moscow.

Dinesman, L. G. (1967). Influence of vertebrates on primary production of terrestrial communities. In: *Secondary productivity of terrestrial ecosystems* (ed. K. Petrusewicz), pp. 261–6. Polish Academy of Sciences, Warsaw.

Dische, Z., Borenfreund, E. & Zelmenis, G. (1965). Changes in lens proteins of rats during aging. *Archives of Ophthalmology*, **55**, 471–83.

Dobrinskij L. N. & Mikhalev, M. V. (1966). K metodike opredelenija vozrastnoj struktury populjacij životnyh. *Akademija Nauk SSSR, Ural'skij Filial, Trudy Instituta Biologii*, **51**, 107–15.

Dolgov, V. A. & Rossolimo, O. L. (1966). Age changes of some structural peculiarities of the skull and baculum in carnivorous mammals. Age determination procedure exemplified by polar fox (*Alopex lagopus* L.). *Zoologiceskie Zhurnal*, **45**, 1075–80. (In Russian with English summary.)

Donaldson, H. H. & King, H. D. (1937). On the growth of the eye in three strains of the Norway rat. *American Journal of Anatomy*, **60**, 203–29.

Drożdż, A. (1965). Wpływ paszy na dojrzewanie samców nornicy rudej (*Clethrionomys glareolus* Schr.). *Zwierz. Lab.*, **3**, 34–45.

Drożdż, A. (1966). Food habits and food supply of rodents in the beech forest. *Acta Therologica*, **11**, 363–84.

Drożdż, A. (1968a). Digestibility and assimilation of natural foods in small rodents. *Acta Theriologica*, **13**, 367–89.

Drożdż, A. (1968b). Food habits and food assimilation in mammals. In: *Methods of ecological bioenergetics* (ed. W. Grodziński & R. Z. Klekowski), pp. 193–206. Polish Academy of Sciences, Warsaw.

Drożdż, A. (1969). Digestibility and utilization of natural foods. In: *Energy flow through small mammal populations* (ed. K. Petrusewicz & L. Ryszkowski), pp. 127–9. Polish Scientific Publishers, Warsaw.

Drożdż, A. (1970). Digestibility and utilization of natural foods in small rodents. In: *Energy flow through small mammal populations* (ed. K. Petrusewicz & L. Ryskowski). Polish Scientific Publishers, Warsaw.

Drożdż, A., Gorecki, A., Grodziński, W. & Pelikan, J. (1971). Bioenergetics

References

of water voles (*Arvicola terrestris* L.) from southern Moravia. *Annales Zoologici Fennici*, **8**, 97–103.

Drummond, D. C. (1966). Rats resistant to warfarin. *New Scientist*, **30**, 771–2.

Drummond, D. C. (1969). Some comments on rat-free towns. *Schriftenreihe des Vereins für Wasser, Boden und Lufthygiene*, **32**, 67–70.

Drummond, D. C. (1970). Variation in the rodent populations in response to control measures. In: *Variations in mammal populations.* Symposia of the Zoological Society of London, **26** (ed. Berry & Southern), pp. 351–67. Academic Press, London.

Drummond, D. C. & Taylor, K. D. (1970). Practical rodent control, pp. 742–768. In: *World food programme storage manual*, Part III. FAO, Rome.

Dub, M. (1971a). Movements of *Microtus arvalis* Pall. and a method of estimating its numbers. *Zoologické Listy*, **20**, 1–14.

Dub, M. (1971b). Remarks on the sex ratio of *Microtus arvalis* (Pall.) caught in pitfall traps. *Zoologické Listy*, **20**, 207–13.

Dubinin, V. B. (1954). K voprosu o faune i ékologii mlekopitayushchikh Khavastskogo raĭona Tashkentskoĭ oblasti Uzbekskoĭ SSR. *Trudy Instituta Zoologii i Parazitologii, Tashkent*, **3**, 38–45.

Dubost, G. G. (1968). Aperçu sur le rythme annuel le réproduction des murides du nord-est du Gabon. *Biologica Gabonica*, **4**, 227–39.

Dudziński, M. L. & Mykytowycz, R. (1961). The eye lens as an indicator of age in the wild rabbit in Australia. *CSIRO Wildlife Research*, **6**, 156–9.

Dufour, B. (1971). Données quantitatives sur la construction du terrier chez *Apodemus sylvaticus* L. *Revue Suisse de Zoologie*, **78**, 568–71.

Dunajeva, T. N. (1955). K izučeniju biologii razmnoženija obyknovennoj burozubki (*Sorex araneus* L.). *Bull. mosk. Obshch. Ispyt. Prir.*, **60**, 27–43. (In Russian.)

Dunaway, P. B. & Kaye, S. V. (1961). Studies of small mammal populations on the radioactive White Oak Lake bed. *Transactions of the North American Wildlife and Natural Resources Conference*, **26**, 167–85.

Dunbar, M. J. (1968). *Ecological development in polar regions.* Prentice-Hall, Englewood Cliffs, New Jersey.

Dunmier, W. W. (1960). An altitudinal survey of reproduction in *Peromyscus maniculatus. Ecology*, **41**, 174–82.

Dymond, J. R. (1947). Fluctuations in animal populations with special reference to those of Canada. *Transactions of the Royal Society of Canada*, Ser. 3, Sect. 5, **41**, 1–34.

Eberhardt, L. L. (1969). Population estimates from recapture frequencies. *Journal of Wildlife Management*, **33**, 28–39.

Ecke, D. H. & Kinney, A. R. (1956). Aging meadow mice, *Microtus californicus*, by observation of molt progression. *Journal of Mammalogy*, **37**, 249–54.

Edwards, R. Y. (1952). How efficient are snap traps in taking small mammals? *Journal of Mammalogy*, **33**, 497–8.

Eibl-Eibesfeldt, I. (1950). Beitrage zur Biologie der Haus- und Ahrenmaus nebst einigen Beobachtungen an anderen Nagern. *Zeitschrift für Tierpsychologie*, **7**, 558.

Eisenberg, J. F. (1963). The behavior of heteromyid rodents. *University of California Publications in Zoology*, **69**, 1–100.

Elder, W. H. & Shanks, C. E. (1962). Age changes in tooth wear and morphology of the baculum in muskrats. *Journal of Mammalogy*, **43**, 144–50.

Eleftheriou, B. E., Bronson, F. H. & Zarrow, M. X. (1962). Interaction of olfactory and other environmental stimuli on implantation in the deermouse. *Science*, **137**, 764.

Elster, H. J. (1953). Ein Beitrag zur Produktionsbiologie des Zooplanktons. Internationale Vereinigung für Theoretische und Angewandte Limnologie Verhandlungen, No. 12, 404–11.

Elton, C. (1942). *Voles, mice and lemmings. Problems of population dynamics.* Oxford University Press, London.

Elton, C. (1953). The use of cats in farm rat control. *British Journal of Animal Behaviour*, **1**, 151–5.

Elton, C., Davis, C. H. S. & Findlay, G. M. (1935). An epidemic among voles (*Microtus agrestis*) on the Scottish border in the spring of 1934. *Journal of Animal Ecology*, **4**, 277–88.

Emlen, J. T., Jr, Hine, R. L., Fuller, W. A. & Alfonzo, P. (1957). Dropping boards for population studies of small mammals. *Journal of Wildlife Management*, **21**, 300–14.

Enders, R. K. (1935). Mammalian life histories from Barro Colorado Island, Panama. *Bulletin of the Museum of Comparative Zoology*, **78**, 383–502.

Erkinaro, E. (1971). The seasonal change of the activity of *Microtus agrestis*. *Oikos*, **12**, 157–63.

Errington, P. L. (1946). Predation and vertebrate population. *Quarterly Review of Biology*, **21**, 144–77, 221–45.

Evans, D. M. (1973). Seasonal variations in the body composition and nutrition of the vole *Microtus agrestis*. *Journal of Animal Ecology*, **42**, 1–18.

Evans, F. C. (1942). Studies of a small mammal population in Bagley Wood, Berkshire. *Journal of Animal Ecology*, **11**, 182–97.

Evans, F. C. & Holdenried, R. (1943). A population study of the beechey ground squirrel in central California. *Journal of Mammalogy*, **24**, 231–60.

Evans, J. V. & King, J. W. (1955). Genetic control of sodium and potassium concentration in the red blood cells of sheep. *Nature, London*, **176**, 171.

Everard, C. O. R. (1964). Some aspects of vertebrate damage to cocoa in West Africa. Proceedings of the Conference on Cocoa Pest WACRI (Nigeria), pp. 114–19.

Faegri, K. & van der Pilj, L. (1966). *The principles of pollination ecology.* Pergamon Press, Oxford.

Faust, B. F., Smith, M. H. & Wray, W. B. (1971). Distances moved by small mammals as an apparent function of grid size. *Acta Theriologica*, **16**, 161–77.

Fedyk, A. (1971). Social thermoregulation in *Apodemus flavicollis* (Melchoir, 1834). *Acta Theriological*, **16**, 221–9.

Fedyk, A. (1974a). Gross body composition of the bank vole during the postnatal development. III. Estimation of the age. *Acta Theriologica*, **19**, 429–40.

Fedyk, A. (1974b). Gross body composition of the bank vole during the

References

postnatal development. II. Differentiation of the seasonal generations. *Acta Theriologics*, **19**, 403–27.

Fenyuk, B. K. & Radchenko, A. G. (1957). Massovoe uvelichenie chislennosti krasnokhvostoĭ peschanki v zapadnoĭ Turkmenii v 1953 g. In: *Voprosȳ ékologii*, vol. 2, pp. 105–16. Kiev.

Ferguson, J. H. & Folk, G. E., Jr (1970). The critical thermal minimum of small rodents in hypothermia. *Cryobiology*, **7**, 44–6.

Fisher, E. W. & Perry, A. E. (1970). Estimating ages of gray squirrels by lensweights. *Journal of Wildlife Management*, **34**, 825–8.

Fisler, G. F. (1971). Age structure and sex ratio in populations of *Reithrodontomys*. *Journal of Mammalogy*, **52**, 653–62.

Fitch, H. S. (1954). Seasonal acceptance of bait by small mammals. *Journal of Mammalogy*, **35**, 39–47.

Fleharty, E. D. & Choate, J. R. (1973). Bioenergetic strategies of the cotton rat, *Sigmodon hispidus*. *Journal of Mammalogy*, **54**, 680–92.

Fleharty, E. D., Krause, M. E. & Stinnett, D. P. (1973). Body composition, energy content and lipid cycles of four species of rodents. *Journal of Mammalogy*, **54**, 426–38.

Fleharty, E. D. & Olson, L. E. (1969). Summer food habits of *Microtus ochrogaster* and *Sigmodon hispidus*. *Journal of Mammalogy*, **50**, 475–86.

Fleming, T. H. (1970). Notes on the rodent faunas of two Panamanian forests. *Journal of Mammalogy*, **51**, 473–90.

Fleming, T. H. (1971). Population ecology of three species of Neotropical rodents. *Miscellaneous Publications of the Museum of Zoology, University of Michigan*, **143**, 1–77.

Fleming, T. H. (1972). Aspects of the population dynamics of three species of opossums in the Panama Canal Zone. *Journal of Mammalogy*, **53**, 619–23.

Fleming, T. H. (1973a). Numbers of mammal species in North and Central American forest communities. *Ecology*, **54**, 555–63.

Fleming, T. H. (1973b). The reproductive cycles of three species of opossum and other mammals in the Panama Çanal Zone. *Journal of Mammalogy*, **54**, 439–55.

Fleming, T. H. (1973c). The number of rodent species in two Costa Rican forests. *Journal of Mammalogy*, **54**, 518–21.

Fleming, T. H. (1974). The population ecology of two species of Costa Rican heteromyid rodents. *Ecology*, **55**, 493–510.

Fleming, T. H., Hooper, E. T. & Wilson, D. E. (1972). Three Central American bat communities: structure, reproductive cycles, and movement patterns. *Ecology*, **53**, 555–69.

Folk, G. E., Jr (1966). *Introduction to environmental physiology*. Lea & Febiger, Philadelphia.

Forbes, R. B. (1966). Studies of the biology of Minnesota chipmunks. *American Midland Naturalist*, **76**, 290–308.

Formozov, A. N. (1946). The snow cover as an environmental factor and its importance in the life of mammals and birds. *Moscow Society of Naturalists, Materials for Fauna and Flora of the USSR, New Series Biology*, **15**, 1–52.

(Trans. W. Pryehodko & W. O. Pruitt, Jr in Occasional Papers No. 1 of the Boreal Institute, University of Alberta, 1963.)

Formozov, A. N., Hodashova, K. S. & Golov, B. A. (1954). Vliaynie gryzunov na rastitelnost pastbishch i senokosov glinistykh polupustyn mezhdurechya Volga, Ural. *Izvestiya Akademii Nauk SSSR*, 331–40. (In Russian.)

Formozov, A. N. & Voronov, A. G. (1939). Deyatel 'nost' grӯzunov na pastbishchakh i senokosnӯkh ugod'yakh Zapadnogo Kazakhstana i ee khozyaĭstvennoe znachenie. *Uchenye Zapiski Moskovskogo Gosudarstven nogo Universitet, Zoologiya*, **20**, 25–63.

Fowle, C. D. & Edwards, R. Y. (1954). The utility of break-back traps in population studies of small mammals. *Journal of Wildlife Management*, **18**, 503–8.

Frank, F. (1952). Umfang, Ursachen und Bekämpfungsmöglichkeiten der Mäusefrasschäden in Forstkulturen. *Nachrichtenblatt des Deutschen Pflanzenschutzdienstes (Braunschweig)*, **4**, 183–9.

Frank, F. (1956). Grundlagen, Möglichkeiten und Methoden der Sanierung von Feldmausplagegebieten. *Nachrichtenblatt des Deutschen Pflanzenschutz-dienstes (Braunschweig)*, **8**, 147–58.

Frank, F. (1957a). The causality of Microtine cycles in Germany. *Journal of Wildlife Management*, **21**, 113–21.

Frank, F. (1957b). Das Fortpflanzungpotential der Feldmaus *Microtus arvalis* (Pallas): eine Spitsenleistung unter den Säugtieren. *Zeitschrift für Säugetierkunde*, **21**, 176–81.

Frank, F. (1965). Grundsätzliche Uberlegungen zur chemischen Bakämpfung der Feldmaus und onderer wühlmausartiger Schadnager nach der Aberkennung des Endrins. *Nachrichtenblatt des Deutschen Pflanzenschutz-dienstes*, **17**, 104–8.

French, N. R. (1965). Radiation and animal populations: problems, progress and projections. *Health Physics*, **11**, 1557–68.

French, N. R. (1969). Radiation sensitivity of rodent species. *Nature, London*, **222**, 1003–4.

French, N. R., Maza, B. G., Hill, H. O., Aschwanden, A. P. & Kaaz, H. W. (1974). A population study of irradiated desert rodents. *Ecological Monographs*, **44**, 45–72.

French, N. R., Tagami, T. Y. & Hayden, P. (1968). Dispersal in a population of desert rodents. *Journal of Mammalogy*, **49**, 272–80.

Friend, M. (1967a). A review of research concerning eye-lens weight as a criterion of age in animals. *New York Fish and Game Journal*, **14**, 152–65.

Friend, M. (1967b). Some observations regarding eye-lens weight as a criterion of age in animals. *New York Fish and Game Journal*, **14**, 91–121.

Friend, M. (1968). The lens technique. Transactions of the 33rd North American Wildlife and Natural Resources Conference, March 1968, Houston, Texas, pp. 279–97. Wildlife Management Institute, Washington.

Fullagar, P. J., Jewell, P. A., Lockley, R. M. & Rowlands, I. W. (1963). The Skomer vole (*Clethrionomys glareolus skomerensis*) and long-tailed field mouse (*Apodemus sylvaticus*) on Skomer Island, Pembrokeshire in 1960. *Proceedings of the Zoological Society of London*, **140**, 295–314.

References

Fuller, W. A. (1967). Winter ecology of lemmings and fluctuations of their populations. *Terre et Vie*, **2**, 97–115. (In French.)

Fuller, W. A., Stebbins, L. L. & Dyke, G. R. (1969). Overwintering of small mammals near Great Slave Lake, Northern Canada. *Arctic*, **22**, 34–55.

Gadgil, M. (1971). Dispersal: population consequences and evolution. *Ecology*, **52**, 253–61.

Gardarsson, A. & Moss, R. (1970). Selection of food by Icelandic ptarmigan in relation to its availability and nutritive value. In: *Animal populations and their food resources* (ed. A. Watson), pp. 47–71. Blackwell Scientific, Oxford.

Gashwiler, J. S. (1970). Further study of conifer seed survival in a western Oregon clearcut. *Ecology*, **51**, 849–54.

Gashwiler, J. S. (1971). Deermouse (*Peromyscus maniculatus*) movement in forest habitat. *Northwest Science*, **45**, 163–70.

Gashwiler, J. S. (1972). Life history notes on the Oregon vole, *Microtus oregoni. Journal of Mammalogy*, **53**, 558–69.

Gębczyńska, Z. (1964). Morphological changes occurring in laboratory *Microtus agrestis* with age. *Acta Theriologica*, **9**, 67–79.

Gębczyńska, Z. (1966). Estimation of rodent numbers in a plot of *Querceto-Carpinetum* forest. *Acta Theriologica*, **11**, 315–28.

Gębczyńska, Z. (1967). Morphologic variability of *Lagurus lagurus* (Pallas, 1773) in laboratory conditions. *Acta Theriologica*, **12**, 533–43.

Gębczyńska, Z. (1970). Bioenergetics of a root vole population. *Acta Theriologica*, **15**, 33–66.

Gębczyński, M., Gorecki, A. & Drożdż, A. (1972). Metabolism, food assimilation and bioenergetics of three species of Dormice (Gliridae). *Acta Theriologica*, **17**, 271–94.

Gentry, J. B. (1968). Dynamics of an enclosed population of pine mice, *Microtus pinetorum. Researches in Population Ecology*, **10**, 21–30.

Gentry, J. B., Golley, F. B. & McGinnis, J. T. (1966). Effect of weather on captures of small mammals. *American Midland Naturalist*, **75**, 526–30.

Gentry, J. B., Golley, F. B. & Smith, M. H. (1968). An evaluation of the proposed International Biological Census method for estimating small mammal populations. *Acta Theriologica*, **13**, 313–27.

Gentry, J. B., Golley, F. B. & Smith, M. H. (1971). Yearly fluctuations in small mammal populations in a southeastern United States hardwood forest. *Acta Theriologica*, **16**, 179–90.

Gentry, J. B. & Odum, E. P. (1957). The effect of weather on the winter activity of old-field rodents. *Journal of Mammalogy*, **38**, 72–7.

Gentry, J. B., Smith, M. H. & Beyers, R. J. (1971*a*). Use of radioactively tagged bait to study movement patterns in small mammal populations. *Annales Zoologici Fennici*, **8**, 17–21.

Gentry, J. B., Smith, M. H. & Beyers, R. J. (1971*b*). Radioactive isotopes in studies of population dynamics of small mammals. In: *Third National Symposium on Radioecology* (ed. D. J. Nelson), pp. 253–9. National Technical Information Service, US Department of Commerce, Springfield, Virginia.

Gentry, J. B., Smith, M. H. & Chelton, J. G. (1971). An evaluation of the octagon census method for estimating small mammal populations. *Acta Theriologica*, **16**, 149–59.

Gentry, J. B. & Stiritz, K. L. (1972). The role of the Florida harvester ant, *Pogonomyrmax badius*, in old-field mineral nutrient relationships. *Environmental Entomology*, **1**, 39–41.

Gessaman, J. A. (1972). Bioenergetics of the snowy owl (*Nyctea scandiaca*). *Arctic and Alpine Research*, **4**, 223–38.

Gessaman, J. A. (1973). Ecological energetics of homeotherms: A view compatible with ecological modeling. Utah State University Monographs Series, **20**, 1–155.

Getz, L. & Ginsberg, V. (1968). Arboreal behaviour of red-back vole, *Clethrionomys gapperi. Animal Behavior*, **16**, 418–24.

Giban, J. (1967). Ways and means of controlling *M. arvalis* in France. *EPPO Publications*, Ser. A, **41**, 107–15.

Giban, J. (1968). Muskrat control through poisoned bait. *EPPO Publications*, Ser. A, **47**, 45–7.

Giban, J. (1970). Expérimentation sur les méthodes de lutte utilisables sur le campagnol des champs, *Microtus arvalis* (Pallas). *EPPO Publications*, Ser. A, **58**, 73–80.

Giban, J. (1972). L'emploi des appats empoisonnés au chlorophacinone pour la déstruction du rat musque. *Bulletin Français de Pisciculture*, **44**, 119–26.

Gibbons, J. W. & Bennett, D. H. (1974). Determination of anuran terrestrial activity patterns by a drift fence method. *Copeia*, **1**, 236–43.

Gintlis, R. V. (1959). A study on the age estimation in *Meriones erythrourus* Gray. *Bjull. MOIP, Otd. biol.*, **64**, 23–30.

Gliwicz, J. (1970). Relation between trappability and age of individuals in a population of the bank vole. *Acta Theriologica*, **15**, 15–32.

Gliwicz, J. (1973). A short characteristic of a population of *Proechimys semispinosus* (Tomes, 1860) – a rodent species of the tropical rain forest. *Bulletin de l'Académie polonaise des Sciences*, Biology Series, **21**, 413–18.

Gliwicz, J., Andrzejewski, R., Bujalska, G. & Petrusewicz, K. (1968). Productivity investigation of an island population of *Clethrionomys glareolus* (Schreber, 1780). *Acta Theriologica*, **23**, 401–13.

Godfrey, G. K. (1955). A field study of the activity of the mole (*Talpa europaea*). *Ecology*, **36**, 678–85.

Godfrey, G. K. (1956). Reproduction in *Talpa europaea* in Suffolk. *Journal of Mammalogy*, **37**, 438–40.

Goertz, J. W. (1970). An ecological study of *Neotoma floridana* in Oklahoma. *Journal of Mammalogy*, **51**, 94–104.

Golley, F. B. (1960). Energy dynamics of a food chain of an old-field community. *Ecological Monographs*, **30**, 187–206.

Golley, F. B. (1961). Interaction of natality, mortality and movement during one annual cycle in a *Microtus* population. *American Midland Naturalist*, **66**, 152–9.

Golley, F. B. (1962). *Mammals of Georgia: a study of their distribution and functional role in the ecosystem*. University of Georgia Press, Athens, Georgia.

References

Golley, F. B. (1967). Methods of measuring secondary productivity in Terrestrial vertebrate populations. In: *Secondary Productivity of Terrestrial ecosystems* (ed. K. Petrusewicz), pp. 99–124. Polish Academy of Sciences, Warsaw.

Golley, F. B. (1973). Impact of small mammals on primary production. In: *Ecological energetics of homeotherms* (ed. J. A. Gessaman), pp. 142–7. Utah State University Press, Logan.

Golley, F. B. & Buechner, H. K. (eds.) (1969). *A practical guide to the study of productivity of large herbivores.* IBP Handbook 7. Blackwell Scientific, Oxford.

Golley, F. B., Pelikan, J. & Ryszkowski, L. (1968). Results of discussion on the IBP Standard-Minimum trapping grid and recommendations for its improvement. *Small Mammal Newsletter,* **2**, 195–8.

Golley, F. B., Wiegert, R. G. & Walter, R. W. (1965). Excretion of orally administered zinc-65 by wild small mammals. *Health Physics,* **11**, 719–22.

Gomez, J. C. (1960). Correlation of a population of roof rats in Venezuela with seasonal changes in habitat. *American Midland Naturalist,* **63**, 177–93.

Goodyear, C. P. & Boyd, C. E. (1972). Elemental composition of large-mouth bass (*Micropterus salmoides*). *American Fisheries Society. Transactions,* **101**, 545–7.

Górecki, A. (1965*a*). Energy values of body in small mammals. *Acta Theriologica,* **10**, 333–52.

Górecki, A. (1965*b*). The bomb calorimeter in ecological research. *Ekologia Polska,* Ser. B, **11**, 145–58.

Górecki, A. (1967). Caloric values of the body in small rodents. In: *Secondary productivity of terrestrial ecosystems* (ed. K. Petrusewicz), pp. 315–21. Polish Academy of Sciences, Warsaw.

Górecki, A. (1968). Metabolic rate and energy budget in the bank vole. *Acta Theriologica,* **13**, 341–65.

Górecki, A. (1971). Metabolism and energy budget in the harvest mouse. *Acta Theriologica,* **16**, 213–20.

Grant, P. R. (1970). A potential bias in the use of Longworth traps. *Journal of Mammalogy,* **51**, 831–5.

Grant, P. R. (1971). The habitat preference of *Microtus pennsylvanicus,* and its relevance to the distribution of this species on islands. *Journal of Mammalogy,* **52**, 351–61.

Gratz, N. G. (1966). A critical review of the currently used acute rodenticides. Seminar of Rodents and Rodent Ectoparasites, WHO/VC/66. 217: 79–87.

Grau, G. A., Sanderson, G. C. & Rogers, J. P. (1970). Age determination of raccoons. *Journal of Wildlife Management,* **34**, 364–72.

Greaves, J. H. (1966). Some laboratory investigations on the toxicity and acceptability of norbormide to wild *Rattus norvegicus* and on feeding behaviour associated with sub-lethal dosing. *Journal of Hygiene, Cambridge,* **64**, 275–85.

Greaves, J. H. (1971). Rodenticides. *Reports on the Progress of Applied Chemistry,* **56**, 465–73.

References

Greaves, J. H., Rennison, B. D. & Redfern, R. (1973). Warfarin resistance in the ship rat in Liverpool. *International Pest Control*, **15**, 17.

Greaves, J. H. & Rowe, F. P. (1969). Responses of confined rodent populations to an ultrasound generator. *Journal of Wildlife Management*, **33**, 407–17.

Greaves, J. H., Rowe, F. P., Redfern, R. & Ayres, P. (1968). Microencapsulation of rodenticides. *Nature, London*, **219**, 402–3.

Grodziński, W. (1961). Metabolism rate and bioenergetics of small rodents from the deciduous forest. *Bulletin de l'Académie polonaise des Sciences*, Cl. II, **9**, 493–9.

Grodziński, W. (1966). Bioenergetics of small mammals from Alaskan taiga forest. *Lynx*, **6**, 51–5.

Grodziński, W. (1968). Energy flow through a vertebrate population. In: *Methods of ecological bioenergetics* (ed. W. Grodziński & R. Z. Klekowski), pp. 239–52. Polish Academy of Sciences, Warsaw.

Grodziński, W. (1971a). Energy flow through populations of small mammals in the Alaskan taiga forest. *Acta Theriologica*, **16**, 231–75.

Grodziński, W. (1971b). Food consumption of small mammals in the Alaskan taiga forest. *Annales Zoologici Fennici*, **8**, 133–6.

Grodziński, W., Bobek, B., Drożdż, A. & Górecki, A. (1970). Energy flow through small rodent populations in a beech forest. In: *Energy flow through small mammal populations* (ed. K. Petrusewicz & L. Ryszkowski), pp. 291–8. Polish Scientific Publishers, Warsaw.

Grodziński, W. & Górecki, A. (1967). Daily energy budgets of small rodents. In: *Secondary productivity of terrestrial ecosystems* (ed. K. Petrusewicz), Vol. 1, pp. 295–314. Polish Academy of Sciences, Warsaw.

Grodziński, W., Górecki, A., Janas, K. & Migula, P. (1966). Effects of rodents on the primary productivity of alpine meadows in Bieszczady mountains. *Acta Theriologica*, **11**, 419–31.

Grodziński, W. & Klekowski, R. Z. (eds.) (1968). *Methods of ecological bioenergetics*. Polish Academy of Sciences, Warsaw.

Grodziński, W., Klekowski, R. Z. & Duncan, A. (eds.) (1975). *Methods for ecological bioenergetics*. IBP Handbook 24. Blackwell Scientific, Oxford.

Grodziński, W., Pucek, Z. & Ryszkowski, L. (1966). Estimation of rodent numbers by means of prebaiting and intensive removal. *Acta Theriologica*, **11**, 297–314.

Grolleau, G. (1971). Comparison de la toxicité de divers raticides anticoagulants a l'égard du campagnol des champs, *Microtus arvalis* (Pallas). *Annalas de Zoologie-Écologie animale*, **3**, 261–7.

Gromadzki, M. & Trojan, P. (1971). Estimation of population density in *Microtus arvalis* (Pall) by three different methods. *Annales Zoologici Fennici*, **8**, 54–9.

Grulich, I. (1967). Zur Methodik der Altersbestimmung des Maulwurfs, *Talpa europaea* L., in der Periode seiner selbständigen Lebensweise. *Zoologické Listy*, **16**, 41–59.

Gutteridge, N. J. A. (1972). Chemicals in rodent control. *Chemical Society Reviews*, **1**, 381–409.

393

References

Habermehl, N. H. (1961). *Die Alterbestimmung bei Haustieren, Pelztieren und beim jagdbaren Wild*, pp. 1–223. Paul Parey, Berlin & Hamburg.

Hadders, G. (1968). Sorksituationen i skogsfröplantagerna. *Institutet för Skogsförbättring, Information*, 4, 1–3.

Haeck, J. (1969). Colonization of the mole (*Talpa europaea* L.) in the Tjssel-meerpolders. *Netherlands Journal of Zoology*, 19, 145–248.

Hagen, B. (1955). Eine neue Methode der Altersbestimmung von Kleinsäugern. *Bonner zoologische Beiträge*, 6, 1–7.

Hagen, B. (1956). Alterbestimmung an einigen Muridenarten. *Zeitschrift für Säugetierkunde*, 21, 39–43.

Haitlinger, R. (1965). Morphological analysis of the Wrocław population of *Clethrionomys glareolus* (Schreber, 1780). *Acta Theriologica*, 10, 243–72.

Hall, E. R. & Kelson, K. R. (1959). *The mammals of North America*. 2 vols. Ronald Press, New York.

Hamar, M., Suteu, G., Sutova, M. & Tuta, A. (1970). Determination of the underground trade and the effectiveness of some gassing methods against *Spalax leucodon* by means of ^{60}C labelling. *EPPO Publications*, Ser. A, 58, 165–70.

Hamar, M. & Sutova, M. (1970). Effective control of the hamster (*Cricetus cricetus* L.) by gassing with polytanol and phostoxin. *EPPO Publications*, Ser. A, 58, 181–4.

Hamar, M., Tuta, A. & Sutova, M. (1971). The applicability of the Standard Minimum method to the estimation of rodent density in agrosystems. *Annales Zoological Fennici*, 8, 45–6.

Hamilton, W. J., Jr (1946). A study of the baculum in some North American Microtinae. *Journal of Mammalogy*, 27, 378–87.

Hammel, H. T., Dawson, T. S., Abrams, R. M. & Anderson, H. T. (1968). Total caloric measurements on *Citellus lateralis* in hibernation. *Physiological Zoology*, 41, 341–57.

Handley, C. O., Jr (1966). Checklist of the mammals of Panama. In: *Ecto-parasites of Panama* (ed. R. L. Wenzel & V. J. Tipton), pp. 753–95. Field Museum of Natural History, Chicago.

Handley, C. O., Jr (1967). Bats of the canopy of an Amazonian forest. *Atas do Simposia Biota Amazonia*, 5, 211–15.

Hanney, P. (1964). The harsh-furred rat in Nyasaland. *Journal of Mammalogy*, 45, 345–58.

Hanney, P. (1965). The Muridae of Malawi (Africa: Nyasaland). *Proceedings of the Zoological Society of London*, 146, 577–633.

Hanney, P. & Morris, B. (1962). Some observations upon the pouched rat in Nyasaland. *Journal of Mammalogy*, 43, 238–48.

Hansen, R. (1957). Development of young varying lemmings (*Dicrostonyx*). *Arctic*, 10, 105–17.

Hansen, R. & Cavender, B. R. (1973). Food intake and digestion by black-tailed prairie dogs under laboratory conditions. *Acta Theriologica*, 18, 191–200.

Hansen, R. M. & Ward, A. L. (1966). Some relations of pocket gophers to rangelands on Grand Mesa, Colorado. *Technical Bulletin of the Agriculture Experiment Station, Colorado State University*, **88**, 17–22.

Hanson, H. C. & Whitman, W. (1938). Characteristics of the major grassland types in northwestern Dakota. *Ecological Monographs*, **8**, 57–114.

Hanson, W. R. (1967). Estimating the density of an animal population. *Journal of Research on the Lepidoptra*, **6**, 203–247.

Hanson, W. R. (1968). Estimating the number of animals: A rapid method for unidentified individuals. *Science*, **162**, 675–6.

Hansson, L. (1967). Index line catches as a basis of population studies on small mammals. *Oikos*, **18**, 261–76.

Hansson, L. (1968). Population densities of small mammals in open field habitats in South Sweden in 1964–1967. *Oikos*, **19**, 53–60.

Hansson, L. (1969a). Home range, population structure and density estimates at removal catches with edge effect. *Acta Theriologica*, **14**, 153–60.

Hansson, L. (1969b). Spring populations of small mammals in central Swedish Lapland in 1964–1968. *Oikos*, **20**, 431–50.

Hansson, L. (1970). Methods of morphological diet microanalysis in rodents. *Oikos*, **21**, 255–66.

Hansson, L. (1970). The standard-minimum method in grassland habitats. In: *Energy flow through small mammal populations* (ed. K. Petrusewicz & L. Ryszkowski), pp. 31–8. Polish Scientific Publishers, Warsaw.

Hansson, L. (1971a). Estimates of the productivity of small mammals in a South Swedish spruce plantation. *Annales Zoologici Fennici*, **8**, 118–26.

Hansson, L. (1971b). Small rodent food, feeding and population dynamics. A comparison between granivorous and herbivorous species in Scandinavia. *Oikos*, **22**, 183–98.

Hansson, L. (1971c). Habitat, food and population dynamics of the field vole, *Microtus agrestis* in South Sweden. *Viltrevy Swedish Wildlife*, **8**, 267–378.

Hansson, L. (1972). Evaluation of the small quadrat method of censusing small mammals. *Annales Zoologici Fennici*, **9**, 184–90.

Hansson, L. (1974). Influence area of trap stations as a function of number of small mammals exposed per trap. *Acta Theriologica*, **19**, 19–25.

Hansson, L. & Grodziński, W. (1970). Bioenergetic parameters of the field vole, *Microtus agrestis* L. *Oikos*, **21**, 76–82.

Happold, D. C. D. (1970). Reproduction and development of the Sudanese jerboa, *Jaculus jaculus butleri* (Rodentia, Dipodidae). *Journal of Zoology, London*, **162**, 505–15.

Harranger, S. (1967). Les campagnols dans l'est de la France. *Phytoma*, **192**, 23–7.

Harrison, J. L. (1951). Reproduction in rats of the subgenus *Rattus*. *Proceedings of the Zoological Society of London*, **121**, 673–94.

Harrison, J. L. (1952). Breeding rhythms of some Selangor rodents. *Bulletin of the Raffles Museum*, **24**, 109–31.

Harrison, J. L. (1954). The natural food of some rats and some other mammals. *Bulletin of the Raffles Museum*, **25**, 157–65.

Harrison, J. L. (1955). Data on the reproduction of some Malayan mammals. *Proceedings of the Zoological Society of London*, **125**, 445–60.

References

Harrison, J. L. (1956). Age and survival of Malayan rats. *Bulletin of the Raffles Museum*, **27**, 5–26.

Harrison, J. L. (1962). The distribution of feeding habits among animals in a tropical rain forest. *Journal of Animal Ecology*, **34**, 53–64.

Harrison, J. L. (1969). The abundance and population density of mammals in Malayan lowland forests. *Malayan Naturalist Journal*, **22**, 174–8.

Harrison, J. L. & Liat, L. B. (1950). Notes on some small mammals of Malaya. *Bulletin of the Raffles Museum*, **23**, 300–9.

Hart, J. S. (1971). Rodents. In: *Comparative physiology of thermoregulation*. Vol. II: *Mammals* (ed. G. C. Whittow), pp. 1–149. Academic Press, New York.

Hart, J. S. & Heroux, O. (1963). Seasonal acclimatization in wild rats (*Rattus norvegicus*). *Canadian Journal of Zoology*, **41**, 711–16.

Hatt, R. T. (1930). The biology of the voles of New York. *Roosevelt Wildlife Bulletin*, **5**, 505–623.

Hawkins, A. E. & Jewell, P. A. (1962). Food consumption and energy requirements of captive British shrews and the mole. *Proceedings of the Zoological Society of London*, **138**, 137–55.

Hayne, D. W. (1949). Two methods for estimating population from trapping records. *Journal of Mammalogy*, **30**, 399–411.

Hayne, D. W. (1950). Apparent home range of *Microtus* in relation to distance between traps. *Journal of Mammalogy*, **31**, 26–39.

Hayne, D. W. & Thompson, D. Q. (1965). Methods for estimating microtine abundance. Transactions of the Thirtieth North American Wildlife and Natural Resources Conference, pp. 393–400.

Hayward, J. S. (1965a). The gross body composition of six geographic races of *Peromyscus*. *Canadian Journal of Zoology*, **43**, 297–308.

Hayward, J. S. (1965b). Microclimate temperature and its adaptive significance in six geographic races of *Peromyscus*. *Canadian Journal of Zoology*, **43**, 341–50.

Healey, M. C. (1967). Aggression and self-regulation of population size in deermice. *Ecology*, **48**, 377–92.

Herman, C. M. (1964). Proceedings of the 2nd bird control seminar, Bowling Green, pp. 112–21.

Herman, R. & Thomas, H. (1963). Observations on the occurrence of pocket gophers in Southern Oregon pine plantations. *Journal of Forestry*, **61**, 527–8.

Hermann, G. (1969). Die gebräuchlichen Rodentizide und ihre Anwendung in einigen Länadern Ost- unf Westeuropas. In: *Rattenbiologie und Rattenbekämpfung* (ed. F. Meinck). *Schriftenreihe des Vereins Für Wasser-, Boden-, und Lufthygiene, Stuttgart*, **32**, 179–96.

Herold, W. (1954). Beobachtungen über den Sitterungseinfluss auf dem Massenwechsel der Feldmaus. *Zeitschrift für Säugetierkunde*, **19**, 86–107.

Herreid, C. F., II & Kessel, B. (1967). Thermal conductance in birds and mammals. *Comparative Biochemistry and Physiology*, **21**, 405–14.

Hildwein, G. (1972). Métabolisme énergétique de quelques mammifères et

References

oiseaux de la foret équatoriale. II. Résultats expérimentaux et discussion. *Arch. Sci. Physiol.*, **26**, 387–400.

Hill, R. W. (1972). Determination of oxygen consumption by use of the paramagnetic oxygen analyzer. *Journal of Applied Physiology*, **33**, 261–3.

Hinds, W. T. & Rickard, W. H. (1973). Correlations between climatological fluctuations and a population of *Phyloithus densicollis* (Horn). *Journal of Animal Ecology*, **42**, 341–51.

Hiraiwa, Y. K., Uchida, T. & Hamajima, F. (1959). Summary: Damages of agricultural products caused by the Norway rat in Sagi-shima, a delta of Nobreoka City, Miyazaki Prefecture. II. Rat control with special reference to the introduction of the Korean weasel as a natural enemy against rats. *Science Bulletin of the Faculty of Agriculture Kyushu University*, **13**, 335–49.

Hodashova, K. S. (1950). Prirodnaya sreda i zhivotnyi mir glinistykh polupustyn Zavolzhya. *Izvestiya Akademii Nauk SSSR*, p. 130.

Hodashova, K. S. (1970). Vozdeistvie pozvonochnykh fitogagov na biologicheskuyu produktivnost' i krugovorot veshchestv v lesostypnykh landshftakh. In: *Sredoabrazuyushchaya deyatelnost' zhivotnykh* (ed. J. A. Isakov), pp. 48–52. Izd. Moskovskogo Univ., Moscow. (In Russian.)

Hoffman, R. S. (1970). Relationships of certain Holarctic shrews, Genus *Sorex. Zeitschrift für Säugetierkunde*, **36**, 193–200.

Hoffman, R. S. & Taber, R. D. (1968). Origin and history of Holarctic tundra ecosystems, with special reference to their vertebrate faunas. In: *Arctic and alpine environments* (ed. H. E. Wright & W. H. Osburn), pp. 143–70. Indiana University Press, Bloomington.

Holdridge, L. R. (1947). Determination of world plant formations from simple climatic data. *Science*, **105**, 367–8.

Holisova, V. (1968). Results of experimental baiting of small mammals with a marking bait. *Zoologické Listy*, **17**, 311–25.

Hood, G. A. (1972). Zinc phosphide – a new look at an old rodenticide for field rodents. Proceedings of the Vth Vertebrate Pest Conference, pp. 85–92.

Hood, G. A., Nass, R. D. & Lindsey, G. D. (1970). The rat in Hawaiian sugarcane. Proceedings of the IVth Vertebrate Pest Conference, pp. 34–7.

Hoslett, S. A. & Imaizumi, Y. H. (1966). Age structure of a Japanese mole population. *Journal of the Mammalogical Society of Japan*, **2**, 151–6.

Houlihan, R. T. (1963). The relationship of population density to endocrine and metabolic changes in the California vole *Microtus californicus. University of California Publications in Zoology*, **65**, 327–62.

Howard, W. E. (1949). Dispersal, amount of inbreeding and longevity in a local population of prairie deermice on the George Reserve, southern Michigan. *Contributions of Laboratory of Vertebrate Biology, University of Michigan*, **43**, 1–52.

Howard, W. E. (1960). Innate and environmental dispersal of individual vertebrates. *American Midland Naturalist*, **63**, 152–61.

Howard, W. E. (1965). Interaction of behavior, ecology, and genetics of introduced mammals. In: *The genetics of colonizing species* (ed. H. G. Baker & G. L. Stebbins), pp. 461–84. Academic Press, New York.

Howard, W. E. (1967a). Biocontrol and chemisterilants. In: *Pest control:*

References

biological, physical and selected chemical methods (ed. W. W. Kilgore &
R. L. Doutt), pp. 343–86. Academic Press, New York.

Howard, W. E. (1967b). Ecological changes in New Zealand due to introduced
mammals. *IUCN Publications*, **9**, 219–40.

Howard, W. E. & Childs, H. E., Jr (1959). Ecology of pocket gophers with
emphasis on *Thomomys bottae mewa. Hilgardia*, **29**, 277–358.

Howard, W. E. & Marsh, R. E. (1969). Mestranol as a reproductive inhibitor
in rats and voles. *Journal of Wildlife Management*, **33**, 403–8.

Howard, W. E., Marsh, R. E. & Cole, R. E. (1968). Food detection by deermice
using olfactory rather than visual cues. *Animal Behavior*, **16**, 13–17.

Howard, W. E., Palmateer, S. D. & Nachman, M. (1968). Aversion to strychnine sulfate by Norway rats, roof rats and pocket gophers. *Toxicology and
Applied Pharmacology*, **12**, 229–41.

Hsia Wu-Ping (1966). Population dynamics of forest region small rodents,
lesser Khiang-an mountains. II. The influences of climatological factors on
the number of rodents. *Acta Zoologica Sinica*, **18**, 8–19.

Hudson, J. W. (1964). Temperature regulation in the round-tailed ground
squirrel, *Citellus tereticaudus. Annales Academiae Scientiarum Fennicae*,
Ser. A, IV, 219–33.

Hunkeler, C. & Hunkeler, P. (1970). Besoins énergétiques de quelques
crocidures (Insectivores) de Cote d'Ivoire. *Terre et Vie*, **24**, 449–56.

Hurter, J., Zürrer, H. & Reuthinger, E. (1966). Rückstandsbestimmungen von
Endrin im Boden, auf Gras und in der Milch. *Zeitschrift für Lebensmittel-
untersuchung und -forschung*, **130**, 20–5.

Hutchinson, E. G. (1938). On the relation between the oxygen deficit and the
productivity and typology of lakes. *International Review of Hydrobiology*,
36, 336–55. I: 277, 281.

Hyvärinen, H. (1969). On the seasonal changes in the skeleton of the common shrew and their physiological background. *Aquilo* (Ser. Zoologica), **7**,
2–32.

Hyvärinen, H. (1972). Seasonal changes in the copper content of the common
shrew, *Sorex araneus*, over a two-year period. *Journal of Zoology, London*,
166, 411–16.

Incerti, G. & Pasquali, A. (1967). Esperimenti sulle attivita' esploratorie in
topi domestici e selvatici (*Mus musculus*). *Istituto Lombardo* (Rend. Sc.) B,
101, 19–46.

Irving, L. (1972). Arctic life of birds and mammals. *Zoophysical Ecology*, **2**,
1–192.

Ishchenko, V. G. (1966). Ispol'zovanie allometricheskikh uravnenii dlya
izucheniya morfologicheskoi differentsiatsii. *Trudy Instituta Biologicheskikh
Sverdlovsk*, **67**, 71.

Ismagilov, M. I. (1961). *Ékologiya landshaftnȳkh grȳzunov Betpak-Dalȳ i
yuzhnogo Pribalkhash'ya*, pp. 1–366. Alma-Ata.

Ivanter, E. V. (1969). A contribution to the study of the mole (*Talpa europaea
europaea* L.) in Karelia. *Uchenye Zapiski Petrozavodskogo Universiteta*,
16, 186–202.

References

Ivanter, E. V. (1973). Method of age determination in *Sicista betulina* (Rodentia, Dipodoidea). *Zoologiceskii Zhurnal*, **52**, 255–7. (In Russian with English summary.)

Iversen, J. A. (1972). Basal energy metabolism of mustelids. *Journal of Comparative Physiology*, **81**, 341–4.

Iverson, S. L. & Turner, B. L. (1972). Natural history of a Manitoba population of Franklin's ground squirrels. *Canadian Field Naturalist*, **86**, 145–9.

Ivlev, V. S. (1945). Biologičeskaja produktivnost' vodoemov. *Uspekhi Souremiennoi Biologii*, **19**, 98–120.

Jackson, M. L. (1958). *Soil chemical analysis*. Prentice-Hall, Englewood Cliffs, New Jersey.

Jackson, W. B. (1951). Food habits of Baltimore, Maryland, cats in relation to rat populations. *Journal of Mammalogy*, **32**, 458–61.

Jackson, W. B. (1962). Reproduction. In: *Pacific island rat ecology* (ed. T. Storer). *Bulletin of the Bishop Museum*, **225**, 1–274.

Jackson, W. B. & Barbehenn, K. R. (1962). Longevity and mortality. In: *Pacific island rat ecology* (ed. T. Storer). *Bulletin of the Bishop Museum*, **225**, 1–274.

Jackson, W. B. & Kaukeinen, D. E. (1972*a*). The problem of anticoagulant rodenticide resistance in the United States. Proceedings of the Vth Vertebrate Pest Conference, pp. 142–8.

Jackson, W. B. & Kaukeinen, D. (1972*b*). Resistance of wild Norway rats in North Carolina to warfarine rodenticide. *Science*, **176**, 1343–4.

Jalkanen, E. (1972). Itse taimissa syyt myyrien tuhoihin. *Metsälehti*, **25**, 1.

Janion, S. M. (1961). Studies on the differentiation of a house mice population according to the occurrence of fleas (*Aphaniptera*). *Bulletin de l'Académie polonaise des Sciences*, Cl. 2, **9**, 501–6.

Janion, S. M., Ryszkowski, L. & Wierzbowska, T. (1968). Estimate of number of rodents with variable probability of capture. *Acta Theriologica*, **13**, 285–94.

Janzen, D. H. (1969). Seed-eaters versus seed size, number, toxicity and dispersal. *Ecology*, **23**, 1–27.

Janzen, D. H. (1970). Herbivores and the number of tree species in tropical forests. *American Naturalist*, **104**, 501–28.

Janzen, D. H. (1971*a*). Seed predation by animals. *Annual Review of Ecology and Systematics*, **2**, 465–92.

Janzen, D. H. (1971*b*). Escape of juvenile *Dioclea megacarpa* (Leguminosae) vines from predators in a deciduous tropical forest. *American Naturalist*, **105**, 97–112.

Janzen, D. H. (1972). Escape in space by *Sterculia apetala* seeds from the bug *Dysdercus fasciatus* in a Costa Rican deciduous forest. *Ecology*, **53**, 350–61.

Jenson, A. G. (1965). Proofing of buildings against rats and mice. Ministry of Agriculture, Fisheries and Food Technical Bulletin No. 12, pp. 1–18. Her Majesty's Stationery Office, London.

Jewell, P. A. (1966). Breeding season and recruitment in some British mammals confined on small islands. In: *Comparative biology of reproduction in mammals* (ed. I. W. Rowlands), pp. 89–116. Zoology Society Symposium No. 15.

References

Johnson, D. R. & Groepper, K. L. (1970). Bioenergetics of North Plains Rodents. *American Midland Naturalist*, **84**, 537–48.

Johnson, D. R. & Maxell, M. H. (1966). Energy dynamics of Colorado pikas. *Ecology*, **47**, 1059–61.

Johnson, E. (1958*a*). Quantitative studies of hair growth in the albino rat. I. Normal males and females. *Journal of Endocrinology*, **16**, 337–50.

Johnson, E. (1958*b*). Quantitative studies of hair growth in the albino rat. II. The effect of sex hormones. *Journal of Endocrinology*, **16**, 351–9.

Johnston, R. F. (1961). Population movements of birds. *Condor*, **63**, 386–9.

Jokela, J. J. & Lorenz, R. W. (1959). Mouse injury to forest planting in the prairie region of Illinois. *Journal of Forestry*, **57**, 21–5.

Jolly, G. M. (1965). Explicit estimates from capture–recapture data with both death and immigration-stochastic model. *Biometrika*, **52**, 225–47.

Jonsgård, A. (1969). Age determination of marine mammals. In: *The biology of marine mammals* (ed. H. F. Andersen), pp. 1–30. Academic Press, New York & London.

Juday, C. (1940). The annual energy budget of an inland lake. *Ecology*, **21**, 438–50.

Juday, C. & Schomer, H. A. (1935). The utilization of solar radiation by algae at different depth in lakes. *Biology Bulletin*, **69**, 75–81.

Judenko, E. (1967). The loss in yield in a crop of sweet corn (*Zea mays* L.) following complete destruction of some plants at an early stage by brown rats (*Rattus norvegicus* Berk.), *Pans*, A, **13**, 412–14.

Junge, C. O., Jr (1963). A quantitative evaluation of the bias in population estimates based on selective samples. International Commission of North Atlantic Fishers, Special Publication 4, pp. 26–8.

Kaczmarski, F. (1966). Bioenergetics of pregnancy and lactation in the bank vole. *Acta Theriologica*, **11**, 409–17.

Kalabukhov, N. I. (1962). Seasonal changes in the organs of mammals as indicators of environmental effects. Symposium Theriologicum, Brno, pp. 156–74.

Kalabukhov, N. I. (1969). *Periodical (seasonal and yearly) changes in the organism of rodents, their causes and consequences.* Izd. Nauka, Leningrad. (In Russian.)

Kale, H. W., II (1972). A high concentration of *Cryptotis parva* in a forest in Florida. *Journal of Mammalogy*, **53**, 216–18.

Kalela, O. (1957). Regulations of reproduction rate in subarctic populations of the vole *Clethrionomys rufocanus* (Sund.). *Annales Academiae Scientiarum Fennicae*, **34**, 1–60.

Kalela, O. (1961). Seasonal change of habitat in the Norwegian lemming, *Lemmus lemmus* (L.). *Suomalainan Tied. Toim.* (*Annales Academiae scientiarum Fennicae*), Ser. A, IV Biology, **55**, 1–72.

Kalela, O., Kilpeläienn, L., Koponen, T. & Tast, J. (1971). Seasonal differences in habitats of the Norwegian lemming, *Lemmus lemmus* (L.), in 1959 and 1960 at Kilpisjärvi, Finnish Lapland. *Annales Academiae Scientiarum Fennicae*, Ser. A, IV Biology, **178**, 1–22.

Kalela, O. & Koponen, T. (1971). Food consumption and movement of the

Norwegian lemming in areas characterized by isolated fells. *Annales Zoologici Fennici*, **8**, 80–4.

Kalela, O., Koponen, T., Lind, E. A., Skaren, U. & Tast, J. (1961). Seasonal change in habitat in the Norwegian lemming, *Lemmus lemmus* (L.). *Annales Academiae Scientiarum Fennicae*, Ser. A, IV Biology, **55**, 1–72.

Kalinowska, A. (1971). Trapping of *Apodemus flavicollis* and *Clethrionomys glareolus* into a double trap. *Acta Theriologica*, **16**, 73–8.

Kandybin, N. V. (1971). (Summary: The theoretical premises of the isolation and use of microorganisms pathogenic for insects and rodents.) *Selskohozjaistnennija biologija*, **6**, 391–6.

Kanervo, V. & Myllymäki, A. (1970). Problems caused by the field vole, *Microtus agrestis* (L.) in Scandinavia. *EPPO Publications*, Ser. A, **58**, 11–26.

Karaseva, E. V., Telitsyn, M., Lapshov, V. A. & Okhotsky, Y. V. (1971). Study of the land vertebrates of Central Yamal. *Byulleten Moskovskogo obshchestva ispytatelei prirody*, **76**, 22–32. (In Russian.)

Karpukhin, I. P. & Karpukhina, N. M. (1971). Eye lens weight as a criterion of age of *Sciurus vulgaris*. *Zoologiceskii Zhurnal*, **50**, 274–7. (In Russian with English summary.)

Karr, J. R. (1971). Structure of avian communities in selected Panama and Illinois habitats. *Ecological Monographs*, **41**, 207–33.

Karr, J. R. (1973). Production, energy pathways, and community diversity in forest birds. In: *Tropical ecological systems* (ed. F. B. Golley & E. Medina), pp. 161–76. Springer-Verlag, Berlin.

Kasatkin, V. I., Rozhkov, A. A. & Meshcherzakova, L. A. (1969). Distribution and changes in the abundance of Muridae in the delta of the Volga River. *Zoologicheskii Zhurnal*, **48**, 746–51.

Kaufman, R. G. & Norton, H. W. (1966). Growth of the porcine eye lens during insufficiences of dietary protein. *Growth*, **30**, 463–70.

Kaufman, R. G., Norton, W. H., Harmon, B. G. & Breindenstein, B. C., (1967). Growth of the porcine eye as an index to chronological age. *Journal of Animal Science*, **26**, 31–5.

Kaufman, D. W., Smith, G. C., Jones, R. M., Gentry, J. B. & Smith, M. H. (1971). Use of assessment lines to estimate density of small mammals. *Acta Theriologica*, **16**, 127–47.

Keiss, R. E. (1969). Comparison of cruption-wear patterns and cementum annuli as age criteria in elk. *Journal of Wildlife Management*, **33**, 175–80.

Keith, J. O., Hansen, R. M. & Ward, A. L. (1959). Effect of 2,4-D on abundance and foods of pocket gophers. *Journal of Wildlife Management*, **23**, 137–45.

Keith, L. B., Meslow, E. C. & Rongstad, O. J. (1968). Techniques for snowshoe hare population studies. *Journal of Wildlife Management*, **32**, 801–12.

Keller, B. L. & Krebs, C. J. (1970). Microtine population biology. III. Reproductive changes in fluctuating populations of *M. ochrogaster* and *M. pennsylvanicus* in southern Indiana, 1965–67. *Ecological Monographs*, **40**, 263–94.

Kelly, J. J., Jr & Weaver, D. F. (1969). Physical processes at the surface of arctic tundra. *Arctic*, **22**, 425–37.

References

Kemp, G. A. & Keith, L. B. (1970). Dynamics and regulation of red squirrel (*Tamiasciurus hudsonicus*) populations. *Ecology*, **51**, 763–79.

Kemper, H. (1968). *Kurzgefasste Geschichte der tierischen Schädlinge, der Schädlingskunde und der Schädlingsbekämpfung*. Duncker & Humblot, Berlin.

Kemper, H. & Kock, T. (1969). Rationale Bekämpfung der Wühlmaus (*Arvicola terrestria* L.). *Nachrichtenblatt des Deutschen Pflanzenschutzdienstes*, **21**, 169–71.

Kendle, K. E., Lazarus, A., Rowe, F. P., Telford, J. M. & Vallance, K. D. (1973). Sterilization of rodent and other pests using a synthetic oestrogen. *Nature, London*, **244**, 105–8.

Keys, J. E., Jr & Van Soest, P. J. (1970). Digestibility of forages by the meadow vole (*Microtus pennsylvanicus*). *Journal of Dairy Science*, **53**, 1502–8.

Khamaganov, S. A. (1968). Information on the reproduction of the brown rat in the Khabarovsk area of the Amur Foreland and in southern Sakhalin. *Izvestiya Irkutskogo gosudarstvennogo Nauch-no-issledovatel'skogo protivochumnogo instituta Sibiri i Dal'nego Vostoka*, **27**, 132–6.

Khodashova, K. S. (1953). Zhiznennye formy gryzunov ravninnogo Kazakhstana i nekotorye zakonomernosti ikh geograficheskogo rasprostraneniya. *Trudy Instituta Geografii Akademii Nauk SSSR, Moskva*, **54**, 33–194.

Khrustselevskii, V. P., Chernonog, N. F., Sharets, A. S. & Dmitryuk, G. Ya. (1963). Landshaftnye osobennosti raspredeleniya i kolebanii chislennosti peschanok v Muyunkumakh. In: *Materialy nauchnoi konferentsii po prirodnoi ochagovosti i profilaktike chumy*, pp. 270–2. Alma-Ata.

Kikkawa, J. (1964). Movement, activity and distribution of the small rodents *Clethrionomys glareolus* and *Apodemus sylvaticus* in woodland. *Journal of Animal Ecology*, **33**, 259–99.

Kim, T. A. (1960). Materialy po ékologii tamariskovoi peschanki pustyni Kyzylkum (*Meriones tamarscinus* Pall.). *Zoologicheskii zhurnal*, **39**, 754–65.

King, J. A. (1955). Social behavior, social organization, and population dynamics in a black-tailed prairie dog town in the Black Hills of South Dakota. *Contributions of Laboratory of Vertebrate Biology, University of Michigan*, **67**, 1–123.

Kirkpatrick, C. M. & Barnett, E. M. (1957). Age criteria in male gray squirrels. *Journal of Wildlife Management*, **21**, 341–7.

Kisiel, D. S. (1972). Effect of two sizes of Sherman traps on success in trapping *Peromyscus* and *Reithrodontomys*. *American Midland Naturalist*, **87**, 551–2.

Kitchings, J. T., Dunaway, P. B. & Storey, J. D. (1969). Uptake and excretion of [134]Cs from fallout simulant and vegetation by cotton rats. *Health Physics*, **17**, 265–77.

Kleiber, M. (1961). *The Fire of life: an introduction to animal energetics*. Wiley, New York.

Klein, D. R. (1962). Rumen contents analysis as an index to range quality. *Transactions of North American Wildlife and Natural Resources Conference*, **27**, 150–62.

Klein, D. R. (1970). Food selection by North American deer and their response to over-utilization of preferred plant species. *Animal populations in relation to their food resources* (ed. A. Watson), pp. 25–46. Blackwell Scientific, Oxford.

Klein, W., Korte, F., Poonavella, N., Weisgerber, I., Kaul, R., Müller, W. & Djirsajai, A. (1967). Über den Metabolismus von Endrin, Chlordan, Heptachlor, Dihydroheptachlor und Telodrin. Abstracta, 6th International Congress of Plant Protection, Vienna.

Klejnenberg, S. E. & Klevezal, G. A. (1962). K metodike opredelenija vozrasta zubatyh kitoobraznyh. *Dokl. AN SSSR*, **145**, 460–2.

Klejnenberg, S. E. & Klevezal, G. A. (1966). Age determination in mammals by the structure of tooth cement. *Zoologiceskii Zhurnal*, **45**, 717–24. (In Russian with English summary.)

Klemm, M. (1958). Die grosse Wühlmaus (*Arvicola terrestris* L.). Verbreitung, Schadgebiete und Auftreten in Deutschland. *Nachrichtenblatt für den Deutschen Pflanzenschutzdienstes*, **12**, 1–9.

Klemm, M. (1960). Beitrag zur Prognose des Auftretens der grossen Wühlmaus (*Arvicola terrestris* L.) in Deutschland. *Zeitschrift für Angewandte Zoologie*, **47**, 129–58.

Kler, V. C. (1927). K metodike issledovanija periodiki rosta metod izodynamičeskih ploskostej. *Zoologicheskii Zhurnal*, **7**, 4.

Klevezal, G. A. (1970). Evaluation of individual peculiarities of the growth in mammals by the layered structures in teeth and bones. *Ontogenez*, **1**, 362–72. (In Russian with English summary.)

Klevezal, G. A. (1972). On the relationship between the growth rate and formation of annual layers in mammal bone. *Zhurnal Obščej Biologii*, **33**, 166–75. (In Russian with English summary.)

Klevezal, G. A. (1973). Some limitations and new possibilities of using layers in tooth and bone tissues for age determination in mammals. *Zoologiceskii Zhurnal*, **52**, 757–65. (In Russian with English summary.)

Klevezal, G. A. & Klejnenberg, S. E. (1967). *Age determination of mammals by layered structure in teeth and bone*, pp. 1–144. Izdatelstvo Nauka, Moscow.

Klevezal, G. A. & Mitchell, E. (1971). Year layers in bones of whalebone whales. *Zoologiceskii Zhurnal*, **50**, 1114–16. (In Russian with English summary.)

Koch, R. C. (1960). *Activation analysis handbook*. Academic Press, New York.

deKock, L. L. & Robinson, A. E. (1966). Observations on a lemming movement in Jamtland, Sweden, in autumn 1963. *Journal of Mammalogy*, **47**, 490–9.

deKock, L. L., Stoddart, D. M. & Kacher, H. (1968). Notes on the behavior and food supply of lemmings (*Lemmus lemmus*, L.) during a peak density in Southern Norway, 1966/67. *Zeitschrift für Tierpsychologie*, **26**, 39–62.

Koford, C. B. (1958). Prairie dog, whitefaces and blue grama. *Wildlife Monographs* No. 3.

Kołodziej, A., Pomianowska, I. & Rajska, E. (1972). Differentiation of contacts between specimens in a *Clethrionomys glareolus* population. *Bulletin de l'Académie polonaise des Sciences*, Cl. 2, **20**, 97–102.

Koponen, T. (1964). The sequence of pelages in the Norwegian lemming,

References

Lemmus lemmus (L.). *Annales Societatis zoologicae botanicae Fennicae Vanamo*, **18**, 260-78.

Koponen, T. (1970). Age structure in sedentary and migratory populations of the Norwegian lemming, *Lemmus lemmus* (L.) at Kilpisjärvi in 1960. *Annales Zoologici Fennici*, **7**, 141-87.

Koponen, T., Kokkonen, A. & Kalela, O. (1961). On a case of spring migration in the Norwegian Lemming. *Annales Academiae Scientiarum Fennicae*, Ser. A, IV Biology, **52**, 1-30.

Koshkina, T. V. (1955). Metod opredelenija vozrasta ryžih polevok i opyt ego primenenija. *Zoologičeskii Zhurnal*, **34**, 631-9.

Koshkina, T. V. (1961). New data on the nutritional habits of the Norwegian lemming. *Byulletin Moskovskogo obshchesteva ispytalelei prirody*, **66**, 15-32.

Koshkina, T. V. (1962). Migrations of *Lemmus lemmus*. *Zoologicheskii Zhurnal*, **41**, 1859-74. (In Russian.)

Koshkina, T. V. (1965). The population density and its significance in regulating the number of the northern red-backed vole (*Clethrionomys rutilus*). *Byulleten' Moskovskogo Obshchestva Ispytatelei Prirody Otdel Biologicheskii*, **70**, 5-20. (In Russian.)

Koshkina, T. V. (1966). On periodic changes in numbers of voles (an example of the Kola Peninsula). *Byulletin Moskovskogo Obshchestva Ispytalelei Prirody*, **71**, 14-26. (In Russian.)

Koshkina, T. V. (1969). Predicting the number of forest voles by ecological indices of their population. *Byulleten' Moskovskogo Obshchestva Ispytatelei Prirody Otdel Biologicheskii*, **74**, 5-16.

Koshkina, T. V. & Khalansky, A. S. (1961). Age variation in the skull of the Norwegian lemming and analysis of the composition of populations. *Byulleten Moskovskogo Obshchestva Ispytalelei Prirody*, **66**, 3-14. (In Russian.)

Koshkina, T. V. & Khalansky, A. S. (1962). On the reproduction of *Lemmus lemmus* on the Kola Penninsula. *Zoologicheskii Zhurnal*, **41**, 604-15. (In Russian.)

Kostial, K. & Momcilovic, B. (1972). The effect of lactation of the absorption of ^{203}Pb and ^{34}Ca in rats. *Health Physics*, **23**, 383-4.

Koval, C. F., Klingbeil, G. C. & Ellarson, R. S. (1970). Meadow mouse control. Cooperative Extension Programs, University of Wisconsin, Fact Sheet 45.

Krebs, C. J. (1964). The lemming cycle at Baker Lake, Northwest Territories, during 1959-62. Arctic Institute of North America, Technical Paper No. 15.

Krebs, C. J. (1966). Demographic changes in fluctuating populations of *Microtus californius*. *Ecological Monographs*, **36**, 23-273.

Krebs, C. J., Gaines, M. S., Keller, B. L., Myers, J. H. & Tamarin, R. H. (1973). Population cycles in small rodents. *Science*, **179**, 35-41.

Krebs, C. J., Keller, B. L. & Myers, J. H. (1971). *Microtus* population densities and soil nutrients in southern Indiana grasslands. *Ecology*, **52**, 660-3.

Krebs, C. J., Keller, B. L. & Tamarin, R. H. (1969). *Microtus* population biology: demographic changes in fluctuating populations of *M. ochrogaster* and *M. pennsylvancius* in southern Indiana. *Ecology*, **50**, 587-607.

Krebs, C. J. & Myers, J. H. (1974). Population cycles in small mammals. *Advances in Ecological Research*, **8**, 268–399.

Kryl'cov, A. I. (1964). Stepnye pestruški i stadnye polevki na severe Kazahstana. *Trudy Naučno-Issledovatel'skogo Instituta Zaščity Rastenij*, **8**, 3–183.

Krylcov, A. J. & Zaleskii, A. N. (1969). Puti uluchsheiya istrebitelnykh meropriyatii protiv malogo suslika. *Trudy Kazakhsk. Instituta Zashchity Rast.*, **10**, 215–21. (In Russian.)

Kubik, J. (1952). Zwergmaus (*Micromys minutus* Pall.) in Naturschutzpark von Białowieża. *Annales Universitatis Mariae Curie-Skłodowska*, Sectio C, **7**, 449–95. (Polish with German and Russian summary.)

Kubik, J. (1965). Biomorphological variability of the population of *Clethrionomys glareolus* (Schreber, 1780). *Acta Theriologica*, **10**, 117–79.

Kucheruk, V. V. (1963). Vozdeistvie travoyadnykh mlekopitayushchikh na produktivnost' travostoya stepi i ikh znachenie v obrazovanii organicheskoi chasti stepnykh pochv. *Transactions of the Moscow Society of Naturalists*, **10**, 157–93. (In Russian.)

Kulik, I. L. (1968). A contribution to the ecology of the water vole (*Arvicola terrestris*) in flood plains of northern rivers. *Zoologicheskii Zhurnal*, **47**, 954–8.

Kuznetzov, G. B. (1970). O royushchey detatielnosti kavkazkogo krota (*Talpa caucasica*). *Zoologicheskii Zhurnal*, **49**, 1254–7. (In Russian.)

Laird, M. (1966). Biological control of rodents. In: *Seminar on rodents and rodent ectoparasites.* WHO/Vector Control/66.217. (Unpublished working document.)

Lamond, D. R. (1959). Effect of stimulation derived from other animals of the same species on oestrous cycles in mice. *Journal of Endocrinology*, **18**, 343–9.

Laurie, C. J. (1946). The reproduction of the house mouse (*Mus musculus*) living in different environments. *Proceedings of the Royal Society*, Ser. B, **133**, 248–81.

LaVal, R. K. (1970). Banding returns and activity periods of some Costa Rican bats. *Southwestern Naturalist*, **15**, 1–10.

Lavrinenko, A. E. & Tarasov, N. S. (1968). Opyt istrebleniya mongol'skikh peschanok v Tuve. *Izvestiya Irkutskogo Nauchno issledovatel'skogo Protivochumnogo Instituta Sibiri i Dal'nogo Vostoka*, **27**, 397–400.

Laws, R. M. (1960). Laminated structure of bones from some marine mammals. *Nature, London*, **187**, 338–9.

Laws, R. M. (1962). Age determination of pinnipeds with special references to growth layers in the teeth. *Zeitschrift für Säugetierkunde*, **27**, 129–46.

Layne, J. N. (1954). The biology of the red squirrel, *Tamiasciurus hudsonicus loquax* (Bangs), in central New York. *Ecological Monographs*, **24**, 227–67.

LeCren, E. D. (1965). A note on the history of mark-recapture population estimates. *Journal of Animal Ecology*, **34**, 453–4.

Lee, J. S. & Lifson, N. (1960). Measurement of total energy and material balance in rats by means of doubly-labeled water. *American Journal of Physiology*, **199**, 238–42.

Leont'ev, A. N. (1954). K ékologii kogtistoĭ peschanki v Buryat-Mongol'skoĭ

References

ASSR. *Izvestiya Irkutskogo Nauchno issledovatel'skogo Protivochumnogo Instituta Sibiri i Dal'nogo Vostoka*, **12**, 137–49.

Leont'ev, A. N. (1962). K izucheniyu populyatsiï mongol'skikh peschanok metodom mecheniya. *Izvestiya Irkutskogo Nauchno issledovatel'skogo Protivochumnogo Instituta Sibiribi i Dal'nogo Vostoka*, **24**, 296–302.

Leont'eva, M. N. (1966). O znachenii gruntovykh vod v ékologii bol'shikh peschanok. *Uchenye Zapiski Gor'kovskogo Universiteta, Seriya Biologischeskaya*, 48–57.

Leopold, J. H. & Calkins, L. (1951). Age changes in the Wistar albino rat eye. *American Journal of Ophthalmology*, **34**, 1735–41.

Lesnyak, A. P. (1959). Kontakt bol'shoï peschanki s drugimi zhivotnymi v chumnom ochage Pribalkhash'ya. In: *X soveshchanie po parasitologischeskim problemam i prirodnoochagovym boleznyam. Tezisy dokladov*, pp. 211–12. Izdatel'stvo Akadademii Nauk SSSR, Moscow.

Lever, R. A. (1962). Rat damage to crops in Fiji and Malaya. *World Crops*, **14**, 236–9.

Liat, L. B. (1966). Abundance and distribution of Malaysian bats in different ecological habitats. *Journal of the Federal Museum*, **11**, 61–76.

Lidicker, W. Z., Jr (1962). Emigration as a possible mechanism permitting the regulation of population density below carrying capacity. *American Naturalist*, **96**, 29–33.

Lidicker, W. Z., Jr (1965). Comparative study of density regulation in confined populations of four species of rodents. *Research in Population Ecology*, **7**, 57–72.

Lidicker, W. Z., Jr (1973). Regulation of numbers in an island population of the California vole; a problem in community dynamics. *Ecological Monographs*, **43**, 271–302.

Lidicker, W. Z., Jr & Anderson, P. K. (1962). Colonization of an island by *Microtus californicus*, analysed on the basis of runway transects. *Journal of Animal Ecology*, **31**, 503–17.

Lidicker, W. Z., Jr & MacLean, S. F., Jr (1969). A method for estimating age in the California vole *Microtus californicus*. *American Midland Naturalist*, **82**, 450–70.

Lifson, N., Gordon, G. B. & McClintock, R. M. (1955). Measurement of total carbon dioxide production by means of $D_2{}^{18}O$. *Journal of Applied Physiology*, **8**, 704–10.

Lifson, N. & Lee, J. S. (1961). Estimation of material balance of totally fasted rats by doubly-labeled water. *Americal Journal of Physiology*, **200**, 85–8.

Lifson, N. & McClintock, R. (1966). Theory and use of turnover rates of body water for measuring energy and material balance. *Journal of Theoretical Biology*, **12**, 46–74.

Lindeman, R. L. (1942). The trophic-dynamic aspect of ecology. *Ecology*, **23**, 399–418.

Linduska, J. P. (1942). Winter rodent populations in field-shocked corn. *Journal of Wildlife Management*, **6**, 353–63.

Linhart, S. B. (1973). Age determination and occurrence of incremental growth lines in the dental cementum of the common vampire bat (*Desmodus rotundus*). *Journal of Mammalogy*, **54**, 493–6.

Linn, I. (1954). Some Norwegian small mammal faunas; a study based on trappings in west and north Norway. *Oikos*, **5**, 1–24.

Linzey, D. W. (1968). An ecological study of the golden mouse, *Ochrotomys nuttalli*, in the Great Smoky Mountains National Park. *American Midland Naturalist*, **79**, 320–45.

Linzey, D. W. & Linzey, A. V. (1967). Maturation and seasonal molts in the golden mouse, *Ochrotomys nuttalli*. *Journal of Mammalogy*, **48**, 236–41.

Littlefield, E. W., Schoomaker, W. J. & Cook, B. (1946). Field mouse damage to coniferous plantations. *Journal of Forestry*, **44**, 745–9.

Longhurst, W. M., Jones, H. M. B. & Repuer, R. E. (1968). A basis for the palatability of deer forage plants. *North American Wildlife Conference. Transactions*, **33**, 181–92.

Lord, R. D., Jr (1959). The lens as an indicator of age in cottontail rabbits. *Journal of Wildlife Management*, **23**, 358–60.

Lord, R. D., Vilches, A. M., Maiztegui, J. I. & Soldini, C. A. (1970). The tracking board: A relative census technique for studying rodents. *Journal of Mammalogy*, **51**, 828–9.

Lotka, A. J. (1956). *Elements of Mathematical Biology*. Dover Publications, New York.

Louarn, H. L. (1971). Determination de l'age par le pesée des cristallins chez quelques éspeces de rongeurs. *Mammalia*, **35**, 636–43.

Louch, C. D. (1956). Adrenalcortical activity in relation to the density and dynamics of three confined populations of *Microtus pennsylvanicus*. *Ecology*, **37**, 701–13.

Lowe, V. P. W. (1967). Teeth as indicators of age with special reference to red deer (*Cervus elaphus*) of known age from Rhum. *Journal of Zoology, London*, **152**, 137–53.

Lowe, V. P. W. (1971). Root development of molar teeth in the bank vole, *Clethrionomys glareolus*. *Journal of Animal Ecology*, **40**, 49–61.

Lozan, M. (1961). Age determination of *Dyromys nitedula* Pall. and of *Muscardinus avellanarius* L. *Zoologiceskii Zhurnal*, **40**, 1740–3.

Ludwick, R. L., Fontenot, J. P. & Mosby, H. S. (1968). Energy metabolism of the eastern gray squirrel. *Journal of Wildlife Management*, **33**, 569–75.

Lund, M. (1964). Resistance to warfarin in the common rat. *Nature, London*, **203**, 778.

Lund, M. (1967). Resistance of rodents to ⌈rodenticides. *World Review of Pest Control*, **6**, 131–8.

Lund, M. (1970). Diurnal activity and distribution of *Arvicola terrestris* L. in an outdoor enclosure. *EPPO Publications*, Ser. A, **58**, 147–58.

Lund, M. (1972a). The need of surface sprays for the control of microtine rodents. Proceedings of the Vth Vertebrate Pest Conference, pp. 22–8.

Lund, M. (1972b). Testing of poisons for control of rodents and moles. Danish Pest Infestation Laboratory Annual Report 1971, pp. 53–61.

MacArthur, R. H. (1969). Patterns of communities in the tropics. *Biological Journal of the Linnean Society*, **1**, 19–30.

References

MacArthur, R. H. & Wilson, E. O. (1967). *The theory of island biogeography.* Princeton University Press, Princeton.

McCarley, H. (1966). Annual cycle, population dynamics and adaptive behavior of *Citellus tridecemlineatus. Journal of Mammalogy,* **47,** 294–316.

McCollum, F. C. (1975). Biochemical variation in populations of *Microtus californicus.* PhD thesis. University of California, Berkeley.

McDonald, P., Edwards, R. A. & Greenhalgh, J. F. D. (1966). *Animal nutrition.* Oliver and Boyd, Edinburgh.

Mackenzie, R. (1972). Public health importance of rodents in South America. *Bulletin of the World Health Organization,* **47,** 161–9.

Maclean, S. F., Fitzgerald, B. M. & Pitelka, F. A. (1974). Population cycles in arctic lemmings: winter reproduction and predation by weasels. *Arctic and Alpine Research,* **6,** 1–12.

McNab, B. K. (1963). A model of the energy budget of a wild mouse. *Ecology,* **44,** 521–32.

McPherson, A. B. & Krull, J. N. (1972). Island populations of small mammals and their affinities with vegetation type, island size and distance from mainland. *American Midland Naturalist,* **88,** 384–92.

Macpherson, A. H. (1965). The origin of diversity in mammals of the Canadian arctic tundra. *Systematic Zoology,* **14,** 153–73.

Macpherson, A. H. (1969). The dynamics of Canadian arctic fox populations. Canadian Wildlife Service Report No. 8.

Maercks, H. (1954). Über den Einfluss der Witterung auf den Massenwechsel der Feldmaus (*Mictorus arvalis* Pallas) in der Wesermarsch. *Nachrichtenblatt des Deutschen Pflanzenschutzdienstes,* **6,** 101–8.

Maher, W. J. (1967). Predation by weasels on a winter population of lemmings, Banks Island, Northwest Territories. *Canadian Field Naturalist,* **81,** 248–50.

Maher, W. J. (1970). The pomarine jaeger as a brown lemming predator in Northern Alaska. *Wilson Bulletin,* **82,** 130–57.

Maksimov, A. A. (1965). Role of agriculture in the formation of the meadow-field type of natural foci of tularemia. In: *Theoretical questions of natural foci of diseases* (ed. B. Rosicky & K. Heyberger), pp. 337–42. Czechoslovakia Academy of Science, Prague.

Manly, B. F. J. (1970). A simulation study of animal population estimation using the capture–recapture method. *Journal of Applied Ecology,* **7,** 13–39.

Manly, B. F. J. & Parr, M. J. (1968). A new method of estimating population size, survivorship and birth rate from capture–recapture data. *Transactions of the Society for British Entomology,* **18,** 81–9.

Maray, A. N., Knizhnikov, V. A. & Karmaeva, A. N. (1966). The effect of calcium in drinking water on the accumulation of ^{226}Ra in the human body. In: *Radioecological concentration processes* (ed. B. Aberg & F. P. Hungate), pp. 333–6. Pergamon Press, Oxford.

Marsden, H. M. & Bronson, F. H. (1964). Estrous synchrony in mice: alteration by exposure to male urine. *Science,* **144,** 1469.

Marsden, W. V. (1964). *The lemming year.* Chatto & Windus, London.

Marsh, R. E. (1968). An aerial method of dispensing ground squirrel bait. *Journal of Range Management,* **21,** 380–4.

References

Marsh, R. E. & Howard, W. E. (1970). Chemosterilants as an approach to rodent control. Proceedings of the IVth Vertebrate Pest Conference, pp. 55–63.

Marten, G. G. (1972). Censusing mouse populations by means of tracking. *Ecology*, **53**, 859–67.

Marten, G. G. (1973). Time patterns of *Peromyscus* activity and their correlations with weather. *Journal of Mammalogy*, **54**, 169–88.

Martin, P. (1971). Movements and activities of the mountain beaver (*Aplodontia rufa*). *Journal of Mammalogy*, **52**, 717–23.

Martinet, L. (1966). Détermination de l'age chez le campagnol des champs (*Microtus arvalis* Pallas) par la pesée du cristallin. *Mammalia*, **30**, 425–30.

Martinet, L. (1967). Cycle saisonnier de reproduction du campagnol des champs *Microtus arvalis*. *Annales de biologie animale, biochimie et biophysique*, **7**, 245–59.

Mathies, J. B., Dunaway, P. B., Schneider, G. & Auerbach, S. I. (1972). *Annual consumption of cesium-137 and cobalt-60 labeled pine seeds by small mammals in an oak-hickory forest.* TM-3912. Oak Ridge National Laboratory. Tennessee.

Maxell, M. H. (1973). Rodent ecology and pronghorn energy relations in the Great Divide Basin of Wyoming. PhD thesis. University of Wyoming.

Mayer, W. V. (1953). A preliminary study of the Barrow ground squirrel *Citellus parryi barrowensis*. *Journal of Mammalogy*, **34**, 334–45.

Maza, B. G., French, N. R. & Aschwanden, A. P. (1973). Home range dynamics in a population of heteromyid rodents. *Journal of Mammalogy*, **54**, 405–25.

Mazák, V. (1963). Notes on the dentition in *Clethrionomys glareolus* (Schreber 1780) in the course of postnatal life. *Säugetierkundliche Mitteilungen*, **11**, 1–11.

Mazurkiewicz, M. (1972). Density and weight structure of populations of the bank vole in open and enclosed areas. *Acta Theriologica*, **17**, 455–65.

Mead, R. A. (1967). Age determination in the spotted skunk. *Journal of Mammalogy*, **48**, 606–16.

Medwecka-Kornaś, A., Łomnicki, A. & Bandoła-Ciołczyk, E. (1973). Energy flow in the deciduous woodland ecosystem, Ispina Project, Poland. In: *Modeling Forest Ecosystems*, pp. 144–50. Report of International Workshop, IBP/PT Oak Ridge, Tennessee.

Mejer, M. N. (1957). On the age variability in *Citellus pygmaeus* Pall. *Zoologiceskii Zhurnal*, **36**, 1393–1402. (In Russian with English summary.)

Melchior, H. R. (1972). Summer herbivory by the brown lemming at Barrow, Alaska. Proceedings of the 1972 IBP Tundra Biome Symposium (ed. J. Brown & S. Bowen), pp. 136–8.

Metzgar, L. H. (1967). An experimental comparison of screech owl predation on resident and transient white-footed mice (*Peromyscus leucopus*). *Journal of Mammalogy*, **48**, 387–91.

Metzgar, L. H. (1971). Behavioral population regulation in the woodmouse, *Peromyscus leucopus*. *American Midland Naturalist*, **86**, 434–48.

Meyer, K. F. (1963). Plague. In: *Diseases transmitted from animals to man*,

409

References

5th edition (ed. T. G. Bull), pp. 527–87. Charles C. Thomas, Springfield, Illinois.

Meylan, A. & Hausser, J. (1973). Chromosomes of *Sorex* of the *araneus-arcticus* group. *Zeitschrift für Säugetierkunde*, **38**, 143–58. (In French.)

Meylan, A. & Morel, J. (1970). Capture et elevage *d'Arvicola terrestris* (L.): premiers résultats. *EPPO Publications*, Ser. A, **58**, 115–24.

Mezhzherin, V. A. (1964). Dehnel's phenomenon and its possible explanation. *Acta Theriologica*, **8**, 95–114. (In Russian.)

Michielsen, N. C. (1966). Intraspecific and interspecific competition in the shrews *Sorex araneus* L. and *S. minutus* L. *Archives Neerlandaises de Zoologie*, **17**, 73–174.

Migula, P. (1969). Bioenergetics of pregnancy and lactation in European common vole. *Acta Theriologica*, **13**, 167–79.

Migula, P., Gano, B., Stepien, Z. & Bagdal, U. (1974). Density estimation and energy flow through small mammal populations in the beech forest of Zaskalskie Gorge (Piening Male). *Acta Theriologica*. (In press.)

Migula, P., Grodziński, W., Jasinski, A. & Musialek, B. (1970). Vole and mouse plagues in southeastern Poland in the years 1945–1967. *Acta Theriologica*, **15**, 233–52.

Millar, J. S. (1970a). The breeding season to reproductive cycle of the western red squirrel. *Canadian Journal of Zoology*, **48**, 471–3.

Millar, J. S. (1970b). Variations in fecundity of the red squirrel, *Tamiasciurus hudsonicus* (Erxleben). *Canadian Journal of Zoology*, **48**, 1055–8.

Miller, R. S. (1964). Ecology and distribution of pocket gophers (Geomyidae) in Colorado. *Ecology*, **45**, 256–72.

Moens, R. (1968). Essai de destruction du rat musqué au moyen d'appats a base de chlopophacinone. *EPPO Publications*, Ser. A, **47**, 49–61.

Mohr, E. (1933). Die postembryonale Entwicklung von *Talpa europaea* L. *Videnskabelige Meddelelser fra Dansk Naturhistorisk Forening i Kjobenhaun*, **94**, 249–72.

Montgomery, S. J., Balph, D. F. & Balph, D. M. (1971). Age determination of Uinta ground squirrels by teeth annuli. *Southwestern Naturalist*, **15**, 400–2.

Morel, J. & Meylan, A. (1970). Une pullulation de campagnols terrestres (*Arvicola terrestris* (L.)) (Mammalia: Rodentia). *Revue Suisse de Zoologie*, **77**, 705–12.

Morhardt, E. J. (1970). Body temperatures of white-footed mice (*Peromyscus* sp.) during daily torpor. *Comparative Biochemistry and Physiology*, **33**, 423–39.

Morris, P. A. (1970). A method for determining absolute age in the hedgehog. *Journal of Zoology, London*, **161**, 277–81.

Morris, P. A. (1971). Epiphyseal fusion in the forefoot as a means of age determination in the hedgehog (*Erinaceus europaeus*). *Journal of Zoology*, **164**, 254–9.

Morris, P. A. (1972). A review of mammalian age determination methods. *Mammal Review*, **2**, 69–104.

Morris, R. D. (1970). The effects of endrin on *Microtus* and *Peromyscus*. I Unenclosed field populations. *Canadian Journal of Zoology*, **48**, 695–708.

Morris, R. D. (1972). The effects of endrin on *Microtus* and *Peromyscus*. II. Enclosed field populations. *Canadian Journal of Zoology*, **50**, 885–96.

Morris, R. F. (1959). Single-factor analysis in population dynamics. *Ecology*, **40**, 580–8.

Morrison, P. R. (1951). An automatic manometric respirometer. *Review of Scientific Instruments*, **22**, 264–7.

Morrison, P. R. & Grodziński, W. (1968). Morrison respirometer and determination of ADMR. In: *Methods of ecological bioenergetics* (ed. W. Grodziński & R. Z. Klekowski), pp. 153–63. Polish Academy of Sciences, Warsaw.

Morrison, P. R., Ryser, F. & Dawe, A. (1959). Studies on the physiology of the masked shrew, *Sorex cinereus*. *Physiological Zoology*, **32**, 256–71.

Morrison, P. R. & Teitz, W. J. (1953). Observations in food consumption in four Alaskan small mammals. *Arctic*, **6**, 52–7.

Moss, R. (1972). Food selection by red grouse (*Lagopus lagopus scoticus* Lath.) in relation to chemical composition. *Journal of Animal Ecology*, **41**, 411–28.

Moulton, C. R. (1923). Age and chemical development in mammals. *Journal of Biological Chemistry*, **57**, 79–97.

Mraz, F. R. (1959). Influence of dietary potassium and sodium on cesium-134 and potassium-42 excretion in sheep. *Journal of Nutrition*, **68**, 655–62.

Mullen, D. A. (1968). Reproduction in brown lemmings (*Lemmus trimucronatus*) and its relevance to their cycle of abundance. *University of California Publications in Zoology*, **85**, 1–24.

Mullen, R. K. (1970). Respiratory metabolism and body water turnover rates of *Perognathus formosus* in its natural environment. *Comparative Biochemistry and Physiology*, **32**, 259–65.

Mullen, R. K. (1971a). Energy metabolism and body water turnover rates of two species of free-living kangaroo rats, *Dipodomys merriami* and *Dipodomys microps*. *Comparative Biochemistry and Physiology*, **39A**, 379–90.

Mullen, R. K. (1971b). Energy metabolism of *Peromyscus crinitus* in its natural environment. *Journal of Mammalogy*, **52**, 633–5.

Mullen, R. K. (1973). The $D_2^{18}O$ method of measuring the energy metabolism of free-living animals. In: *Ecological energetics of homeotherms* (ed. J. Gessaman). Utah State University Monograph Series, **20**, 1–155.

Mullen, R. K. & Chew, R. M. (1973). Estimating the energy metabolism of free-living *Perognathus formosus*: a comparison of direct and indirect methods. *Ecology*, **54**, 631–7.

Murio Di A. & Lorito, N. (1969). L'arvicola *Pitymys savii* un pericoloso nemico per la cultura del tabacco. *Tabacco*, **73**, 11–17. (In Italian.)

Murray, B. M. & Murray, D. F. (1969). Notes on mammals in Alpine areas of the Northern St. Ellias Mountains, Yukon Territory and Alaska. *Canadian Field Naturalist*, **83**, 331–8.

Muul, I. (1968). Behavioral and physiological influences on the distribution of the flying squirrel, *Glaucomys volans*. *Miscellaneous Publications of the Museum of Zoology, University of Michigan*, **134**, 1–66.

Myers, J. H. & Krebs, C. J. (1971). Genetic, behavioral, and reproductive

References

attributes of dispersing field voles *Microtus pennsylvanicus* and *Microtus ochrogaster*. *Ecological Monographs*, **41**, 53–78.

Myers, K. & Gilbert, N. (1968). Determination of age of wild rabbits in Australia. *Journal of Wildlife Management*, **32**, 841–9.

Myllymäki, A. (1964). Om vattensorkens populations ekologi. NJF Kongress, Sektion 4, Copenhagen (Mimeograph).

Myllymäki, A. (1967). Peltomyyrän tuhot ja torjunta. (Summary: Damage caused by field voles on garden plants, field crops and forest trees in Finland.) *Maatalous ja Koetomiminta*, **21**, 183–94.

Myllymäki, A. (1969). Productivity of a free-living population of the field vole, *Microtus agrestis* (L.). In: *Energy flow through small mammal populations* (ed. K. Petrusewicz & L. Ryszkowski), pp. 255–66. Polish Scientific Publishers, Warsaw.

Myllymäki, A. (1969–70). Trapping experiments on the water vole, *Arvicola terrestris* (L.), with the aid of the isotope technique. In: *Energy flow through small mammal populations* (ed. K. Petrusewicz & L. Ryskowski), pp. 39–55. Polish Scientific Publishers, Warsaw.

Myllymäki, A. (1970). Population ecology and its application to the control of the field vole, *Microtus agrestis* (L.). *EPPO Publications*, Ser. A, **58**, 27–48.

Myllymäki, A. (1974*a*). Experience from an unsuccessful removal of a semi-isolated population of *Arvicola terrestris* (L.). International Symposium on Species and Zoogeography of European Mammals, Brno. (In press.)

Myllymäki, A. (1974*b*). *Erfarenheter av programmerad sorkbekämpning på fröplantager*. Nordisk Bekämpningsmedelskonferens, Ellivuori. (In press.)

Myllymäki, A., Aho, J., Lind, E. & Tast, J. (1962). Behavior and daily activity of the Norwegian lemming, *Lemmus lemmus* (L.) during autumn migration. *Annales Zoologici Societatis Zoologicae – Botanicae Fennicae Vanamo*, **24**, 1–31.

Myllymäki, A. & Paasikallio, A. (1972). The detection of seed-eating small mammals by means of P-32 treatment of spruce seed. *Aquilo, Seria Zoologica*, **13**, 21–4.

Myllymäki, A., Paasikallio, A. & Hakkinen, A. (1971). Analysis of a standard trapping of *Microtus agrestis* (L.) with triple isotope marking outside the quadrat. *Annales Zoologici Fennici*, **8**, 22–34.

Myllymäki, A., Paasikallio, A., Pankakoski, E. & Kanervo, V. (1971). Removal experiments on small quadrats as a means of rapid assessment of the abundance of small mammals. *Annales Zoologici Fennici*, **8**, 177–85.

Myrcha, A. (1968). Caloric value and chemical composition of the body of the european hare. *Acta Theriologica*, **13**, 65–70.

Myrcha, A., Ryszkowski, L. & Walkowa, W. (1969). Bioenergetics of pregnancy and lactation in the white mouse. *Acta Theriologica*, **14**, 161–6.

Myrcha, A. & Walkowa, W. (1968). Changes in the caloric value of the body during the postnatal development of white mice. *Acta Theriologica*, **13**, 391–400.

Mysterud, I. (1970). Hypotheses concerning characteristics and causes of population movements in Tengmaim's owl (*Aegolius funereus* (L.). *Norwegian Journal of Zoology*, **18**, 47–74.

Mystkowska, E. & Sidorowicz, J. (1961). Influence of the weather on captures of micromammalia. II. Insectivora. *Acta Theriologica*, **5**, 263–73.

Nabholz, J. V. (1973). Small mammals and mineral cycling on three Coweeta watersheds. MS Thesis. University of Georgia, Athens. (Unpublished.)

Naumov, N. P. (1934). Opredelenie vozrasta belki (*Sciurus vulgaris*). *Uč. zap. Mosk. Univ.*, **2**.

Naumov, N. P. (1940). The ecology of the hillock mouse, *Mus musculus hortulanus*. *Journal of the Institute of Evolutionary Morphology*, **3**, 33–77 (In Russian.)

Naumov, N. P. (1948). *Ocerki sravnitelnoi ekologii mysevidnykh gryzunov*. Akademiia Nauk SSSR, Moscow.

Naumov, N. P. (1954). Tipy̆ poselenii gry̆zunov i ikh ékologicheskoe znachenie. *Zoologicheskii Zhurnal*, **33**, 268–90.

Naumov, N. P. (1967). Struktura populyatskiĭ i dinamika chislennosti nazemny̆kh pozvonochny̆kh. *Zoologicheskii Zhurnal*, **66**, 1470–83.

Naumov, N. P. (1971). Prostranstvennaya struktura vida u mlekopitayushchikh. *Zoologicheskii Zhurnal*, **50**, 965–81.

Naumov, N. P., Dmitriev, P. P. & Lobachev, V. S. (1970). Izmeneniya v biotsenozakh Priaral'skikh Karakumov pri istreblenii bol'shikh peschanok. *Zoologicheskii Zhurnal*, **49**, 1758–66.

Naumov, N. P., Gibet, L. A. & Shatalova, S. P. (1969). Dynamics of sex ratio with respect to changes in number of mammals. *Zhurnal Obshchei Biologii*, **30**, 673–80.

Naumov, N. P., Lobachev, V. C., Dmitriev, P. P. & Smirin, B. M. (1972). *Prirodnỹi ochag chumỹ v Priaral'skikh Karakumakh*, pp. 1–405. Izdatel'stvo gosudarstvennogo Universiteta, Moscow.

Neal, B. R. (1968). The ecology of small rodents in the grassland community of the Queen Elizabeth National Park, Uganda. PhD Thesis. University of Southampton.

Negus, N. C., Gould, E. & Chipman, R. K. (1961). Ecology of the rice rat, *Oryzomys palustrus* (Harlan), on Breton Island, Gulf of Mexico, with a critique of the social stress theory. *Tulane Studies in Zoology*, **8**, 94–123.

Negus, N. C. & Pinter, A. J. (1965). Litter sizes of *Microtus montanus* in the laboratory. *Journal of Mammalogy*, **46**, 434–7.

Nel'zina, E. N. & Medvedev, S. I. (1962). Éntomotsenoz gnezda malogo suslika na territorii Severnogo Kazakhstana. *Zoologicheskii Zhurnal*, **41**, 217 20.

Ness, J. & Dugdale, R. C. (1959). Computation of production for populations of aquatic midge larvae. *Ecology*, **40**, 425–30.

Newman, J. R. (1971). Energy flow of a secondary consumer (*Sorex cinereus*) in a salt marsh community. PhD Thesis. University of California, Davis.

Newsome, A. E. (1969*a*). A population study of house mice temporarily inhabiting a South Australian wheatfield. *Journal of Animal Ecology*, **38**, 341–60.

Newsome, A. E. (1969*b*). A population study of house mice permanently inhabiting a reed bed in South Australia. *Journal of Animal Ecology*, **38**, 361–78.

References

Newsome, A. E. (1971). The ecology of house mice in cereal haystacks. *Journal of Animal Ecology*, **40**, 1–16.

Newsome, A. E. (1975). Outbreaks of rodents in semi-arid to arid Australia: causes, preventions and evolutionary considerations. In: *Rodents in desert environments* (ed. I. Prakash), Monographicae Biologicae, Junk Publishers.

Newson, R. (1963). Differences in numbers, reproduction and survival between two neighboring populations of bank voles (*Clethrionomys glareolus*). *Ecology*, **44**, 110–20.

Nicholson, I. A., Paterson, I. S. & Currie, A. (1970). A study of vegetational dynamics; selection by sheep and cattle in *Nardus* pasture. In: *Animal populations in relation to their food resources* (ed. A. Watson), pp. 129–43. Blackwell Scientific, Oxford.

Nikodemusz, E. (1973). A klorfacinon taxocitasa a mizei pockon (*Microtus arvalis* Pall.) (Summary: The toxicity of chlorophacione against the field vole (*Microtus arvalis* Pall.).) *Növényvépelem*, **9**, 59–61.

Nishiwaki, M., Ohsumi, S. & Kasuya, T. (1961). Age characteristics in the sperm whale mandible. *Norsk hvalfangst-tidende*, **12**, 499–506.

Nosek, J., Kozuch, O. & Chemla, J. (1972). Contribution to the knowledge of home range in common shrew *Sorex araneus* L. *Oecologia, Berlin*, **9**, 59–63.

Novikov, G. A. & Petrov, O. V. (1953). K ekologii podzemnoi polevki v lesostepnykh dubravakh. *Zoologicheskii Zhurnal*, **32**, 130–9.

Nurgel'dÿev, O. N. (1969). *Ékologiya mlekopotayushchikh ravninnoĭ Turkmenii*. Izdatel'stvo Ylym, Ashkhabad.

Nyholm, E. S. (1968). Ecological observations on the snow hare (*Lepus timidus* L.) on the islands of Krunnit and in Kuusamo. *Suomen Riista*, **20**, 15–31.

Odum, E. P. (1959). *Fundamentals of ecology*. W. B. Saunders, Philadelphia and London.

Odum, E. P., Connell, C. E. & Davenport, L. B. (1962). Population energy flow of three primary consumer components of old-field ecosystems. *Ecology*, **43**, 88–95.

Ognev, S. I. (1928–50). *Mammals of the USSR and adjacent countries*. 7 vols. (Trans. A. Birron & Z. S. Cole, 1962–4, Israel Program for Scientific Translation.)

Okhotina, M. V. (1966). *Dal' nevostočnyj krot i egoprom ysel*, pp. 1–136. Izdatelstvo Nauka, Moscow. (In Russian.)

Okulova, N. M., Aristova, V. A. & Koshkina, T. V. (1971). Effects of population density upon the size of home range of small rodents in the west Siberian Taiga. *Zoologicheskii Zhurnal*, **50**, 908–15.

Olszewski, J. L. (1968). Role of uprooted trees in the movements of rodents in forests. *Oikos*, **19**, 99–104.

Ophof, A. J. & Langeveld, D. W. (1969). Warfarin – resistance in the Netherlands. *Shriftenreihe des Vereins für Wasser, Boden und Lufthygiene*, **32**, 39–47.

Opuszynski, K. & Trojan, P. (1963). Distribution of burrows and elements of the population structure of small forest rodents. *Ekologia Polska*, Ser. A, **11**, 339–52.

Östbye, E. & Semb-Johansson, A. (1970). The eye lens as an age indicator in the Norwegian lemming (*Lemmus lemmus* L.). *Nytt magasin for Zoologi*, **18**, 239–43.

Otero, J. G. & Dapson, R. W. (1972). Procedures in the biochemical estimation of age of vertebrates. *Researches in Population Ecology*, **13**, 152–60.

Ovchinnikova, S. L. (1969). Some specific ecological features of the Russian mole rat (*Spalax microphthalmus*) in the Chernozem zone. *Zoologicheskii Zhurnal*, **48**, 1564–70.

Owen, R. (1840–5). *Odontography or a treatise of the comparative anatomy of the teeth, their physiological relations, mode of development and microscopic structure in the vertebrate animals.* London.

Packard, R. L. (1968). An ecological study of the fulvous harvest mouse in eastern Texas. *American Midland Naturalist*, **79**, 68–88.

Paine, R. T. (1971). The measurement and application of the calorie to ecological problems. *Annual Review of Ecology*, **2**, 145–63.

Palmen, E. (ed.) (1971). Proceedings of the IBP meeting on secondary productivity in small mammal populations. *Annales Zoologici Fennici*, **8**, 1–185.

Panteleev, P. A. (1967). Outbreaks of *Arvicola terrestris* L. and their relation to solar activity cycles. *Zhurnal Obshchei Biologii*, **28**, 649–57.

Panteleev, P. A. (1968). *Populyatsionnaya ekologiya vodyanio polebki i mery bor'by.* Nauka, Moscow.

Parsons, P. A. (1963). Migration as a factor in natural selection. *Genetica*, **33**, 184–206.

Patric, E. F. (1970). Bait preference of small mammals. *Journal of Mammalogy*, **51**, 179–82.

Pavlenen, V. A. (1971). The white hare (*Lepus timidus* L., 1758), In: *Mammals of the Yamal and arctic Urals* (ed. S. S. Schwarts), pp. 75–106. Academy of Science, USSR. (In Russian.)

Pavlov, A. N. (1959). Osobennosti razmnozheniya peschanok poludennoĭ i grebenshchikovoĭ v usloviyakh severo-zapadnogo Prikaspiya. *Zoologicheskii Zhurnal*, **38**, 1876–85.

Pavlovskiĭ, E. N. (1964). *Prirodnaya ochagovost' transmissivnȳkh Sologneĭ*, pp. 1–211. Izdatel'stvo Akademii Nauk SSSR, Moscow.

Pearson, A. (1966). Population dynamics and disease. *Proceedings of the Royal Society of Medicine*, **59**, 55–6.

Pearson, A. M. (1962). Activity patterns, energy metabolism, and growth rates of the voles *Clethrionomys rufocanus* and *C. glareolus* in Finland. *Annales Societatis Zoolog.-bottanicae Fennicae Vanamo*, **24**, 1–58.

Pearson, O. P. (1960). The oxygen consumption and bioenergetics of harvest mice. *Physiological Zoology*, **33**, 152–60.

Pearson, O. P. (1963). History of two local outbreaks of feral house mice. *Ecology*, **44**, 540–9.

Pearson, O. P. (1964). Carnivore–mouse predation: an example of its intensity and bioenergetics. *Journal of Mammalogy*, **45**, 177–88.

Pearson, O. P. (1966). The prey of carnivores during one cycle of mouse abundance. *Journal of Animal Ecology*, **35**, 217–33.

Pearson, O. P. (1967). La estructura por edades y la dinamica reproductiva en

References

una poblacion de ratones de campo, *Akodon azarae*. *Physis Revista de la Asociacion Argentina de Ciencias Naturales*, **27**, 53–8.

Pearson, O., Binsztein, N., Boiry, L., Busch, C., Di Pace, M., Gallopin, G., Penchaszadeh, P. & Piantanida, M. (1968). Estructura social, distribucion espacial y composicion por edades de una poblacion de tuco-tucos (*Ctenomys talarum*). *Inv. Zool. Chilenas*, **13**, 47–80.

Peiper, R. (1963). Production and chemical composition of arctic tundra vegetation and their relation to the lemming cycle. PhD Thesis. University of California, Berkeley.

Pelikan, J. (1967). The estimation of population density in small mammals. In: *Secondary productivity of terrestrial ecosystems* (ed. K. Petrusewicz), pp. 267–73.

Pelikan, J. (1970). Testing and elimination of the edge effect in trapping small mammals. In: *Energy flow through small mammal populations* (ed. K. Petrusewicz & L. Ryszkowski), pp. 57–61. Polish Scientific Publishers, Warsaw.

Pelikan, J. (1971). Calculated densities of small mammals in relation to quadrat size. *Annales Zoologici Fennici*, **8**, 3–6.

Pelikan, J. & Holisova, V. (1969). Movements and home ranges of *Arvicola terrestris* on a brook. *Zoologické Listy*, **18**, 207–24.

Perry, A. E. & Herreid, C. F. (1969). Comparison of the toothwear and lens-weight methods of age determination in the guano bat, *Tadarida brasiliensis mexicana*. *Journal of Mammalogy*, **50**, 357–60.

Peterson, P. J. (1971). Unusual accumulations of elements by plants and animals. *Science Progress, Oxford*, **59**, 505–26.

Petrides, G. A. (1951). Notes on age determination in squirrels. *Journal of Mammalogy*, **32**, 111–12.

Petrides, G. A. & Stewart, P. G. (1970). Determination of energy flow in small mammals using ^{51}chromium. In: *Energy flow through small mammal populations* (ed. K. Petrusewicz & L. Ryszkowski), pp. 131–42. Polish Scientific Publishers, Warsaw.

Petrusewicz, K. (1957). Investigation of experimentally induced population growth. *Ekologia Polska*, Ser. A, **5**, 281–301.

Petrusewicz, K. (1963). Population growth induced by disturbance in the ecological structure of the population. *Ekologia Polska*, Ser. A, **11**, 87–125.

Petrusewicz, K. (1966a). Dynamics, organization and ecological structure of population. *Ekologia Polska*, Ser. A, **14**, 413–36.

Petrusewicz, K. (1966b). Production vs. turnover of biomass and individuals. *Bulletin de l'Académie polonaises des Sciences*. Biology Series, **15**, 621–5.

Petrusewicz, K. (1967a). Concepts in studies on the secondary productivity of terrestrial ecosystems. In: *Secondary productivity of terrestrial ecosystems* (ed. K. Petrusewicz), pp. 17–50. Polish Academy of Sciences, Warsaw.

Petrusewicz, K. (1967b). Suggested list of more important concepts in productivity studies (definitions and symbols). In: *Secondary productivity of terrestrial ecosystems* (ed. K. Petrusewicz), pp. 51–8. Polish Academy of Sciences, Warsaw.

416

References

Petrusewicz, K. (ed.) (1967c). *Secondary productivity of terrestrial ecosystems.* 2 vols. Polish Academy of Sciences, Warsaw.

Petrusewicz, K. (1968). Calculation of the number of individuals born by a population. *Bulletin de l'Académie polonaise des Sciences*, Cl. II, **16**, 545–53.

Petrusewicz, K. (1969). Estimation of number of new-born animals. In: *Energy flow through small mammals populations* (ed. K. Petrusewicz & L. Ryszkowski), pp. 181–6. Polish Scientific Publishers, Warsaw.

Petrusewicz, K. (1970). Dynamics and production of the hare population in Poland. *Acta Theriologica*, **26**, 413–45.

Petrusewicz, K. & Andrezejewski, R. (1962). Natural history of a free-living population of house mice (*Mus musculus* L.) with particular reference to grouping within the population. *Ekologia Polska*, Ser. A, **10**, 85–122.

Petrusewicz, K., Andrezejewski, R., Bujalska, G. & Gliwicz, J. (1968). Productivity investigation of an island population of *Clethrionomys glareolus* (Schreber, 1780). IV. Production. *Acta Theriologica*, **13**, 435–45.

Petrusewicz, K., Andrezejewski, R., Bujalska, G. & Gliwicz, J. (1969). The role of spring, summer and autumn generation in the productivity of a free-living population of *Clethrionomys glareolus* (Schreber, 1780). In: *Energy flow through small mammal populations* (ed. K. Petrusewicz & L.Ryskowski), pp. 235–45. Polish Scientific Publishers, Warsaw.

Petrusewicz, K., Bujalska, G., Andrezejewski, R. & Gliwicz, J. (1971). Productivity processes in an island population of *Clethrionomys glareolus*. *Annales Zoologici Fennici*, **8**, 127–32.

Petrusewicz, K. & Macfadyen, A. (1970). *Productivity of terrestrial animals: Principles and methods.* IBP Handbook 13. Blackwell Scientific, Oxford.

Petrusewicz, K., Markov, G., Gliwicz, J. & Christov, L. (1972). A population of *Clethrionomys glareolus pirineus* on the Vithosa Mountains, Bulgaria. IV. Production. *Acta Theriologica*, **17**, 437–53.

Petrusewicz, K. & Ryszkowski, L. (eds.) (1969/70). *Energy flow through small mammal populations.* Polish Scientific Publishers, Warsaw.

Petrusewicz, K. & Walkowa, W. (1968). Contribution of production due to reproduction to the total production of the population and individual. *Bulletin de l'Académie polonaise des Sciences*, Cl. II, **16**, 439–42.

Petryszyn, Y. & Fleharty, E. D. (1972). Mass and energy of detritus clipped from grassland vegetation by the cotton rat (*Sigmodon hispidus*). *Journal of Mammalogy*, **53**, 168–75.

Petticrew, B. G. & Sadleir, R. M. F. S. (1970). The use of index trap lines to indicate population numbers of deermice (*Peromyscus maniculatus*) in a forest environment in British Columbia. *Canadian Journal of Zoology*, **48**, 385–9.

Péwé, T. L. (1957). Permafrost and its effect on life in the north. In: *Arctic biology* (ed. H. P. Hansen), pp. 27–65. Oregon State University Press, Corvallis.

Pianka, E. R. (1970). On *r*- and *K*-selection. *American Naturalist*, **104**, 592–7.

Pielowski, Z. (1962). Untersuchungen über die ökologie der Kreuzotter (*Vipera berus* L.). *Zoologische Jahrbücher (Systematik)*, **89**, 479–500.

417

References

Pilarska, J. (1969). Individual growth curve and food consumption by European hare *Lepus europaeus* Pallas, 1778, in laboratory conditions. *Bulletin de l'Académie polonaise des Sciences*, Cl. II, **17**, 299–305.

Pilj, L. van der (1968). *Principles of dispersal in higher plants*. K. Springer, Berlin.

Pingale, S. V. (1966). Economic importance of sylvan or field rodents. In: *Seminar on rodents and rodent ectoparasites*. WHO/Vector control 66.217. (Unpublished working document.)

Pirie, A. & Heyningen, R. van (1965). *Biochemistry of the eye*. Charles C. Thomas, Springfield, Illinois.

Pitelka, F. A. (1957a). Some characteristics of microtine cycles in the Arctic. In: *Arctic biology* (ed. H. P. Hansen), pp. 153–84. Oregon State University Press, Corvallis.

Pitelka, F. A. (1957b). Some aspects of population structure in the short-term cycle of the brown lemming in Northern Alaska. *Cold Spring Harbor Symposia on Quantitative Biology*, **22**, 237–51.

Pitelka, F. A. (1959). Numbers, breeding schedule, and territoriality in pectoral sandpipers of northern Alaska. *Condor*, **61**, 233–64.

Pitelka, F. A. (1973). Cyclic pattern in lemming populations near Barrow, Alaska. In: *NARL 25th Anniversary Symposium* (ed. W. Gunn). Arctic Institute of North America (In press).

Pitelka, F. A. & Schultz, A. M. (1964). The nutrient-recovery hypothesis for arctic microtine rodents. In: *Grazing in terrestrial and marine environments* (ed. D. Crisp), pp. 55–68. Blackwell Scientific, Oxford.

Pitelka, F. A., Tomich, P. Q. & Treichel, G. W. (1955). Ecological relations of jaegers and owls and lemming predators near Barrow, Alaska. *Ecological Monographs*, **25**, 85–117.

Pjastolova, O. A. (1964). Spetsificheskie osobennosti vozrastnoi struktury populyatsii polevki-ékonomki na Kraïnem severnom predele ee rasprostraneniya. In: *Souremennÿe problemy izucheniya dinamiki chislennosti populyatsiï zhivotnykh*, pp. 81–2. Moscow.

Pjastolova, O. A. (1971). Age structure of subarctic populations of *Microtus middendorffi* and *M. oeconomus*. *Annales Zoologici Fennici*, **8**, 72–4.

Pjastolova, O. A. (1972). The role of rodents in energetics of biogeocoenoses of forest tundra and southern tundra. Proceedings of the IV International Meeting on the Biological Productivity of Tundra (ed. F. W. Wielgolasky & T. Rosswall). Leningrad.

Platt, F. B. W. & Rowe, J. (1964). Damage by the edible dormouse (*Glis glis* L.) at Wendover forest (Chilterns). *Quarterly Journal of Forestry*, **65**, 228–33.

Plehanov, P. (1933). The determination of age in seals. *Sovetskii Sever*, **4**, 111–14. (In Russian.)

Plessis, du, S. S. (1972). Ecology of blesbok with special reference to productivity. *Wildlife Monographs*, No. 30.

Poirier, F. E. (1972). *Primate socialization*. Random House, New York.

Pollitzer, R. (1954). *Plague*. WHO, Geneva.

References

Polyakov, I. J. (1958). Methods of forecasting and warning on the appearance of pests and diseases in crops. In: IX International Conference on Quarantine and Plant Protection against Pests and Diseases.

Polyakov, I. J. (1959). Forecasting of rodent populations. Report of the International Conference of Harmful Mammals and their Control, pp. 39–47, Paris.

Pomeroy, L. R. (1970). The strategy of mineral cycling. *Annual Review of Ecology and Systematics*, **1**, 171–90.

Ponugaeva, A. G. (1960). *Fizjologiceskie isseldovanija instinktov u mlekopitajuscick*. Moskova, Lenningrad.

Porter, W. P. & Gates, D. M. (1969). Thermodynamic equilibria of animals with environment. *Ecological Monographs*, **39**, 245–70.

Poulet, A. R. (1972*a*). Ecologie des populations de *Taterillus pygargus* (Cuvier) (Rongeurs, Gerbillides) du Sahel Senegalais. PhD Thesis. Université de Paris.

Poulet, A. R. (1972*b*) Recherches écologiques sur une savane sahelienne du ferlo septentrional, Senegal: les mammifères. *Terre et Vie*, **26**, 440–72.

Prakash, I. (1971). Eco-toxicology and control of the Indian Desert gerbil, *Meriones hurrianae* (Jordan): VIII. Body weight, sex ratio and age structure in the population. *Journal of the Bombay Natural History Society*, **8**, 717–25.

Prakash, I. & Kametkar, L. R. (1969). Body weight, sex and age factors in a population of the northern palm squirrel, *Funambulus pennanti* Wroughton. *Journal of the Bombay Natural History Society*, **66**, 99–115.

Prakash, I., Kametkar, L. R. & Purohit, K. G. (1968). Home range and territoriality of the northern palm squirrel, *Funambulus pennanti* Wroughton. *Mammalia*, **32**, 603–11.

Prakash, I. & Rana, B. D. (1970). A study of field population of rodents in the Indian Desert. *Zeitschrift für Angewandte Zoologie*, **57**, 129–36.

Provo, M. M. (1962). The role of energy utilization habitat selection temperature and light in the regulation of a *Sigmodon* population. PhD Thesis. University of Georgia, Athens.

Pruitt, W. O., Jr (1966). Ecology of terrestrial mammals. In: *Environment of the Cape Thompson Region, Alaska* (ed. N. J. Wilmovsky & J. N. Wolfe), pp. 519–63. US Atomic Energy Commission.

Pruitt, W. O., Jr (1970). Some ecological aspects of snow. In: *Ecology of the subarctic regions*, pp. 83–99. UNESCO, Ecology and Conservation Series.

Prychodko, W. (1951). Zur Variabilität der Rötelmaus (*Clethrionomys glareolus*) in Bayern. *Zoologische Jahrbücher* (*Systematik*), **80**, 482–506.

Prychodko, W. (1958). Effect of aggregation of laboratory mice (*Mus musculus*) on food intake at different temperatures. *Ecology*, **39**, 499–503.

Pucek, Z. (1960). Sexual maturation and variability of the reproductive system in young shrews (*Sorex* L.) in the first calendar year of life. *Acta Theriologica*, **3**, 269–96.

Pucek, M. (1967). Changes in the weight of some internal organs of Micromammalia due to fixing. *Acta Theriologica*, **12**, 545–53.

Pucek, Z. (1969). Trap response and estimation of numbers of shrews in removal catches. *Acta Theriologica*, **14**, 403–26.

419

References

Puček, Z. (1970). Seasonal and age changes in shrews as an adaptive process. *Symposia of the Zoological Society of London*, **26**, 189–207. Academic Press, London.

Puček, A. & Olszewski, J. (1971). Results of extended removal catches of rodents. *Annales Zoologici Fennici*, **8**, 37–44.

Puček, Z., Ryszkowski, L. & Zejda, J. (1968). Estimation of average length of life in bank vole, *Clethrionomys glareolus* (Schreber, 1780). In: *Energy flow through small mammal populations* (ed. K. Petrusewicz & L. Ryszkowski), pp. 187–201. Polish Scientific Publishers, Warsaw.

Puček, Z. & Zejda, J. (1968). Technique for determining age in the red-backed vole, *Clethrionomys glareolus* (Schreber, 1780). *Small Mammals Newsletters*, **2**, 51–60.

Pulliainen, E. (1972). Nutrition of the arctic hare (*Lepus timidus*) in north-eastern Lapland. *Annales Zoologici Fennici*, **9**, 17–22.

Pulliam, H. R., Barrett, G. W. & Odum, E. P. (1969). Bioelimination of tracer ^{65}Zn in relation to metabolic rates in mice. In: *Symposium on radioecology* (ed. D. J. Nelson & F. C. Evans), pp. 725–30.

Quast, J. C. & Howard, W. E. (1953). Comparison of catch of two sizes of small-mammal live traps. *Journal of Mammalogy*, **34**, 514–15.

Quay, W. B. & Quay, J. F. (1956). The requirements and biology of the collared lemming, *Dictostonys torquatus* Pallas, 1778, in captivity. *Säugetierkündliche Mitteilungen*, **4**, 174–80.

Radda, A. (1969). Studies on the home range of *Apodemus flavicollis* (Melchior 1834). *Zoologické Listy*, **18**, 11–22.

Radvanyi, A. (1966). Destruction of radio-tagged seeds of white spruce by small mammals during summer months. *Forest Science*, **12**, 307–15.

Radvanyi, A. (1972). Protecting coniferous seeds from rodents. Proceedings of the Vth Vertebrate Pest Conference, pp. 29–35.

Radwan, M. A. (1963). Protecting forest trees and their seed from wild mammals. US Forest Service Research Paper PNW-6. Portland, Oregon.

Radwan, M. A. (1970). Destruction of conifer seed and methods of protection. Proceedings of the IVth Vertebrate Pest Conference, pp. 77–82.

Rahm, U. (1967). Les Murides des environs du Lac Kivu et des régions voisines (Afrique Centrale) et leur écologie. *Revue Suisse Zoologie*, **74**, 439–519.

Rahm, U. (1970). Note sur la reproduction des sciurides et murides dans la Forêt équatoriale au Congo. *Revue Suisse Zoologie*, **77**, 635–46.

Rall', Yu. M. (1938–9). Vvedenie v ékologiyu poludennykh peschanok (*Pallasiomys meridianus* Pall.). *Vestnik Microbiologii, Epidemiologii i Parazitologii*, **17**, 320–58.

Ramsey, P. R. & Briese, L. A. (1971). Effects of immigrants on the spatial structure of a small mammal community. *Acta Theriologica*, **16**, 191–202.

Randolph, J. C. (1973). The ecological energetics of a homeothermic predator, the short-tailed shrew. *Ecology*, **54**, 1166–87.

Ranson, R. M. (1934). The field vole (*Microtus*) as a laboratory animal. *Journal of Animal Ecology*, **3**, 70–6.

420

References

Ranson, R. M. (1941). Pre-natal and infant mortality in laboratory populations of voles (*Microtus agrestis*). *Proceedings of the Zoological Society of London*, III A, 45–57.

Rasweiler, J. J., IV (1973). The care and management of bats as laboratory animals. In: *The biology of bats* (ed. W. A. Wimsatt), vol. 3. Academic Press, New York.

Raun, G. C. (1966). A population of woodrats (*Neotoma micropus*) in southern Texas. *Bulletin of the Texas Memorial Museum*, **11**, 1–62.

Rautapää, J., Siltanen, H., Valta, A. L. & Mattinen, V. (1972). DDT, lindane and endrin in some agricultural soils in Finland. *Journal of the Scientific Agricultural Society of Finland*, **44**, 199–206.

Reid, V. H., Hansen, R. M. & Ward, A. L. (1966). Counting mounds and earth plugs to census mountain pocket gophers. *Journal of Wildlife Management*, **30**, 327–44.

Reig, O. A. (1970). Ecological notes on the fossorial octodont rodent *Spalacopus cyanus* (Molina). *Journal of Mammalogy*, **51**, 592–601.

Reimov, R. (1972). *Opȳt ėkologicheskogo i morfo-fiziologicheskogo analiza fauny mlekopitayushchikh Yuzhnogo Priaral'ya*, pp. 1–411. Izdatelstvo Karakalpakstan.

Rennison, B. D., Hammond, L. E. & Jones, G. L. (1968). A comparative trial of norbormide and zinc phosphide against *Rattus norvegicus* on farms. *Journal of Hygiene, Cambridge*, **66**, 147–58.

Report of the International Conference of Rodents and Rodenticides. *EPPO Publications*, Ser. A, **41**,

Reynolds, H. G. (1960). Life history notes on Merriam's kangaroo rats in southern Arizona. *Journal of Mammalogy*, **41**, 48–58.

Richens, V. B. (1967). The status and use of gophacide. *Proceedings of the IIIrd Vertebrate Pest Conference*, pp. 118–25.

Richmond, M. & Conaway, C. H. (1969). Induced ovulation and oestrus in *Microtus ochrogaster*. *Journal of Reproduction and Fertility*, Suppl. 6, 357–76.

Rieck, W. (1962). Analyse von Feldhasenstrecken nach dem Gewicht der Augenlinse. *Ricerche di Zoologia Applicata alla Caccia*, Suppl. (Transactions of the Vth Congress of the International Union of Game Biologists, Bolonia), **4**, 21–9.

Robson, D. S. & Regier, H. A. (1964). Sample size in Petersen mark–recapture experiments. *Transactions of the American Fisheries Society*, **93**, 215–26.

Rockstein, M. & Hrachovec, J. P. (1963). Biochemical criteria for senescence in mammalian structures. II. Age changes in the chemical composition of rat liver and muscle, water content, defatted dry weight, total lipids and various lipid compounds. *Gerontologia*, **7**, 30–43.

Roe, H. S. J. (1967). Seasonal formation of laminae in the ear plug of the fin whale. *Discovery Reports*, **35**, 1–30.

Roff, D. A. (1973a). On the accuracy of some mark–recapture estimators. *Oecologia*, **12**, 15–34.

Roff, D. A. (1973b). An examination of some statistical tests used in the analysis of mark–recapture data. *Oecologia*, **12**, 35–54.

421

References

Romejs, B. (1953). *Mikroskopičeskaja tehnika*, pp. 1–719. Izdatel'stvo Inno-strannoj Literatury, Moscow.

Rongstad, O. J. (1966). A cottontail rabbit lens-growth curve from southern Wisconsin. *Journal of Wildlife Management*, **30**, 114–21.

Roseberry, J. L. & Klimstra, W. D. (1970). Productivity of white-tailed deer on Crab Orchard National Wildlife Refuge. *Journal of Wildlife Management*, **34**, 23–8.

Rosenzweig, M. L. (1973). Habitat selection experiments with a pair of coexisting heteromyid rodent species. *Ecology*, **54**, 111–17.

Rosenzweig, M. L. & Sterner, P. W. (1970). Population ecology of desert rodent communities: body size and seed husking as a basis for heteromyid coexistence. *Ecology*, **51**, 17–224.

Rosevear, D. S. (1969). *The rodents of West Africa*. British Museum Publication No. 677. British Museum, London.

Rossolimo, O. L. (1958). Opredelenie vozrasta nutrii po kraňiologičeskim priznakam. *Nauč. dokl. Vysšej školy, Biol. nauki.*, **4**.

Rotshil'd, E. V. (1957). Gruppirovki rastitel'nosti, kak pokazateli otnositel'nogo vozrasta poselenii bol'skikh peschanok. In: *Nauchnaya konferentsiya po prirodnoĭ ochagovosti i épidemiologii osobo opasnȳkh infektsionnykh zabolevaniĭ*, pp. 346–50. Saratov.

Rotshil'd, E. V. (1967). Matodika srednemas-shtabnogo kartirovaniya poselelenii bol'shikh peschanok pri izuchenii prirodnoĭ ochagovosti chumȳ. *Zoologicheskii Źhurnal*, **45**, 151–63.

Rotshil'd, E. V. (1968). *Azotolyubivaya rastitel'nost' pustȳni i zhivotnȳe*, pp. 1–202. Izdatel'stvo Moskovskogo gosudarstvennogo Universiteta, Moscow.

Rowe, F. P., Greaves, J. H., Redfern, R. & Martin, A. D. (1970). Rodents and current research. Proceedings of the IVth Vertebrate Pest Conference (ed. R. H. Dana), pp. 126–8. University of California, Davis.

Rowe, F. P. & Redfern, R. (1965). Toxicity tests on suspected warfarin-resistant house mice (*Mus musculus* L.). *Journal of Hygiene, Cambridge*, **63**, 417–25.

Rowe, F. P. & Taylor, K. D. (1970). Rodent biology. In: *World food programme storage manual*, Part I, pp. 136–67. FAO, Rome.

Rudenchik, Yu. V. (1959). K ekologii i rasprostraneniyu krasnokhvostoĭ peschanki v zapadnȳkh Kȳzȳlkumakh. *Trudȳ sredne-Aziatskogo Nauchno-isseledovatel'skogo Protivochumnogo Instituta, Alma-Ata*, **5**, 75–80.

Rudd, R. L. (1965). Weight and growth in Malaysian rain forest mammals. *Journal of Mammalogy*, **46**, 588–94.

Rupp, R. S. (1966). Generalized equation for the ratio method of estimating population abundance. *Journal of Wildlife Management*, **30**, 523–6.

Ryan, G. E. & Jones, E. L. (1972). *A report on the mouse plague in the Murrumbidgee and Coleambally irrigation areas, 1970*. NSW Department of Agriculture, New South Wales.

Rybář, P. (1969). Ossification of bones as age criterion in bats (Chiroptera). *Práce a Studie, Přír.-Pardubice*, **1**, 115–36. (In Czech. with English summary.)

References

Rybář, P. (1970). The methods of age determination in mammals. *Práce a Studie, Přir-Pardubice*, **2**, 129–55. (In Czech. with English summary.)

Rybář, P. (1971). On the problems of the practical use of ossification of bones as an age criterion in bats (Microchiroptera). *Práce a Studie, Přir.-Pardubice*, **3**, 97–121. (In Czech. with English summary.)

Ryszkowski, L. (1966). The space organization of nutria (*Myocaster coypus*) populations. *Symposia of the Zoological Society of London*, **18**, 259–65.

Ryszkowski, L. (1967). Short cut methods for the estimation of mean length of life in small mammal populations. In: *Secondary productivity of terrestrial ecosystems* (ed. K. Petrusewicz), pp. 283–95. Polish Academy of Sciences, Warsaw.

Ryszkowski, L. (1969a). Estimates of consumption of rodent populations in different pine forest ecosystems. In: *Energy flow through small mammal populations* (ed. K. Petrusewicz & L. Ryszkowski), pp. 281–9. Polish Scientific Publishers, Warsaw.

Ryszkowski, L. (1969b). Operation of the Standard–Minimum method. In: *Energy flow through small mammal populations* (ed. K. Petrusewicz & L. Ryszkowski), pp. 13–34. Polish Scientific Publishers, Warsaw.

Ryszkowski, L. (1971a). Estimation of small rodent density with the aid of coloured bait. *Annales Zoologici Fennici*, **8**, 8–13.

Ryszkowski, L. (1971b). Reproduction of bank voles (*Clethrionomys glareolus*) and survival of juveniles in different pine forest ecosystems. *Annales Zoologici Fennici*, **8**, 85–91.

Ryszkowski, L., Andrzejewski, R. & Petrusewicz, K. (1966). Comparison of estimates of numbers obtained by the methods of release of marked individuals and complete removal of rodents. *Acta Theriologica*, **11**, 329–41.

Ryszkowski, L., Gentry, J. B. & Smith, M. H. (1971). Proposals to test the density estimation techniques for small mammals living in temperate forests. *Small Mammal Newsletter*, **5**, 40–52.

Ryszkowski, L., Goszczynski, J. & Truszkowski, J. (1973). Trophic relationships of the common vole in cultivated fields. *Acta Theriologica*, **18**, 125–65.

Ryszkowski, L. & Truszkowski, J. (1970). Survival of unweaned and juvenile bank voles under field conditions. *Acta Theriologica*, **15**, 223–32.

Ryszkowski, L., Wagner, C. K., Goszczynski, J. & Truszkowski, J. (1971). Operation of predators in a forest and cultivated fields. *Annales Zoologici Fennici*, **8**, 160–8.

Sadleir, R. M. F. S. (1965). The relationship between agonistic behaviour and population changes in the deermouse, *Peromyscus maniculatus* (Wagner). *Journal of Animal Ecology*, **34**, 331–52.

Sadleir, R. M. F. S. (1970). Population dynamics and breeding of the deermouse (*Peromyscus maniculatus*) on Burnaby Mountain, British Columbia. *Syesis*, **3**, 67–74.

Saint Girons, M. Ch. (1966). Le rythme circadien de l'activité chez les mammifères holarctiques. *Mémoires du Muséum national d'Histoire naturelle (Zoologique)*, **40**, 101–87.

Sater, J. E. (ed.) (1969). *The arctic basin*. Arctic Institute of North America, Washington, DC.

423

References

Sawby, S. W. (1973). An evaluation of radioisotopic methods of measuring free-living metabolism. In: *Ecological energetics of homeotherms* (ed. J. A. Gessaman), Chapter 8. Utah State University Press Monographs, Series 20, Logan.

Sawicka-Kapusta, K. (1970). Changes in the gross body composition and caloric value of the common vole during their postnatal development. *Acta Theriologica*, **15**, 67–79.

Schamurin, V. F., Polozova, T. G. & Khodachek, A. E. (1972). Plant biomass of main plant communities at the Tareya station (Taimyr). In: *Tundra Biome.* Proceedings of the IV International Meeting on the Biological Productivity of Tundra (ed. F. E. Wielgolaski & T. Rosswall), pp. 163–81. Leningrad.

Schendel, R. R. (1940). Life history notes on *Sigmodon hispidus texianus* with special emphasis on populations and nesting habits. MS Thesis. Oklahoma State University of Agriculture and Applied Science, Stillwater.

Schindler, U. (1955). Eine neue wirksame Methode zur Bekämpfung der Erdmaus (*Microtus agrestis* L.). *Allgemeine Forstzeitschrift*, **10**, 384–7.

Schindler, U. (1956). Erdmausbekämpfung mit Insektiziden. *Zeitschrift für Angewandte Zoologie*, **43**, 407–23.

Schindler, U. (1959). Zur Erdmaus-Prognose. *Anzeiger für Schädlingskunde*, **32**, 101–6.

Schindler, U. (1970). Erfolgskontrolle praxisüblicher Bekämpfungen der Erdmaus (*Microtus agrestis* L.) und der Rötelmaus (*Clethrionomys glareolus* Schreb) in forstlichen verjungungen mit Hilfe der Lebendfang-Methode. *Zeitschrift für Pflanzenkraukh und Pflanzenschutz*, **77**, 76–82.

Schindler, U. (1972). Massenwechsel der Erdmaus, *Microtus agrestis* L., in Süd-Niedersachsen von 1952–1971. *Zeitschrift für Angewandte Zoologie*, **59**, 189–204.

Schmidt-Nielsen, K. (1972*a*). Locomotion: Energy cost of swimming, flying and running. *Science*, **177**, 222–8.

Schmidt-Nielsen, K. (1972*b*). *How animals work.* Cambridge University Press, London.

Schnabel, Z. E. (1938). The estimation of the total fish population of a lake. *American Mathematical Monthly*, **43**, 348–52.

Schultz, A. M. (1964). The nutrient-recovery hypothesis for Arctic microtine cycles. II. Ecosystem variables in relation to Arctic microtine cycles. In: *Grazing in terrestrial and marine environments* (ed. D. J. Crisp). Blackwell Scientific, Oxford.

Schultz, A. M. (1969). A study of an ecosystem: the arctic tundra. In: *The ecosystem concept in natural resource management* (ed. G. VanDyne), pp. 77–93. Academic Press, New York.

Sealander, J. A., Jr (1952). The relationship of nest protection and huddling to survival of *Peromyscus* at low temperature. *Ecology*, **33**, 63–71.

Sealander, J. A. (1972). Circum-annual changes in age, pelage characteristics and adipose tissue in the northern red-backed vole in interior Alaska. *Acta Theriologica*, **17**, 1–24.

424

Seber, G. A. F. (1962). The multi-sample single recapture census. *Biometrika*, **49**, 339–50.

Sella, L. D. (1973). Trace elements in the cotton rat, *Sigmodon hispidus*, on the Piedmont of Georgia. MS Thesis. University of Georgia, Athens.

Semenov-Tyan-Shanskii, O. I. (1970). Population cycles in the forest voles (*Clethrionomys*). *Byulleten' Moskovskogo Obshchestva Ispytatelei Prirody Otdel Biologicheskii*, **75**, 11–26.

Semeonoff, R. & Robertson, F. W. (1968). A biochemical and ecological study of plasma esterase polymorphism in natural populations of the field vole,· *Microtus agrestis* L. *Biochemical Genetics*, **1**, 205–27.

Semizorova, I. & Popper, J. (1971). Vorbericht über die Möglichkeit, das Röntgengerät zur Bestimmung des Alters der Rötelmaus (*Clethrionomys glareolus* Schreb.) anzuwenden. *Lynx*, **12**, 75–84. (In Czech. with German summary.)

Sergeant, D. (1967). Age determination of land mammals from annuli. *Zeitschrift für Säugetierkunde*, **32**, 297–300.

Shakirzyanova, M. S. (1954). Norovye moskity Kazakhstana i ikh vozmozhnaya rol' v peredache vistseral'nogo leĭshmanioza v Kzyl-Ordinskoĭ oblasti. In: *Prirodnaya ochagovost' zaraznykh boleznei v Kazakhstrane*, pp. 105–12. Alma-Ata.

Sharp, H. F. (1967). Food ecology of the rice rat, *Oryzomys palustris* (Harlan), in a Georgia salt marsh. *Journal of Mammalogy*, **48**, 557–63.

Shaw, T. H., Hsia, W. P. & Lee, T. C. (1959). Age and growth in the red-backed vole, *Clethrionomys rutilus*, from Dailing, Lesser Khingan Mountains. *Acta Zoologica Sinica*, **11**, 57–66. (In Chinese with English summary.)

Shcheglova, A. I. (1962). Bol'shaya peschanka kak predstavitel' zhiznennoi formy pustyni. *Voprosy Ékologii*, **6**, 174–6.

Sheng, H. P. & Huggins, R. A. (1971). Growth of the beagle: changes in chemical composition. *Growth*, **35**, 369–76.

Sheppe, W. (1965). Island populations and gene flow in the deer mouse,. *Peromyscus leucopus*. *Evolution*, **19**, 480–95.

Sheppe, W. (1966a). Habitat restriction by competitive exclusion in the mice *Peromyscus* and *Mus*. *Canadian Field Naturalist*, **81**, 81–98.

Sheppe, W. (1966b). Determinants of home range in the deer mouse, *Peromyscus leucopus*. *Proceedings of the California Academy of Science*, **34**, 377–418.

Sheppe, W. (1972). The annual cycle of small mammal populations on a Zambian floodplain. *Journal of Mammalogy*, **53**, 445–60.

Shiranovich, P. I. (1968). Causes for the fall in the number of little souslik (*Citellus pygmaeus*) in the semi-desert of North-Western Caspian foreland. *Zoologicheskii Zhurnal*, **47**, 1539–48.

Shmal'gauzen, J. J. (1946). *Problemy darwinizma*, pp. 1–528. Sovietskaja Nauka, Moscow.

Shubin, I. G. & Bekenov, A. (1971). Ecological characteristics of the tamanisk gerbil in the Zaisan basin. *Ekologiya*, **2**, 97–8.

Shuyler, H. R. (1972). Rodents in the tropics: their effects and control. *PNS*, **18**, 445–51.

References

Shvarts, S. S. (1962). Izuchenie korrelystsii morfologicheskikh osobennosteĭ grÿzunov so skorost'yu ikh rosta v svyazi s nekotorÿmi voprosami vnutrividovoĭ sistematiki. Voprosÿ vnutrividovoĭ izmenchivosti mlekopitayushchikh. *Trudÿ Instituta Biologicheskikh Sverdlovsk*, **29**, 5–14.

Shvarts, S. S. (1963). *Ways of adaptation of terrestrial vertebrates to the conditions of the subarctic*, vol. 1. *Mammals*. Institute of Biology, Academy of Sciences of the USSR, Ural Branch. (Trans. E. Issakoff & W. Fuller, Boreal Institute, University of Alberta.)

Shvarts, S. S. (1968). Printsip optimal'nogo fenotipa (k teorii stabiliziruyushchego otbora). *Zh. obshch. Biol.*, **34**, 12–24.

Shvarts, S. S. (1969). *Évolyutsionnoya ékologiya zhivotnÿkh*. Ural Branch of Academy of Science, Sverdlovsk.

Shvarts, S. S. (ed.) (1971). *Mammals of the Yamal and arctic Urals*. Academy of Sciences of the USSR, Ural Science Central Department of the Institute of Ecology. (In Russian.)

Shvarts, S. S., Bolshakov, V. N., Olenov, V. G. & Pjastolova, O. A. (1969). Population dynamics of rodents from northern and mountainous geographical zones. In: *Energy flow through small mammals populations* (ed. K. Petrusewicz & L. Ryszkowski), pp. 205–20. Polish Scientific Publishers, Warsaw.

Shvarts, S. S., Pokrowski, A. V., Istchenko, V. G., Olenjey, V. G., Ovtschinnikova, N. A. & Pjastolova, O. A. (1964). Biological peculiarities of seasonal generation of rodents with special reference to the problem of senescence in mammals. *Acta Theriologica*, **8**, 11–43.

Shvarts, S. S., Smirnov, V. S. & Dobrinskiĭ, L. N. (1968). *Metod morfofiziologicheskikh indikatorov v ékologii nazemnÿkh pozvonochnÿkh*. Sverdlovsk.

Sidorowicz, J. (1960a). Influence of the weather on capture of micromammalia. I. Rodents (Rodentia). *Acta Theriologica*, **4**, 139–58.

Sidorowicz, J. (1960b). Problems of the morphology and zoogeography of representatives of the genus *Lemmus* Link 1795 from the Palearctic. *Acta Theriologica*, **4**, 53–77.

Sidorowicz, J. (1964). Comparison of the morphology of representatives of the genus *Lemmus* Link from Alaska and the Palearctic. *Acta Theriologica*, **8**, 217–25.

Simkiss, K. (1967). *Calcium in reproductive physiology. A comparative study of vertebrates*. Van Nostrand Reinhold, New York.

Simpson, G. G., Roe, A. & Lewontin, R. C. (1960). *Quantitative zoology*, pp. 1–440. Harcourt Brace & Co., New York-Burlingame.

Singh, S. (1967). The rodent problem in India. *Farmer and Parliament* (*New Delhi*), **2**, 7–10.

Skoczen, A. (1966). Age determination, age structure and sex ratio in mole, *Talpa europaea* Linnaeus 1758, populations. *Acta Theriologica*, **11**, 523–36.

Skvorcova, V. K. & Utehin, V. D. V. (1969). Vliyanie royushchei deyatelnosti slepusha (*Spalax microphthalmus*) na vidovoy sostav i produktivnost travoyanistykh fitocenozov lesostepi. *Biogeografia*, **3**, 7–10. (In Russian.)

Slonim, A. D. (1961). *Basis of general ecological physiology in mammals*. Moscow. (In Russian.)

References

Slonim, A. D. & Shcheglova, A. I. (1963). Fiziologicheskie osobennosti mlekopitayushchikh pustȳn' Srednei Azii. In: *Prirodnȳe usloviya, zhivotnovod stovo ikormovaya baza pustȳn'*, pp. 139–51. Ashkhabad.

Sludskiĭ, A. A. (ed.) (1969). *Mlekopitayushchie Kazakhstana*, pp. 83–158. Alma-Ata.

Smirin, V. M. (1963). Itogi kartirovaniya poselenii bol'shikh peschonok v severnȳkh Kȳzȳlkumakh i sovremennoĭ del'te Sȳrdari. In: *Materialȳ nauchnoĭ konferentsii po prirodnoĭ ochagovosti chumȳ*, pp. 220–2. Alma-Ata.

Smirnov, V. S. & Shvarts, S. S. (1957). Sezonnȳe izmeneniya otnosti-tel'nogo vesa nadpochechnikov u mlekopitayushchikh v prirodnykh usloviyakh. *DAN SSSR*, **115**, 1193–6.

Smirnov, V. S. & Shvarts, S. S. (1959). Sravnitel'naya ekologo-fiziologicheskaya kharateristika ondatry v lesostepnykh i pripolyornykh reĭonakh. Voprosy akklimatizatsiĭ mlekopitayushchikh na Urale. *Trudȳ Instituta Biologicheskikh Sverdlovsk*, **18**, 91–137.

Smirnov, V. S. & Tokmakova, S. G. (1972). Influence of consumers on natural phytocenoses' production variation. In: Tundra Biome. Proceedings of the IV International Meeting on the Biological Productivity of Tundra (ed. F. E. Wielgolaski & T. Rosswall), pp. 122–7. Leningrad.

Smirnov, V. S. & Tokmakova, S. G. (1971). Preliminary data in the influence of different numbers of voles upon the forest tundra vegetation. *Annales Zoologici Fennici*, **8**, 154–6.

Smirnov, V. S., Pavlenko, T. A. & Pokrovskii, A. V. (1971). Method of analyzing the age structure of a population of the small five-toed jerboa. *Ekologiya*, **2**, 88–90.

Smith, C. F. & Aldous, S. E. (1947). The influence of mammals and birds in retarding artificial and natural reseeding of coniferous forests in the United States. *Journal of Forestry*, **45**, 361–9.

Smith, F. E. (1954). Quantitative aspects of population growth. In: *Dynamics of growth processes* (ed. E. Boell). Princeton University Press, Princeton.

Smith, F. L. (1968). Rat damage to coconuts in the Gilbert and Ellice Islands. In: Rodents as Factors in Disease and Economic loss. *Asia-Pacific Interchange Proceedings, Honolulu*, **513**, 55.

Smith, G. C., Kaufman, D. W., Jones, R. M., Gentry, J. B. & Smith, M. H. (1971). The relative effectiveness of two types of snap traps. *Acta Theriologica*, **16**, 240–4.

Smith, H. D., Jorgensen, C. D. & Tolley, H. D. (1972). Estimation of small mammals using recapture methods: partitioning of estimator variables. *Acta Theriologica*, **17**, 57–66.

Smith, M. H. (1968*a*). A comparison of different methods of capturing and estimating numbers of mice. *Journal of Mammalogy*, **49**, 455–62.

Smith, M. H. (1968*b*). Dispersal of the old-field mouse, *Peromyscus polionotus*. *Bulletin of the Georgia Academy of Science*, **26**, 45–51.

Smith, M. H. (1971). Food as a limiting factor in the population ecology of *Peromyscus polionotus* (Wagner). *Annales Zoologici Fennici*, **8**, 109–12.

Smith, M. H. (1974). Seasonality in mammals. In: *Phenology and seasonality modeling* (ed. H. Leith), pp. 149–62. Springer-Verlag, New York.

427

References

Smith, M. H. & Blessing, R. (1969). Trap response and food availability. *Journal of Mammalogy*, **50**, 368–9.

Smith, M. H., Blessing, R., Chelton, J. G., Gentry, J. B., Golley, F. B. & McGinnis, J. T. (1971). Determining density for small mammal populations using a grid and assessment lines. *Acta Theriologica*, **16**, 105–25.

Smith, M. H., Carmon, J. L. & Gentry, J. B. (1972). Pelage color polymorphism in *Peromyscus polionotus*. *Journal of Mammalogy*, **53**, 824–33.

Smith, M. H., Gentry, J. B. & Golley, F. B. (1969–70). A preliminary report on the examination of small mammal census methods. In: *Energy flow through small mammal populations* (ed. K. Petrusewicz & L. Ryskowski), pp. 25–9. Polish Scientific Publishers, Warsaw.

Smith, M. H., Gentry, J. B. & Pinder, J. (1974). Annual fluctuations in small mammal populations in an eastern hardwood forest. *Journal of Mammalogy*, **55**, 231–4.

Smith, M. H. & McGinnis, J. T. (1968). Relationships of latitude, altitude and body size to litter size, and annual production of offspring in *Peromyscus*. *Research in Population Ecology*, **10**, 115–26.

Smith, R. W. (1967). The control of rats in coconuts using rat blocks. *Oleagineaux*, **22**, 159–60.

Smith, P. (1883). Diseases of crystalline lens and capsule. I. On the growth of the crystalline lens. *Transactions of the Ophthalmology Society of the United Kingdom*, **3**, 79–99.

Smyth, M. (1966). Winter breeding in woodland mice, *Apodemus sylvaticus*, and voles, *Clethrionomys glareolus* and *Microtus agrestis*, near Oxford. *Journal of Animal Ecology*, **35**, 471–85.

Smyth, M. (1968). The effects of the removal of individuals from a population of bank voles *Clethrionomys glareolus*. *Journal of Animal Ecology*, **37**, 167–83.

Smythe, N. (1970). Relationships between fruiting seasons and seed dispersal methods in a neotropical forest. *American Naturalist*, **104**, 25–35.

Snigirevskaya, E. M. (1955). Dannye po pitaniyu i kolebaniyam chislennosti zheltogoloi myshy v Zhigulyakh. *Zoologicheskii Zhurnal*, **34**, 432–40. (In Russian.)

Snyder, D. P. (1956). Survival rates, longevity, and population fluctuations in the white-footed mouse, *Peromyscus leucopus*, in southeastern Michigan. *Miscellaneous Publications of the Museum of Zoology, University of Michigan*, **95**, 1–33.

Solomonov, N. G. (1970). Direction of average brood size variability among certain mammals of Yakutia. *Ekologiya*, **5**, 89–92.

Southern, H. N. & Laurie, E. M. O. (1946). The house mouse (*Mus musculus*) in corn ricks. *Journal of Animal Ecology*, **15**, 134–49.

Southern, H. N. & Lowe, V. P. W. (1968). The pattern of distribution of prey and predation in tawny owl territories. *Journal of Animal Ecology*, **37**, 75–98.

Southwick, C. H. (1955a). The population dynamics of confined house mice supplied with unlimited food. *Ecology*, **36**, 212–25.

References

Southwick, C. H. (1955*b*). Regulatory mechanisms of house mouse populations: social behavior affecting litter survival. *Ecology*, **36**, 627–34.

Sowls, L. K. (1948). The Franklin ground squirrel, *Citellus franklinii*, (Sabine), and its relationship to nesting ducks. *Journal of Mammalogy*, **29**, 113–37.

Speller, S. W. (1972). Biology of *Dicrostonyx groenlandicus* on Truelove Lowland, Devon Island, N.W.T. In: *Devon Island IBP Project, high arctic ecosystem* (ed. L. Bliss), pp. 257–71. University of Alberta.

Spencer, P. R. (1955). The effects of rodents on reforestation. *Proceedings of the Society of American Foresters* 1955, pp. 125–8.

Spitz, F. (1965). Controle par piégeage des essais de destruction de campagnols en plein champ. *Phytiatrie-Phytopharmacie*, **14**, 3–8.

Spitz, F. (1967). The causes and dynamics of population explosion of *Microtus arvalis*: a survey of current French research. *EPPO Publications*, Ser. A, **41**, 97–105.

Spitz, F. (1968). Interactions entre la végétation épigée d'une Luzernière et les populations enclose ou non enclose de *Microtus arvalis* (Pallas). *Terre et Vie*, **3**, 274–306.

Spitz, F. (1970). Demographie de *Microtus arvalis* Pallas dans l'ouest de la France et techniques de surveillance. *EPPO Publictions*, Ser. A, **58**, 61–4.

Spitz, F. (1972). Demographie du campagnol des champs, *Microtus arvalis* en Vendee. PhD Thesis. University of Paris.

Stark, N. (1973). *Nutrient cycling in a Jeffrey Pine forest ecosystem.* Institute for Microbiology, University of Montana, Missoula.

Startin, L. (1969). Body composition of the kangaroo rat (*Dipodomys merriami*). In: *Physiological systems in semi-arid environments* (ed. C. C. Hoff & M. L. Riedesel), pp. 35–44. University of New Mexico Press.

Steel, R. G. D. & Torrie, J. H. (1960). *Principles and procedures of statistics.* McGraw-Hill Co., New York.

Stein, G. H. W. (1961). Beziehungen zwischen Bestandsdichte und Vermehrung bei der Waldspitzmaus, *Sorex araneus*, und weiteren Rotzahnspitzmäusen. *Zeitschrift für Säugetierkunde*, **26**, 1–64.

Stein, G. H. W. & Reichstein, H. (1957). Über ein neues Verfahren zur Bestimmung der Bestandsdichte bei Feldmäusen, *Microtus arvalis* Pallas. *Nachrichtenblatt für den Deutschen Pflanzenschutzdienst*, **11**, 149–54.

Stenmark, A. (1963). Guventering au sorkskadoruas ebonomiska betydelse for tradgardsodlingen under 1962. *Vaxtskyddsnotiser*, **27**, 24–9. (In Swedish.)

Stickel, L. F. (1948). The trap line as a measure of small mammal populations. *Journal of Wildlife Management*, **12**, 153–61.

Stickel, L. F. (1954). A comparison of certain methods of measuring ranges of small mammals. *Journal of Mammalogy*, **35**, 1–15.

Stoddart, D. M. (1967). A note on the food of the Norway lemming. *Proceedings of the Zoological Society of London*, **151**, 211–13.

Stoddart, D. M. (1970). Individual range, dispersion and dispersal in a population of water voles (*Arvicola terrestr s* (L.)). *Journal of Animal Ecology*, **39**, 403–25.

Stoddart, D. M. (1971). Breeding and survival in a population of water voles. *Journal of Animal Ecology*, **40**, 487–94.

References

Stones, R. C. & Hayward, C. L. (1968). Natural history of the desert woodrat, *Neotoma lepida*. *American Midland Naturalist*, **80**, 458–76.

Storer, T. I. (ed.) (1962). Pacific island rat ecology. *Bulletin of the Bishop Museum*, **225**, 1–274.

Straka, F. (1970). Rolyata na obyknovenata polevka pri bdeshchoto razvitie na trevnofurazhnoto proizvodstvo. *Selskostopanska Nauka*, **9**, 31–5. (In Bulgarian.)

Strecker, R. L. (1954). Regulatory mechanisms in house-mouse populations: the effect of limited food supply on an unconfined population. *Ecology*, **35**, 249–53.

Summerlin, C. T. & Wolfe, J. L. (1973). Social influences on trap response of the cotton rat, *Sigmodon hispidus*. *Ecology*, **54**, 1156–9.

Sviridenko, P. A. (1940). Pitanie myshevidnykh gryzunov i znachenie ikh v problemievozobnovleniya lesa. *Zoologicheskii Zhurnal*, **19**, 680–703. (In Russian.)

Sviridenko, P. A. (1957). *Zapasanie korma zhivotnymi.* Izvestiya Akademii USSR, Kiev. (In Russian.)

Syuzyumova, L. M. (1973). Vnutrividovӳe osobennosti reaktsii tkanevoï nesovmestimosti u polevok. In: *Eksperimental'nӳe issledovaniya problemӯ vida*, pp. 41–73. Sverdlovsk.

Szabik, E. (1973). Age estimation of roe-deer from different hunting grounds of south-eastern Poland. *Acta Theriologica*, **18**, 223–36.

Taber, R. D. (1969). Criteria of sex and age. In: *Wildlife management techniques* (ed. R. H. Giles, Jr), pp. 359–401. The Wildlife Society, Washington DC.

Tahon, J. (1969). Nonrelated criteria between food consumption by small mammals and waste caused to vegetation. In: *Energy flow through small mammal populations* (ed. K. Petrusewicz & L. Ryszkowski), pp. 159–65. Polish Scientific Publishers, Warsaw.

Tamarin, R. H. & Krebs, C. J. (1969). *Microtus* population biology. II. Genetic changes at the transferrin locus in fluctuating populations of two vole species. *Evolution*, **23**, 183–211.

Tamarin, R. H. & Malecha, S. R. (1971). The population biology of Hawaiian rodents: demographic parameters. *Ecology*, **52**, 383–94.

Tamarin, R. H. & Malecha, S. R. (1972). Reproductive parameters in *Rattus rattus* and *Rattus exulans* of Hawaii, 1968 to 1970. *Journal of Mammalogy*, **53**, 513–28.

Tanaka, R. (1956). On differential response to live traps of marked and unmarked small mammals. *Annotationes Zoologicae Japonenses*, **29**, 44–51.

Tanaka, R. (1963*a*). Examination of the routine census equation by considering multiple collisions with a single-catch trap in small mammals. *Journal of Japanese Ecology*, **13**, 16–21.

Tanaka, R. (1963*b*). On the problem of trap-response types of small mammal populations. *Research in Population Ecology*, **5**, 139–46.

Tanaka, R. (1966). A possible discrepancy between the exposed and the whole population depending on range-size and trap-spacing in vole populations. *Research in Population Ecology*, **8**, 93–101.

Tanaka, R. (1968). Analysis of molar-wear amount for age determination in cage-reared stocks of the brown rat. *Japanese Journal of Zoology*, **15**, 377–86.

Tanaka, R. (1970). A field study of the effect of prebaiting on censusing by the capture–recapture method in a vole population. *Research in Population Ecology*, **12**, 111–25.

Tanaka, R. (1972). Investigation into the edge effect by use of capture–recapture data in a vole population. *Research in Population Ecology*, **13**, 127–51.

Tanaka, R. & Kanamori, M. (1967). New regression formula to estimate the whole population for recapture–addicted small mammals. *Research in Population Ecology*, **9**, 83–94.

Tanaka, R. & Kanamori, M. (1969). Inquiry into effects of prebaiting on removal census in a vole population. *Research in Population Ecology*, **11**, 1–13.

Tanton, M. T. (1965*a*). Acorn destruction potential of small mammals and birds in British woodlands. *Quarterly Journal of Forestry*, **59**, 230–4.

Tanton, M. T. (1965*b*). Problems of live-trapping and population estimation for the wood mouse, *Apodemus sylvaticus* (L.). *Journal of Animal Ecology*, **34**, 1–22.

Tanton, M. T. (1969). The estimation and biology of populations of the bank vole (*Clethrionomys glareolus* (Schr.)) and wood mouse (*Apodemus sylvaticus* (L.)). *Journal of Animal Ecology*, **38**, 511–29.

Tarkowski, A. K. (1957). Studies on reproduction and prenatal mortality of the common shrew (*Sorex araneus* L.). II. Reproduction under natural conditions. *Annales Universitatis Mariae Curie – Sklodowska*, Sectio C, **10**, 117–244.

Tast, J. (1966). The root vole, *Microtus oeconomus* (Pallas), as an inhabitant of seasonally flooded land. *Annales Zoologici Fennici*, **3**, 127–71.

Tast, J. (1968). Influence of the root vole, *Microtus oeconomus* (Pallas), upon the habitat selection of the field vole, *Microtus agrestis* (L.) in Northern Finland. *Annales Academiae Scientarum Fennicae*, A, IV Biologica, **136**, 1–23.

Tast, J. & Kalela, O. (1971). Comparisons between rodent cycles and plant production in Finnish Lapland. *Annales Academiae Scientiarum, Fennicae*, IV Biologica, **186**, 1–9.

Taylor, C. R., Caldwell, S. L. & Rowntree, V. J. (1972). Running up and down hills: some consequences of size. *Science*, **178**, 1096–7.

Taylor, C. R., Schmidt-Nielsen, K. & Raab, J. L. (1970). Scaling of energetic cost of running to body size in mammals. *American Journal of Physiology*, **219**, 1104–7.

Taylor, D. (1970). Growth, decline, and equilibrium in a beaver population at Sagehen Creek, California. PhD Thesis. University of California, Berkeley.

Taylor, J. C. (1970). The role of arboreal rodents on their habitat and man. *EPPO Publications*, Ser. A, **58**, 217–20.

Taylor, J. C., Lloyd, H. G. & Shillito, J. F. (1968). Experiments with warfarin for grey squirrel control. *Annals of Applied Biology*, **61**, 312–21.

References

Taylor, J. M. & Horner, B. E. (1971). Reproduction in the Australian tree rat *Conilurus penicillatus* (Rodentia: Muridae). *CSIRO (Commonwealth Scientific and Industrial Research Organization) Wildlife Research*, **16,** 1–9.

Taylor, K. D. (1968). An outbreak of rats in agricultural areas of Kenya in 1962. *East-African Agricultural and Forestry Journal*, **34,** 66–77.

Taylor, K. D. (1971). Assessment of losses due to rodents. In: *Crop loss assessment methods* (ed. L. Chiarappa). FAO manual on the evaluation and prevention of losses by pests, diseases and weeds.

Taylor, K. D. (1972). The rodent problem: *Outlook on Agriculture*, **7,** 60–7.

Taylor, K. D., Shorten, M., Lloyd, H. G. & Courtier, F. A. (1971). Movements of the grey squirrel as revealed by trapping. *Journal of Applied Ecology*, **8,** 123–46.

Telle, H. J. (1962). Rattenfrei Städte und Kreise in Niedersachsen. *Schädlingsbekämpfer*, **14,** 158–9.

Terborgh, J. (1973). Faunal equilibria and the preservation of tropical forest faunas. In: *Tropical ecological systems* (ed. F. Golley & E. Medina), pp. 369–80. Springer-Verlag, New York.

Terman, C. R. (1968). Population dynamics. In: *Biology of* Peromyscus (*Rodentia*) (ed. J. A. King). Special Publication No. 2, American Society of Mammalogists.

Tertil, R. (1972). The effect of behavioral thermoregulation on the daily metabolism of *Apodemus agrarius* (Pall.). *Acta Theriologica*, **17,** 295–313.

Tesh, R. & Arata, A. (1967). Bats as laboratory animals. *Laboratory Science*, **4,** 106–12.

Teshima, H. (1970). Rodent control in the Hawaiian sugar industry. Proceedings of the IVth Vertebrate Pest Conference, pp. 38–40.

Tevis, L. (1956). Effect of a slash burn on forest mice. *Journal of Wildlife Management*, **20,** 405–9.

Thiers, R. E. (1957). Contamination in trace element analysis and its control. *Methods of Biochemical Analysis*, **5,** 273–335.

Thomas, J. R., Casper, H. R. & Bever, W. (1964). Effects of fertilizers on the growth of grass and its use by deer in the Black Hills of South Dakota. *Agronomy Journal*, **56,** 223–6.

Thompson, D. Q. (1955*a*). The 1953 lemming emigration at Point Barrow, Alaska. *Arctic*, **8,** 37–45.

Thompson, D. Q. (1955*b*). The ecology and dynamics of the brown lemming (*Lemmus trimucronatus*) at Point Barrow, Alaska. PhD Thesis. University of Missouri.

Thompson, D. Q. (1955*c*). The role of food and cover in population fluctuations of the brown lemming at Point Barrow, Alaska. *Transactions of the North American Wildlife Conference*, **20,** 166–74.

Tietjen, P., Halvorson, C. K., Hegdal, P. L. & Johnson, M. (1967). 2,4-D herbicide, vegetation and pocket gopher relationships, Black Mesa, Colorado. *Ecology*, **48,** 634–43.

Tikhomirov, B. A. (1959). *Relationship of the animal world and the plant cover of the tundra*. Botanical Institute, Academy of Sciences, USSR. (Trans. E. Issakoff & T. W. Barry, Boreal Institute, University of Alberta.)

432

References

Tikhomirov, E. A. & Klevezal, G. A. (1964). Metody opredelenija vozrasta nekotorykh lastonogih. In: *Opredelenie vozrasta promyslovyh lastonogih i racional'noe ispol'zovanie morskih mlekopitajuščih* (ed. S. E. Kleinenberg), pp. 5–20. Izdatelstvo Nauka, Moscow. (In Russian).

Toktosunov, A. T. (1973). Morfofiziologicheskaya differentsiatsiya amfibiï i mlekopitayushchikh v usloviyakh Tyan'-Shanya. PhD Thesis. Sverdlovsk.

Trojan, P. (1969). Energy flow through a population of *Microtus arvalis* (Pall.) in an agrocoenosis during a period of mass occurrence. In: *Energy flow through small mammal populations* (ed. K. Petrusewicz & L. Ryszkowski), pp. 267–79. Polish Scientific Publishers, Warsaw.

Trojan, P. (1969). An ecological model of the costs of maintenance of *Microtus arvalis* (Pall.). In: *Energy flow through small mammal populations* (ed. K. Petrusewicz & L. Ryszkowski), pp. 113–22. Polish Scientific Publishers, Warsaw.

Trojan, P. & Wojciechowska, B. (1967). Resting metabolism rate during pregnancy and lactation in the European common vole, *Microtus arvalis* (Pall.). *Ekologia Polska*, Ser. A, **15**, 811–17.

Trojan, P. & Wojciechowska, B. (1969). Ecological model and tables of the daily costs of maintenance (DEB) of *Microtus arvalis* (Pall.). *Ekologia Polska*, Ser. A, **17**, 313–42.

Truková, E. (1966). Methods of age determination in muskrat, *Ondatra zibethica* Linnaeus 1758. *Lynx*, **6**, 165–72. (In Czech. with English summary.)

Tryon, C. A. & Snyder, D. P. (1973). Biology of the eastern chipmunk, *Tamias striatus*: life tables, age distributions, and trends in population numbers. *Journal of Mammalogy*, **54**, 145–68.

Tupikova, N. V. (1964). Izučenie razmnoženija i vozrastnogo sostava populjacii melkih mlekopitajuščih. In: *Metody izučenija prirodnyh očagov boleznej čeloveka* (ed. P. A. Petriščeva & N. T. Olsufev), pp. 154–91. Medgiz, Moscow.

Tupikova, N. V., Sidorova, G. A. & Konovalova, E. A. (1968). A method of age determination in Clethrionomys. *Acta Theriologica*, **13**, 99–115.

Tupikova, N. V., Sidorova, G. A. & Konovalova, E. A. (1970). A guide for age determinations in red-backed voles. *Fauna i Ekologija Gryzunov*, **9**, 160–7. (In Russian with English summary.)

Turček, F. S. (1956). Quantitative experiments on the consumption of tree seeds by mice of the species *Apodemus flavicollis*. *Annales Societatis zoolog.-bottanicae Fennicae Vanamo*, **10**, 50–9.

Turček, F. J. (1960). Über Rötelmausschäden in den slowakischen Wäldern im Jahre 1959. *Zeitschrift für Angewandte Zoologie*, **47**, 449–56.

Turner, F. B. (1970). The ecological efficiency of consumer populations. *Ecology*, **51**, 741–2.

Turner, G. T., Hansen, R. M., Reid, V. H., Tietjen, H. P. & Ward, A. L. (1973). Pocket gophers and Colorado mountain rangeland. Colorado State University Experiment Station (Fort Collins) Bulletin 54S.

Ullrey, D. E., Youatt, W. G., Johnson, H. E., Fay, L. D., Purser, D. B., Schoepke, B. L. & Magee, W. T. (1971). Limitations of winter aspen browse for the white-tailed deer. *Journal of Wildlife Managment*, **35**, 732–43.

433

References

Usuki, H. (1966). Studies of the shrew mole (*Urotrichus talpoides*). I. Age determination, population structure and behavior. *Journal of the Mammalogical Society of Japan*, **3**, 24–9.

Utrecht-Cock, C. N. van (1965). Age determination and reproduction of female fin whales, *Balaenoptera physalus* (Linnaeus, 1758) with special regard to baleen plates and ovaries. *Bijdragen Tot de Dierkunde*, **35**, 39–87.

Valentine, G. L. & Kirkpatrick, R. L. (1970). Seasonal changes in reproductive and related organs in the pine vole, *Microtus pinetorum*, in southwestern Virginia. *Journal of Mammalogy*, **51**, 553–60.

van der Lee, S. & Boot, L. M. (1955). Spontaneous pseudo-pregnancy in mice. *Acta physiologica et pharmacologica néerlandica*, **4**, 442–4.

Van Soest, P. J. (1966). Non-nutritive residues: a system of analysis for replacement of crude fiber. *Journal of the Association of Official Agricultural Chemists*, **49**, 546.

Van Valen, L. (1971). Group selection and the evolution of dispersal. *Evolution*, **25**, 591–8.

Van Vleck, D. B. (1968). Movements of *Microtus pennsylvanicus* in relation to depopulated areas. *Journal of Mammalogy*, **49**, 92–103.

Van Vleck, D. B. (1969). Standardization of *Microtus* home range calculation. *Journal of Mammalogy*, **50**, 69–80.

Varley, G. C. (1967). The effects of grazing by animals on plant productivity. In: *Secondary productivity of terrestrial ecosystems* (ed. K. Petrusewicz), pp. 773–8. Polish Academy of Sciences, Warsaw.

Varley, G. C. & Gradwell, G. R. (1962). The effect of partial defoliation by caterpillars on the timber production of oak trees in England. International Congress of Entomology, Vienna, pp. 211–14.

Varšavskij, S. N. & Krylova, K. T. (1948). Osnovnyje principy opredelenija vozrasta myševidnyh gryzunov. I. Myši-Murinae. *Fauna i Ekologija Gryzunov*, **3**, 179–90.

Vasil'ev, S. V., Efimov, V. I. & Zarkhidze, V. A. (1963). Osobennosti razmnozheniya bol'shoĭ i krasnokhvostoĭ peschanok na Krasnovodskom poluostrove. *Trudy̆ Vsesoyuznogo Nauchno-issledovatel'skogo Instituta Zashchity Rastenii*, **18**, 96–122.

Viktorov, L. V. (1967). Opredelenie vozrasta obyknovennoj burozubki (*Sorex araneus* L.). *Bjulleten' Moskovskogo Obščestva Ispitatelej Prirody, Otdel biologičeskij*, **6**, 151–2. (In Russian.)

Vinokurov, A. A., Orlov, V. A. & Okhotsky, Y. V. (1972). Population and faunal dynamics of vertebrates in tundra biocenoses (Taimyr). In: Tundra Biome. Proceedings of the IV International Meeting on the Biological Productivity of Tundra (ed. F. E. Wielgolaski & T. Rosswall), pp. 187–9. Leningrad.

Vitala, J. (1971). Age determination in *Clethrionomys rufocanus* (Sundevall). *Annales Zoologici Fennici*, **8**, 63–7.

Vogl, R. J. (1967). Wood rat densities in Southern California manzanita chaparral. *Southwestern Naturalist*, **12**, 176–9.

Voronov, N. P. (1953). Iz nablyudenii nad royushchei deyatelnostyu gryzunov v lesu. *Pochvovedenie*, **10**, 61–74. (In Russian.)

References

Wade, P. (1958). Breeding season among mammals in the lowland rain-forest of northern Borneo. *Journal of Mammalogy*, **39**, 429–33.

Wagner, C. K. (1968). Relationship between oxygen consumption, ambient temperature, and excretion of 32-phosphorous in laboratory and field populations of cotton rats. MS Thesis. University of Georgia, Athens, USA.

Wagner, C. K. (1970). Oxygen consumption, ambient temperature and excretion of phosphorus-32 in cotton rats. *Ecology*, **51**, 311–17.

Walkowa, W. (1967). Production due to reproduction and to body growth in a confined mouse population. *Ekologia Polska*, Ser. A, **15**, 819–22.

Walkowa, W. (1970). Operation of compensation mechanisms in exploited populations of white mice. In: *Energy flow through small mammal populations* (ed. K. Petrusewicz & L. Ryszkowski), pp. 247–53. Polish Scientific Publishers, Warsaw.

Walkowa, W. (1971). The effect of exploitation on the productivity of laboratory mouse populations. *Acta Theriologica*, **16**, 19–25.

Walkowa, W. & Petrusewicz, K. (1967). Net production of confined mouse populations. In: *Secondary productivity of terrestrial ecosystems* (ed. K. Petrusewicz), pp. 335–48. Polish Academy of Sciences, Warsaw.

Wallin, L. (1971). Spatial pattern of trappability of two populations of small mammals. *Oikos*, **22**, 221–4.

Wang, L. C. H., Jones, D. L., MacArthur, R. A. & Fuller, W. A. (1973). Adaptation to cold: energy metabolism in an atypical lagomorph, the arctic hare (*Lepus arcticus*). *Canadian Journal of Zoology*, **51**, 841–6.

Ward, A. L. & Hansen, R. M. (1962). Pocket gopher control with the burrow-builder in forest nurseries and plantations. *Journal of Forestry*, **60**, 42–4.

Wasilewski, W. (1952). Morphologische Untersuchungen über *Clethrionomys glareolus glareolus* Schreb. *Annales Universitatis Mariae Curie-Sklodowska*, Sectio C, **7**, 120–211. (In Polish with German summary.)

Wasilewski, W. (1956a). Untersuchungen über die morphologische Veränderlichkeit der Erdmaus (*Microtus agrestis* Linne). *Annales Universitatis Mariae Curie-Sklodowska*, Sectio C, **9**, 261–305.

Wasilewski, W. (1956b). Untersuchungen über die Veranderlichkeit des *Microtus eoconomus* Pall. in Biatowieża Nationalpark. *Annales Universitatis Mariae Curie-Sklodowska*, Sectio C, **9**, 355–86.

Watson, A. (1956). Ecological notes on the lemmings *Lemmus trimucronatus* and *Dicrostonyx groenlandicus* in Baffin Island. *Journal of Animal Ecology*, **25**, 289–302.

Watts, C. H. S. (1968). The foods eaten by wood mice (*Apodemus sylvaticus*) and bank voles (*Clethrionomys glareolus*) in Wytham Woods, Berkshire. *Journal of Animal Ecology*, **37**, 25–41.

Watts, C. H. S. (1969). The regulation of wood mouse (*Apodemus sylvaticus*) numbers in Wytham Woods, Berkshire. *Journal of Animal Ecology*, **38**, 285–304.

Webb. W. L. (1965). Small mammal populations on islands. *Ecology*, **46**, 479–88.

Webb, R. E. & Horsfall, F., Jr (1967). Endrin resistance in the pine mouse. *Science*, **156**, 1762.

435

References

Weinbren, M. P., Weinbren, B. M., Jackson, W. B. & Villella, J. B. (1970). Studies on the roof rat (*Rattus rattus*) in the El Verde forest. In: *A tropical rain forest* (ed. H. T. Odum & R. F. Pigeon), pp. E169–81. US Atomic Energy Commission.

Welch, J. F. (1967). Review of animal repellants. Proceedings of the IIIrd Vertebrate Pest Conference, pp. 36–40.

West, N. E. (1968). Rodent influenced establishment of ponderosa pine and bitter-bush seedlings in central Oregon. *Ecology*, **49**, 1009–11.

Wheeler, G. G. & Calhoun, J. B. (1967). *Programs and procedures of the International Census of Small Mammals (ICSM)*. US Department of Health, Education, and Welfare, National Institute of Mental Health, Bethesda, Maryland.

Wheeler, G. G. & Calhoun, J. B. (1968). Manual for conducting ICSM census category 04 (octagon census and assessment traplines). *International Census of Small Mammals Manual, No. 4.* Parts 1 and 2, Edition 1.

Whitaker, J. O., Jr (1963). A study of the meadow jumping mouse, *Zapus hudsonius* (Zimmerman) in central New York. *Ecological Monographs*, **33**, 215–54.

Whitaker, J. O., Jr (1966). Food of *Mus musculus, Peromyscus maniculatus baridi*, and *Peromyscus leucopus* in Vigo County, Indiana. *Journal of Mammalogy*, **47**, 473–86.

White, F. A. (1968). *Mass spectrometry in science and technology.* Wiley & Sons Inc., New York.

White, L. (1967). Problems in county-wide rodent control programming. Proceedings of the IIIrd Vertebrate Pest Conference, pp. 7–10.

White, L. (1972). The Oregon ground squirrel in Northeastern California: its adaptation to a changing agricultural environment. Proceedings of the Vth Vertebrate Pest Conference, pp. 82–4.

Whittaker, R. H. & Feeny, P. D. (1971). Allelochemics: chemical interactions between species. *Science*, **171**, 757–70.

Whitten, W. K. (1966). Pheromones and mammalian reproduction. *Advances in Reproductive Physiology*, **1**, 155–77.

Whitten, W. K., Bronson, F. H. & Greenstein, J. A. (1968). Estrus-inducing pheromone of male mice: transport by movement of air. *Science*, **161**, 584–5.

WHO (1972). Report of informal discussions on wildlife rabies in Europe. VPH/73.3. (Unpublished.)

WHO (1973). *Weekly epidemiological record*, **48**, 305–12.

Wiegert, R. G. (1961). Respiratory energy loss and activity patterns in the meadow vole, *Microtus pennsylvanicus pennsylvanicus. Ecology*, **42**, 245–53.

Wiegert, R. G. & Evans, F. C. (1967). Investigations of secondary productivity in grassland. In: *Secondary productivity in terrestrial ecosystems* (ed. K. Petrusewicz), pp. 499–518. Polish Academy of Sciences, Warsaw.

Wiegert, R. G. & Owen, D. F. (1971). Trophic structure, available resources and population density in terrestrial vs. aquatic ecosystems. *Journal of Theoretical Biology*, **30**, 69–81.

References

Wiener, G., Field, A. C. & Wood, J. (1969). The concentration of minerals in the blood of genetically diverse groups of sheep I. Copper concentration at different seasons in Blackface, Cheviot, Welsh Mountain and crossbred sheep at pasture. *Journal of Agricultural Science, Cambridge*, **72**, 93–101.

Wiener, J. G. & Smith, M. H. (1972). Efficiency of four types of traps in sampling small mammal populations. *Journal of Mammalogy*, **53**, 868–73.

Wiens, J. (1973). Pattern and process in grassland bird communities. *Ecological Monographs*, **43**, 237–70.

Wijngaarden, A. van (1954). Biologie en bestrijding van de woelrat, *Arvicola terrestris* (L.) in Netherlands. *Medeling Plantenziektenkundige Dienst te Wageningen*, **123**, 1–147.

Wijngaarden, A. van (1957). The rise and disappearance of continental vole plague zones in the Netherlands. *Verslagen van Landbouwkundige Onderzoekingen*, **63**, 1–21.

Wijngaarden, A. van & Morzer Bruijns, M. F. (1961). De hermelijnen, *Mustela erminea* L., van Terschelling. *Lutra*, **3**, 35–42.

Williams, G. C. (1966). Natural selection, the costs of reproduction, and a refinement of Lock's principle *American Naturalist*, **100**, 687–92.

Williams, T. C. & Williams, J. M. (1970). Radio tracking of homing and feeding flights of a neotropical bat, *Phyllostomus hastatus*. *Animal Behavior*, **18**, 302–9.

Willis, E. O. (1973). The loss of birds from a tropical forest preserve. Paper presented at the Second International Symposium on Tropical Ecology, Caracas, Venezuela.

Wilson, D. E. (1973). A trophic comparison of bat faunas. *Systematic Zoology*, **22**, 14–29.

Wilson, D. E. & Janzen, D. H. (1972). Predation on *Scheelea* palm seeds by bruchid beetles: seed density and distance from parent palm. *Ecology*, **53**, 954–9.

Winberg, G. G. (1934). Opyt izučenija fotosinteza i dychanija v vodnoj masse ozera. K. voprosu o balanse organičeskogo veščestva. Soobšč. 1. *Trudy limnol. St. v Kosine*, **18**, 5–24.

Winberg, G. G. (ed.) (1971). *Methods for the estimation of production of aquatic animals*. Academic Press, London and New York.

Wirtz, W. O., II (1972). Population ecology of the Polynesian rat, *Rattus exulans*, on Kure Atoll, Hawaii. *Pacific Science*, **26**, 433–64.

Wise, D. H. (1967). Home range of a wandering shrew, *Sorex vagrans*, by tracking. *Journal of Mammalogy*, **48**, 490–2.

Wood, B. J. (1969). Population studies on the Malaysian wood rat (*Rattus tiomanicus*) in oil palms, demonstrating an effective new control method and assessing some older ones. *Planter*, **45**, 547–55.

Wood, D. H. (1970). An ecological study of *Antechinus stuartii* (Marsupialia) in a south-east Queensland rain forest. *Australian Journal of Zoology*, **18**, 185–207.

Wood, D. H. (1971). The ecology of *Rattus fuscipes* and *Melomys cervinipes* (Rodentia: Muridae) in a south-east Queensland rain forest. *Australian Journal of Zoology*, **19**, 371–92.

References

Wodzicki, K. (1969a). Preliminary report on damage to coconuts and the ecology of the Polynesian rat (*Rattus exulans*) in the Tokelau Islands. *Proceedings of the Ecological Society of New Zealand*, **16**, 7–12.

Wodzicki, K. (1969b). Results of second Tokelau Island survey. WHO Information Circular VBC/VC/69, **4**, 85.

Wojciechowska, B. (1969). Fluctuations in numbers and intra-population relations in *Microtus arvalis* (Pall.) population in agrocenose. In: *Energy Flow Through Small Mammal Populations* (ed. K. Petrusewicz & L. Ryszkowski), pp. 75–9. Polish Scientific Publishers, Warsaw.

Wolf, L. Y. (1966). Control of field rodents. Seminar on Rodents and Rodent Ectoparasites. WHO/VC/66, **217**, 107–111.

Wrangel, H. F. (1940). Beitrage zur Biologie der Rötelmaus *Clethrionomys glareolus* (Schreber, 1780). *Zeitschrift für Saügertierkunde*, **14**, 52–93.

Wunder, B. A. (1970). Energetics of running activity in Merriam's chipmunk, *Eutamias merriami*. *Comparative Biochemistry and Physiology*, **33**, 821–36.

Wunder, B. A. (1975). A model for estimating metabolic rate of active or resting mammals. *Journal of Theoretical Biology*, **49**, 345–54.

Wunder, B. A. & Morrison, P. R. (1974). Red squirrel metabolism during incline running. *Comparative Biochemistry and Physiology*, **48A**, 153–63.

Wynne-Edwards, V. C. (1962). *Animal dispersion in relation to social behaviour*. Hafner Publishing Co., New York.

Yerger, R. W. (1955). Life history notes on the eastern chipmunk, *Tamias striatus lysteri* (Richardson), in central New York. *American Midland Naturalist*, **53**, 312–23.

Yoshida, H. (1970). Small mammals of Mount Kiyomizu, Fukoka Prefecture. I. Ecological distribution of the small mammals. *Journal of the Mammalogy Society of Japan*, **5**, 8–14.

Young, H., Neess, J. & Emlen, J. T., Jr (1952). Heterogeneity of trap response in a population of house mice. *Journal of Wildlife Management*, **16**, 169–80.

Yudin, B. S. (1962). Ecology of the shrews (genus *Sorex*) of western Siberia. *Transactions of the Biological Institute, Siberian Branch, Academy of Sciences of the USSR*, **8**, 33–134. (In Russian.)

Zejda, J. (1961). Age structure in populations of the bank vole, *Clethrionomys glareolus* Schreber 1780. *Zoologické Listy*, **10**, 249–64.

Zejda, J. (1962). Winter breeding in the bank vole, *Clethrionomys glareolus* Schreb. *Zoologické Listy*, **11**, 309–21.

Zejda, J. (1966). Litter size in *Clethrionomys glareolus* (Schreber, 1780). *Zoologické Listy*, **15**, 193–204.

Zejda, J. (1971). Differential growth of three cohorts of the bank vole, *Clethrionomys glareolus* Schreb. 1780. *Zoologické Listy*, **20**, 229–46.

Zejda, J. (1972). Movements and individual home ranges in a population of the water vole (*Arvicola terrestris* L.) on a pond. *Zoologické Listy*, **21**, 97–113.

Zejda, J. & Holisova, V. (1970). On the prebaiting of small mammals in the estimation of their abundance. *Zoologické Listy*, **19**, 103–18.

Zejda, J. & Holisova, V. (1971). Quadrat size and the prebaiting effect in trapping small mammals. *Annales Zoologici Fennici*, **8**, 14–16.

Zeljda, J. & Pelikan, J. (1969). Movements and home ranges of some rodents in lowland forests. *Zoologické Listy*, **18**, 143–62.

Zhukov, A. B. (1949). Dubravy USSR i sposoby ikh vosstanovleniya. In: *Dubravy SSSR, I Vses. nauchno issled*, pp. 30–352. Institut lesn.-khoz 28.

Zhukov, V. V. (1973). Immunologicheskie vzaimootnosheniya nekotorȳkh form v podsemeĭstve microtinae, In: *Éksperimental'nye issledovaniya problemȳ vida*, pp. 74–94. Sverdlovsk.

Zimina, R. P. (1970). Royushchaya deyatelnost' melkikh mlekoptiayushchikh i ikh landshaftoobrazuyushchaya rol. In: *Sredoobrazyushchaya deyatelnost' zhivotnykh* (ed. J. A. Isakov), pp. 74–5. Moskowskogo Univ., Moscow. (In Russian.)

Zimina, R. P., Pogodina, G. S. & Urushadze, T. F. (1970). Landshaftoobrazuyushchaya rol surkov v aridnykh vysokogoryakh Tyan-Shanya i Pamira. *Materialy k poznanyu fauny i flory SSSR (MOIP)*, **45**, 177–91. (In Russian.)

Zimmerman, E. G. (1972). Growth and age determination in the thirteen-lined ground squirrel, *Spermophilus tridecemlineatus*. *American Midland Naturalist*, **87**, 314–25.

Zimmermann, K. (1937). Die märkische Rötelmaus. Analyse einer population. *Märkische Tierwelt*, **3**, 24–40.

Zinsser, H. (1967). *Rats, lice and history*. Bantam Books Inc., New York.

Zippin, C. (1956). An evaluation of the removal method of estimating animal populations. *Biometrics*, **12**, 163–89.

Zippin, C. (1958). The removal method of population estimation. *Journal of Wildlife Management*, **22**, 82–90.

Index

acclimation and acclimatization, 191-2
accuracy: in density estimates, 34, 35-6; in model for energy flow, 204; in production estimates, 154, 165-8
Acomys cahirinus, 289, 292, 296, 339
Acomys spp., may carry plague, 351
activity of animals: in energy balance, 180, 181, 193, 195-6, 197, 198; high cost of, in small mammals, 7; weather and patterns of, 36
adrenal glands: hypertrophy of, in populations under stress; 135, 138, relative weight of, 130, 135, 136
aestivation, 6, 33
Aethomys, 279
Aethomys chrysophilus, A. kaiseri, 292
age of animal: in calculations of production, 161, 165; and morpho-physiological indices, 134; and serological characteristics, 356; at sexual maturity (tropical rodents), 286-7; and trappability, 37
age of animal, criteria of, 55-6, 69-72, 251; body size, 56; bone growth and ossification, 56-60; bone marrow, 69; chemical composition, 64-5, 207, 208; moult and type of pelage, 65-6; tooth succession and development, 60-1; tooth wear, 66-8; weight of eye lens, 61-4; yearly rings in hoofs, horns, etc., 68
age structure of population: dispersal and, 109; of lemmings, 252, 260, 261
Agouti paca, 278, 279
agriculture, *see* cultivated land, crops
Akodon azarae, 76-7, 296
Allactaga elater, 84-5
allometric indices, 144, 145
Alopex lagopus, 266
Alticola lemminus, A. roylei, 76-7
Alticola strelzowi, 230
aluminium, body content of, 207, 208, 211, 212
Ammospermophilus leucurus, 212
Antechinus stuartii, 285, 290, 294
anticoagulants, for rodent control, 312, 335, 343-4, 345; resistance to, 337, 344
Aplodentia rufa, 76-7
Apodemus agrarius, 39, 76-7, 187; criteria of age of, 61, 67, 70; morpho-physiological indices for, 131, 132
Apodemus argenteus, 76-7
Apodemus flavicollis, 76-7, 132, 187; burrowing by, 236; criterion of age of,

Apodemus flavicollis, cont.
60; density estimates for, 31, 37, 39; dispersal of, 109; as pest, 231, 232; production estimates for, 158, 170, 171; storage of food by, 233
Apodemus spp., as pests, 313
Apodemus sylvaticus, 76-9, 114; burrowing by, 236, 237
appendix, relative length of, 147
arboreal animals: density estimates for, 33; in tropical ecosystems, 278-80
arctic ecosystems, 243-6, 267; lemmings in, 247-53; mammals other than lemmings in, 253-5; nutrition and energetics in, 255-62; population dynamics in, 246-7; predators in, 266-7; small mammals and soil in, 264-6; small mammals and vegetation in, 262-4
arsenic, body content of, 208
arthropods: carrying disease, 350; parasitic, 299, 300
Artibeus jamaicensis, 275, 284
Artibeus tituratus, 275
Arvicanthus niloticus, 316
Arvicanthus spp., may carry plague, 351
Arvicola spp., carriers of tularemia, 354
Arvicola terrestris, 78-9, 187, 258; burrowing by, 236; dispersal of, 110, 119; as pest, 313; prediction of population of, 330; storage of food by, 233
Asio flammeus, 266
assimilation, 11, 184-7; in calculations of energy flow, 19, 183, 188, 190, 199; feeding trials to determine, 178
assimilation/consumption efficiency, 21
Atherurus, 279
Azotobacter, rodent faeces and growth of, 235

bacterial diseases, carried by small mammals, 350, 356, 358
baculum: size and shape of, as criteria of age, 58
baiting: with poison, 322, 337, 343, 348; of traps for density estimates, 28-9
Bandicota bengalensis, 315, 339
Bandicota indica, 315
Bandicota spp., may carry plague, 351
barium, body content of, 208, 211, 212
Bassaricyon gabbii, 282
bats, 3, 4, 184, 280; diseases carried by 351, 358; pollination by, 282, 283-4, 297;

Small mammals

bats, cont.
relative abundance of, 274, 275; in tropical ecosystems, 270, 271, 272, 296, 297
Beamys major, 289
bioenergetics, 19, 20
biomass: expressed in terms of an element, 205; storage of energy as, 19; in tropical ecosystems, 275–8, 297; turnover of, 18, 19
birds: in burrows of *Rhombomys*, 305; energy needed for tropical and temperate communities of, 281
birth: season of, and morpho-physiological indices, 134
birth rate, in calculations of reproduction, 155
Blarina brevicauda, 82–3, 184, 212; density estimates for, 39, 40; as seed-eater, 233
Blarina spp., as pests, 314
body composition, 206; determinations of, 178–9, 195, 205–6; interspecific differences in, 209; intraspecific differences in, 207–9
body shape, and metabolic rate, 190
body weight: and age, 166, 261; and body composition, 207, 210, 211; in calculation of production, 162, 163–4, 292–4; and energy cost of activity, 193; and food consumption, 214; and metabolic rate, 180–2, 190, 198, 258; seasonal changes in, 139, 140
bone marrow: change in colour of, with age, 69
bones: incremental layers in, as criteria of age, 58–60; of limbs and tail, fusion of epiphyses to, as criterion of age, 57–8
boron: balance of, over month, 218–19, 220, and over year, 216–17; body content of, 207, 211, 212
breeding season: and density estimates, 34, 37; metabolic rate in, 145; for microtine rodents, 250, 253, 254, 261; and morpho-physiological indices, 139; in tropical ecosystems, 286–7, 289–90
burrowing, effects of: in deserts, 300, 301, 302, 303, 304, 305, 309; on habitat, 226, 236–8, 239; on soil, 265–6; on vegetation, 263–5
burrowing animals, *see* fossorial mammals

calcium: balance of, over month, 218–19, 220, and over year, 216–17; body content of, 207, 208, 211, 212; may be limiting element, 215
Callosciurus, 279
Calomys callosus, 355

calorimetry: for calculations of energy flow, (direct) 176, (indirect) 177–8; for determination of energy value of body content, 178
Caluromys, 279
Capreolus capreolus, 66
carnivores, utilization of food energy by, 184–5, 188, 189, 190
Carollia perspicillata, C. subrufa, 275
carrying capacity of an area, 120; emigration preventing population from reaching, 121–5; emigration stopping population growth at, 120–1
Castor canadensis 125, 127, 187
cats, for rodent control, 324, 336, 346
census techniques: behavioural response to, 25, 37–8; factors affecting, 33–7, 38–40; trapping methods for, 26–33, 43–5
chemical repellents, for rodent control, 318–19, 336
chemosterilants, for rodent control, 325–6, 336, 347
Chiropodomys gliroides, 287, 295
Chiroptera, *see* bats
Citellus fulvus, 147, 301
Citellus leucurus, 86–7
Citellus parryii, see *Spermophilus undulatus*
Citellus pygmaeus, 59, 86–7, 300–1; burrowing by, 237–8; as pest, 228
Citellus relictus, 146–7
Citellus spp.: and composition of plant community, 230; in desert areas, 300; may carry plague, 351; as pests, 313, 314, 315
Citellus undulatus, 86–7, 233
claws, yearly rings of growth in, 68
Cleithrionomys gapperi, 39, 78–9, 187; criterion of age of, 58; as seed-eater, 232–3
Cleithrionomys glareolus, 78–81, 187, 253; burrowing by, 236; carrier of rabies? 355; criteria of age of, 58, 60, 61; density estimates for, 31, 37, 39; dispersal of, 108, 109, 110, 114, 115; growth rates of, 114, 166; growth–survival curve for, 160; on islands, 119; low survival rate of, 75; morpho-physiological indices for, 133, 136, 137; as pest, 228, 313; production estimates for, 170, 171; reproduction by, 155, 156, 157, 158, 166, 169; reproduction of, and energy requirements, 191
Cleithrionomys rufocanus, 80–1, 246, 249, 250; criterion of age of, 61; habitats of, 253, 254
Cleithrionomys rutilus, 80, 246, 249, 250; criterion of age of, 60–1; energy budget for, 259; habitats of, 254; production estimate for, 170

Small mammals

Mastomys natalensis, 275, 289, 291, 293; as carrier of disease, 354; as pest, 316
Mastomys spp., may carry plague, 351
Melomys cervinipes, 287, 288, 290, 295
Meriones blackleri, 301
Meriones erythrourus, 68, 148, 303
Meriones hurrianae, 84–5
Meriones meridianus, 302
Meriones persicus, 301, 357
Meriones spp., in desert areas, 300; may carry plague, 351–2
Meriones tamariscinus, 61, 84–5, 235, 301, 302
Meriones unguiculatus, 301–2
metabolic rate, 6, 7; average daily (ADMR), 177, 179, 180, 258; average daily, for arctic small mammals, 259, 261; average daily, model for energy budget based on, 181–2, 193–9; basal (BMR), 179, 180; in breeding season, 145; estimations of, in the field, 201–3; resting (RMR), 179, 180; resting, for arctic small mammals, 261; resting, model for energy budget based on, 180–1, 197, seasonal change in, 302
Metachirus, 279
mice: methods of catching, for density estimates, 27; *see also Mus, etc.*
Micromys minutus, 39
Microsorex hoyi, 184
microtine rodents, 76–82; breeding seasons of, 250; diets of, 250; immunological divergence in subfamily Microtinae of, 150; metabolic rates of, 180; predators and, 266; survival and reproductive rates in, 92, 94, 95, 97, 101
Microtus, see also Pitymys
Microtus agrestis, 6, 80–1, 187; changes in gene frequency in, 126; control of, in Finnish seed orchards, 332–3, 334; epidemic among, 324; as pest, 228, 312, 321, 328; prediction of outbreaks of, 330; production estimate for, 171; reproduction of, 116, 157, 158; spread of habitats for, 311, 317
Microtus arvalis, 80–3, 187; burrowing by, 236, 237; carrier of rabies? 355; control of, in Hungary, 332; criteria of age of, 57, 63; density estimates for, 167, 329; DNA of, 151–2; energy budget for, 261; habitats of, 240, 254; morpho-physiological indices for, 141; as pest, 228, 312, 313, 321, 322, 324, 327–8; prediction of outbreaks of, 330; production estimate for, 171; reproduction estimate for, 158; storage of food by, 233
Microtus californicus, 82–3, 258; changes in gene frequency in, 127; criterion of age of, 65; dispersal of, 108, 111–12, 112–13, 124; on islands, 118

Microtus drummondii, 314
Microtus gregalis, 82–3, 254; in arctic, 246; criterion of age of, 57; morpho-physiological indices for, 140, 141, 145, 148
Microtus guentheri, 312
Microtus juldeschi, 82–3
Microtus middendorffi, 82–3, 246, 254; morpho-physiological indices for, 140, 141
Microtus miurus, 246, 253, 256
Microtus montanus, 58
Microtus nivalis, 233
Microtus ochrogaster, 82–3; changes in gene frequency in, 5, 126; dispersal of, 107, 108, 110, 113, 116; on islands, 118, and in enclosures, 119, 123–4
Microtus oeconomus, 82–3, 187, 250, 254; in arctic, 145; densities of, 249; diet of, 256; dispersal of, 110, 114, 117; DNA of, 151–2; energy budget for, 259, 261; habitats of, 253, 254; morpho-physiological indices for, 130, 134, 135; as pest, 230, 313
Microtus pennsylvanicus, 82–3, 187; changes in gene frequency in, 5, 126; criterion of age of, 58; dispersal of, 107, 108, 110, 113, 117; energy budget for, 261; on islands, 118, and in enclosures, 119, 123–4; as pest, 314, 337; production estimates for, 171, 331; reproduction estimate for, 158
Microtus pinetorum, 39, 82–3, 113–14, 124, 337
Microtus socialis, 233, 237, 330; as pest, 230, 312
Microtus spp.: in arctic, 253; as carriers of tularemia, 354; dispersal of, 125, 127; effect of, on habitats, 230; litter size of, 155; metabolic rates of, 258; as pests, 314, 322
migration: of birds, 107; of populations, 104–5; *see also* emigration, immigration
mineral elements: analysis of animals for, 205–6; balance of, over year, 215–18; body contents of, 211–12; consumption and egestion of, 210, 213–15; cycling of, 227, 228, 235–6, 238, 265; interspecific differences in body contents of, 209; intraspecific differences in body contents of, 207–9; limiting, 215; prediction of amounts of, in standing crops, 210
molybdenum, body content of, 207, 211, 212
morpho-physiological indices, 129–31; dynamic standards for, 131, 134
mortality: frustrated dispersal increases rate of, 119; gross, usually taken to include emigration, 103, 106, 108–9;

Small mammals